Aspects of Mechanism and Organometallic Chemistry

Herbert C. Brown

A Volume in Honor of Professor Herbert C. Brown

Aspects of Mechanism and Organometallic Chemistry

Edited by
James H. Brewster

Purdue University
West Lafayette, Indiana

PLENUM PRESS • NEW YORK AND LONDON

Library of Congress Cataloging in Publication Data

Main entry under title:

Aspects of mechanism and organometallic chemistry.

Part of the proceedings of a symposium in honor of Professor Herbert C. Brown,
held at Purdue University, West Lafayette, Indiana, May 5—6, 1978.
Includes index.
1. Chemical reaction, Conditions and laws of — Congresses. 2. Organometallic chemistry — Congresses. 3. Brown, Herbert Charles, 1912-   I. Brewster, James Henry,
1922-   II. Brown, Herbert Charles, 1912-   III. Purdue University, Lafayette,
Ind.

QD501.A834   547'.05   78-21684
ISBN-13: 978-1-4684-3395-1   e-ISBN-13: 978-1-4684-3393-7
DOI: 10.1007/978-1-4684-3393-7

Proceedings of a symposium in honor of Professor Herbert C. Brown
held at Purdue University, West Lafayette, Indiana, May 5—6, 1978

© 1978 Plenum Press, New York
Softcover reprint of the hardcover 1st edition 1978
A Division of Plenum Publishing Corporation
227 West 17th Street, New York, N.Y. 10011

PREFACE

In May of 1978, several hundred of the friends, colleagues and
former students of Professor Herbert C. Brown gathered on the campus
of Purdue University to note his formal retirement, to honor him for
his past contributions to chemistry and to wish him continued success
in research. It was a time of reunion and recollection, a time for
looking back and giving recognition to a lifetime of accomplishment.
There was the ceremony of a banquet, presided over with inimitable
wit by Professor Derek Davenport, and the dedication of the Herbert
C. Brown Archives, with addresses by Dr. Alfred Bader, of Aldrich
Chemicals, and Dr. Alan Schriesheim, of Exxon. There was the publication of a book of the personal reminiscences of students and postdoctoral colleagues - "Remembering HCB." But it was also a time
for looking at the present and into the future with a set of scientific lectures, mainly by former students or associates, who described their current or projected research activities. That is
what this book is about.

The papers, some of which are expanded versions of the lectures,
fall into two broad groups - some deal with the interplay of structure and mechanism, the others deal with the use of organometallics
in synthesis. It is, perhaps, no accident that these are the two
main areas of H. C. Brown's research interest.

In the lead-off paper, Edward Peters (with M. Ravindranathan
and H. C. Brown) reviews the use of the mechanistic probe of increasing the electron demand as a means of detecting the donation of
election density by other parts of the molecule to a developing cationic center. This method allows magnitudes to be assigned to such
effects as conjugation, participation and neighboring group effects;
it appears to provide decisive answers to questions about the extent
to which different kinds of carbonium ion can be considered to be
"non-classical."

Leon Stock considers the trifluoromethyl group as an electron
acceptor and shows that while it acts mainly through field and inductive effects in uncharged ground state molecules, it can act as

a strong acceptor of electron spin and charge, *via* delocalization or hyperconjugation, in electron-rich transition states and radical anions.

A particular class of radical anions, the semidiones, was discovered by Glen Russell and his group. In a wide-ranging review, he covers the use of the semidione spin label as a means of detecting the components of conformational and molecular equilibria, particularly with highly unsaturated or strained substances.

The concept of homomorphs, developed by H. C. Brown, is used by Shelton Bank to relate the properties of arylated anions and isoelectronic amines. Steric effects, influencing the ability of the aromatic rings to conjugate with unshared electrons provoke interesting differences in kinetic and thermodynamic acidity of certain hydrocarbons. This provides yet another example of the far-reaching impact of Brown's analytical treatment of steric effects.

The more intimate details of mechanisms, especially those subject to catlysis by transient or fragile species, may be difficult to discern. Milorad Rogić and his coworkers have made major progress in understanding how copper complexes induce the oxidative cleavage of o-benzoquinones, catechols and even phenols. They suggest that their work may provide a basis for understanding the way in which certain oxygenase enzymes work.

In a paper providing something of a bridge to those dealing with synthetic applications of organometallics, Frederick Jensen considers the stereochemistry of the S_E2 cleavage of tetra-alkytin compounds by bromine. With chiral trineopentyl-*sec*-butylin, clean stereochemical results are obtained only when radical reactions are suppressed; in methanol inversion occurs, while in carbon tetrachloride the retention pathway dominates. The factors which can shift this delicately balanced system one way or the other are considered.

The reduction of carbonyl compounds by diborane was first studied by H. C. Brown when he was a student of Professor H. I. Schlesinger at Chicago and the development of hydride reducing agents has continued as an important part of his research. This area is now under commercial development at Aldrich-Boranes and their progress in the development of convenient, stable and selective reducing agents is summarized by Clinton Lane.

George Kabalka calls attention to the many ways in which the boron chemistry developed by Professor Brown at Purdue can be used for the introduction of radionuclides into organic compounds biological importance. He outlines briefly work in progress or projected along these lines.

Mark Midland discusses further areas of boron chemistry, namely the rearrangement of alkyl groups from boron "ate" species and reductions by trialkylboranes, in which the hydrogen atom comes from a carbon atom. These reactions can be used in asymmetric synthesis, the first allowing formation of chiral allenic boranes (and, thus, chiral allenes) and the second, the reduction of aldehydes to primary deutero alcohols of high enantiomeric purity.

In his years at Purdue, George Zweifel oversaw the research of many of Professor Brown's doctoral students and played a major role in the development of hydroboration and of the chemistry of boranes. He has gone on to work in the area of hydroalumination and has found a number of useful reactions, especially of the alkenylalanes. Here he compares their chemistry with that of the alkenylboranes and draws attention to their utility in the synthesis of unsaturated compounds.

Organomercurials are available by many routes, including direct mercuration and displacement of boron from vinylboranes. Richard Larock has found that unsaturated mercurials can be coupled with palladium compounds as catalysts. With molar amounts of palladium compounds, such mercurials add to olefins to give organopalladiums. Carboxylation with carbon monoxide can also be achieved. With rhodium compounds as catalysts, replacement by methyl or acyl groups can be achieved; unsaturated aldehydes can by cyclized to cyclanones. The mechanism of the reactions probably include transmetalation steps.

Going a step further, Ei-ichi Negishi considers in particular the synthetic problem of cross-couplings of organometallics with halides. Nickel and palladium catalysts can serve to bring many of these reactions about with minimal homo-coupling and maximal retention of stereochemical integrity. Alkenylzirconium compounds cross-couple well with nickel or palladium catalysts. Finally it is found that the cross-coupling of alkenylalanes under the influence of a palladium catalyst can be further improved by addition of small amounts of zinc compounds, to give a double metal catalysis.

This, then, is in some sense a partial record of the scientific part of the symposium. Not all of the speakers were able to contribute chapters to this volume but those who did should be thanked for their cooperation and, in most cases, their promptness. Those who have fleshed out their lectures into full reviews deserve special thanks for making this more than just a record of an event. I wish also to thank those who helped make the symposium itself a success, especially Dorothy Murphy of the Chemistry Department office. The three typists, whose skill and patience contributed so much to the production of this book were Sue Baker, Roberta Molander and Joy Anne Shulke. A final word of thanks should go to those whose financial aid made it possible for us to get the speakers to and from

Lafayette – the School of Science at Purdue, E. I. du Pont de Nemours
and Company and Union Carbide Corporation.

James H. Brewster
West Lafayette, Indiana
August 7, 1978

CONTENTS

---

* Senior author

THE MECHANISTIC PROBE OF INCREASING ELECTRON DEMAND IN THE STUDY
OF SOLVOLYTIC REACTIONS

Edward N. Peters, M. Ravindranathan and Herbert C. Brown

Union Carbide Corporation, One River Road, Bound Brook,

New Jersey, and Richard B. Wetherill Laboratory, Purdue

University, West Lafayette, Indiana

INTRODUCTION

Among the most striking developments in the history of physical organic chemistry have been methods for the quantitative correlation of rate and equilibrium constants in terms of relative free energy changes.[1] One of the best known and most useful of these is an equation proposed by Hammett, which relates equilibrium and rate constants for the reactions of meta- and para-substituted benzene derivatives.[1,2] The Hammett relationship is based on the fact that as the substituent is varied, the logarithms of the rate constants for aromatic side-chain reactions are linearly related to one another. Thus, a rate or equilibrium constant for a compound is related to the rate or equilibrium constant for the unsubstituted compound in the same process through two parameters, ρ and σ. In the case of rate constants, the relationship is as shown in equation 1, where k_0 is the rate constant for the

$$\log \frac{k}{k_0} = \rho\sigma \tag{1}$$

unsubstituted compound. Of these two parameters, the substituent constant, σ, is characteristic only of the substituent on an aryl group, while ρ is a reaction constant and is a function of the reaction being considered. Thus the reaction constant, or ρ value, is a measure of the sensitivity of a particular reaction to the substituents.

1

Table I

Effect of Increasing the Electron
Demand on 2-Aryl-2-propyl Chlorides[a]

Substituent on 2-aryl	σ^+	k_1 x 10^5 (sec^{-1}) at 25°	Relative Rates
p-CH$_3$O-	-0.778	41,700	3360
p-CH$_3$S-	-0.604	6,860	553
p-CH$_3$-	-0.311	322	26.0
H-	0.0	12.4	1.0
p-Cl-	0.114	3.78	0.305
p-CF$_3$-	0.612	0.0206	0.00106
m,m'-(CF$_3$)$_2$-[b]	1.04	2.35 x 10^{-4}	1.90 x 10^{-5}
m,m'-Cl$_2$,p-CN-[c]	1.22	3.62 x 10^{-5}	2.92 x 10^{-6}

[a] Solvolysis in 90 percent acetone. Data from reference 3a unless otherwise noted.

[b] Data calculated from 2 x σ^+ for m-CF$_3$-.

[c] Data from P. G. Gassman and A. F. Fentiman, Jr., *Tetrahedron Lett.*, 1021 (1970).

The Hammett equation fails when a full positive charge is developed on the aromatic side chain, for example in the solvolysis of benzyl derivatives. However, Brown and co-workers developed an analogous expression which overcame these shortcomings.[3] In the Hammett-Brown relationship (eq 2), σ^+ represents the resonance and inductive contributions of a substituent on an aryl group in a reaction where a full positive charge is developed during the course of the reaction. The use of σ^+ constants has

$$\log \frac{k}{k_0} = \rho^+ \sigma^+$$

$$(2)$$

been found to be generally applicable to a number of systems.[3c]
In particular, several workers have investigated the rates of
solvolysis of various tertiary benzylic systems, 1. Here it is

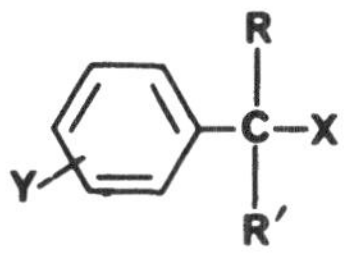

1

possible to introduce both activating and deactivating substituents, Y, into the aromatic ring and, hence, vary the electron
demand at the reaction site over a wide range while maintaining
the steric effects around that site essentially constant. For
example, from the rates of solvolysis of 2-aryl-2-propyl chlorides
(1; X=Cl, R=R'=CH$_3$), the spread in rate constants is 10^9 (Table
I).

The diagnostic probe of increasing the electron demand has
proven to be a very powerful technique in physical organic
chemistry. Its use in the study of solvolytic reactions will be
reviewed here. Most of the available rate data is for p-nitrobenzoates in 80 percent acetone; exceptions will be noted. Most
kinetic data are compared at 25°.

ARYLALKYL, ARYLCYCLOALKYL, AND ARYLPOLYCYCLOALKYL SYSTEMS

Arylalkyl Systems. Aryldialkylcarbinyl systems have been
examined by use of the probe of increasing the electron demand.[4,5]
Simple alkyl groups such as methyl, ethyl, and isopropyl do not
have a significant effect on ρ^+ (80 percent acetone). For
example, aryldimethylcarbinyl (2), aryldiethylcarbinyl (3), and
arylisopropylmethylcarbinyl (4) p-nitrobenzoates (OPNB) have
similar ρ^+ values of -4.72, -4.52, and -4.76, respectively.[4]

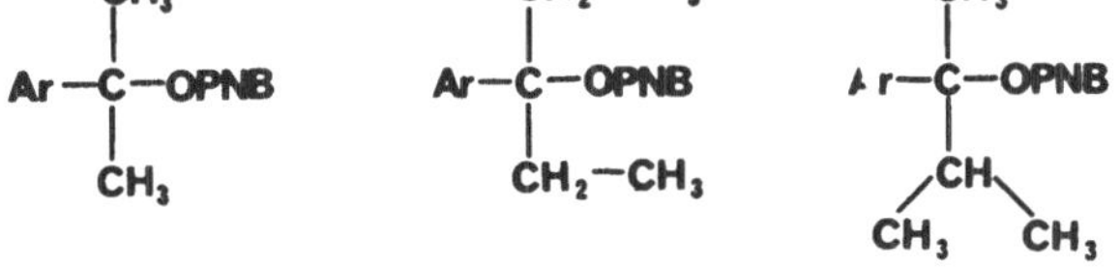

	2	3	4
ρ^+	-4.72	-4.52	-4.76

The large negative and essentially constant ρ^+ values indicate that the alkyl groups do not differ significantly in their hyperconjugative stabilization of the cationic center; had they done so, significant differences in the electronic demand on the aryl group would have resulted.

On the other hand, Tanida and Matsumura found a decrease in ρ^+ in their study of the solvolysis, in 70 percent acetone, of a series of aryldialkylcarbinyl p-nitrobenzoates with bulky alkyl groups.[5] For aryldimethylcarbinyl (2), aryl-tert-butylneopentyl-carbinyl (5), aryldineopentylcarbinyl (6), and aryl-di-tert-butylcarbinyl (7) p-nitrobenzoates the ρ^+ values at 25° are −4.60, −3.87, −3.14, and −1.53, respectively.[5,6] Thus

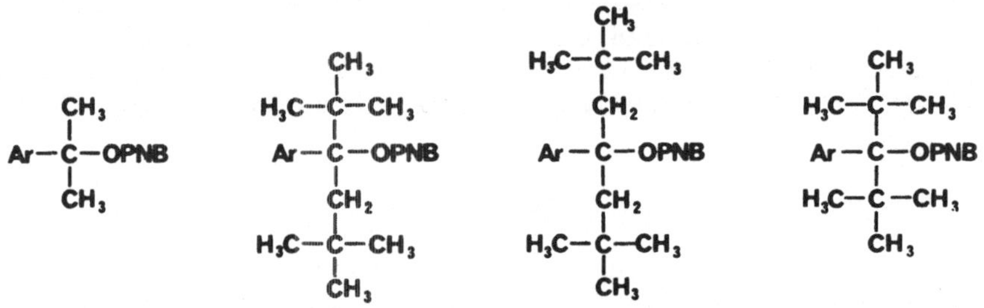

increasingly bulky groups produce steric effects which result in progressively less negative ρ^+ values.

Arylcycloalkyl Systems. Alicyclic compounds exhibit considerable changes in reactivity with changes in the number of carbon atoms in the ring. This is true both for S_N2 and S_N1 reactions, as well as for the reactions of cyclic ketones.[7-13] The effect of ring size on chemical reactivity has been accounted for in terms of I-strain - the increase in internal strain in a cyclic structure resulting from alterations in bond angles and conformations which accompany a change in the coordination number of a ring atom in the course of the reaction.[7] In this way the inertness of cyclopropane and cyclobutane derivatives to solvolysis (in systems where rearrangements are not involved) is attributed to the additional strain involved in forming an sp^2 carbonium carbon atom, with a preferred angle of 120°, from an sp^3 carbon atom in a small ring. The inertness of cyclohexyl derivatives in solvolytic reactions is attributed to the additional strain accompanying the formation of an sp^2 carbonium atom in the nicely

staggered, essentially strain-free cyclohexyl system. On the
other hand, the greater reactivity of cyclopentyl, cycloheptyl,
and cyclooctyl systems is attributed to the relief of nonbonded
interaction accompanying the transformation of a tetrahedral ring
atom into one with only three ligands.

The probe of increasing the electron demand was applied to
arylalicyclic systems.[4] In comparison to the arylaliphatic system
(3), cyclopropyl and cyclobutyl derivatives had large negative
ρ^+ values. The l-aryl-l-cyclobutyl p-nitrobenzoates (8) gave a
ρ^+ of -4.91[4] and (from the data of DePuy and co-workers[14]) l-aryl-
l-cyclopropyl tosylates (9) -5.15.[4] Solvolysis of l-aryl-l-
cyclopropyl dinitrobenzoates gave a ρ^+ of -5.19.[15] Thus, the

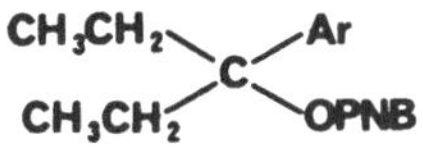

	3	**8**	**9**
ρ^+	-4.52	-4.91	-5.15

difficulty of transforming a ring atom from sp^3 to sp^2 causes
the developing cationic center to become highly electron demanding; this results in a relatively large electron supply from the
aryl group and produces a relatively large negative ρ^+ value.

The more reactive l-aryl-l-cyclopentyl derivatives (10) exhibited a ρ^+ value of -3.82, whereas the less reactive l-aryl-l-
cyclohexyl derivatives (11) showed a value of -4.60. Similarly,

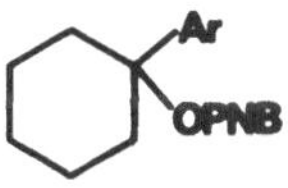

	10	**11**
ρ^+	-3.82	-4.60

the relatively more reactive cycloheptyl (12) and cyclooctyl (13)
derivatives gave ρ^+ values close to that for the cyclopentyl
system. Bond opposition forces in the cyclopentyl, cycloheptyl,
and cyclooctyl systems result in less negative values of ρ^+
because the strain in the ground state is partially dissipated in
the transition state and this strain relief reduces the energy of

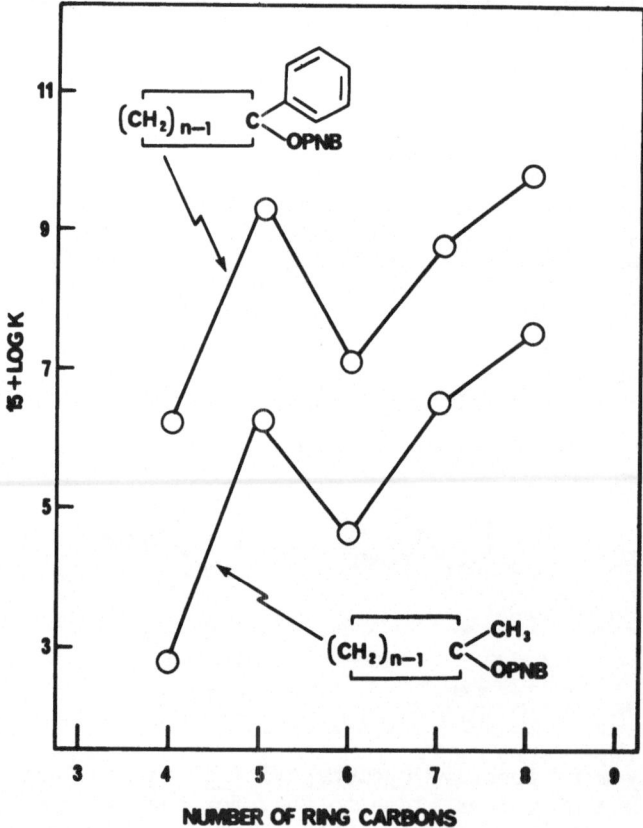

12   **13**

ρ^+ -3.87 -3.83

activation. According to the Hammond postulate,[16] the transition
state of such systems will not be so far along on the reaction
coordinate. The development of electron deficiency in the transition state will be less, and the center will make a smaller demand
on the aryl system for stabilization.[13] A similar explanation
was previously suggested by Schleyer for the results of Tanida on
the solvolysis of several 1-aryl-1-cycloalkyl chlorides.[17]

Figure 1. Effect of ring size on the rate of solvolysis of
1-phenyl- and 1-methyl-1-cycloalkyl p-nitrobenzoates in 80% acetone.

The question arises as to whether this parallelism between
the value of ρ^+ for the 1-aryl-1-cycloalkyl systems and their
reactivities is fortuitous, arising from differences in hindrance
to rotation of the aryl group at the developing cationic center.
Accordingly, the 1-methyl-1-cycloalkyl p-nitrobenzoates were
examined. A comparison with 1-phenyl-1-cycloalkyl p-nitrobenzoates appears in Figure 1.[4] The results reveal the same pattern
of reactivity. The parallelism of the observed ρ^+ values with
the relative reactivities of these ring systems, previously
accounted for in terms of I-strain effects, is remarkable. It
suggests that both the relative reactivities and the relative
values of ρ^+ may have their origin in I-strain.

    Arylpolycycloalkyl Systems. Tanida and Tsushima studied the
solvolysis of 7-aryl-7-norbornyl (14) and 2-aryl-2-adamantyl
(15) chlorides.[17] A very large negative ρ^+ value of -5.64 was
found for 14; this was attributed in part to increasing bond
angle strain during solvolysis. The adamantyl system (15)

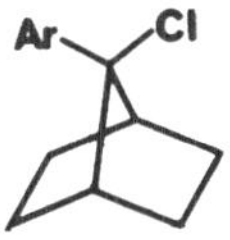

ρ^+

14

-5.64

exhibited a ρ^+ of -4.83, similar to the ρ^+ of -4.65 for 1-aryl-1-
cyclohexyl chloride (16).

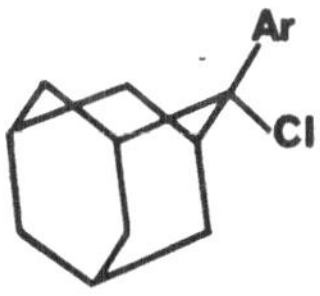

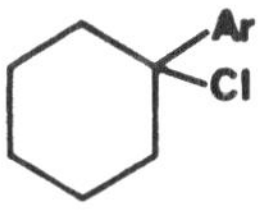

15   16

ρ^+ -4.83 -4.65

It is evident that the strain accompanying the formation of
an sp^2 center in the 7-norbornyl, cyclopropyl, or cyclobutyl
systems must result in a large demand by that center for additional electron density. In these systems the low rates of ionization
and the large negative values of ρ^+, i.e., -5.64, -5.15, and
-4.91, are readily accounted for. Similarly, the formation of an

sp^2 center in the nicely staggered strain-free adamantyl system results in considerable strain. Here, also, the slow rate of ionization and the large (-) value of ρ^+, -4.83, are readily accounted for. In these systems, the higher energies of activation for the reactions cause the transition state to lie further along the reaction coordinate. It will be more ionic, closer to the cationic intermediate. The higher electron deficiency makes a greater demand on the aryl group and its substituents for electronic stabilization.

NEIGHBORING GROUP EFFECTS

Groups near to developing cationic centers can have enormous effects on rates of solvolysis.[18] Such neighboring groups can act _via_ conjugation or participation.

The effect of conjugation of the reaction center with an adjacent double bond or benzene ring (π-conjugation) is pronounced. The observed rate enhancements in these allylic and benzylic systems may be attributed to stabilization of the transition state by π-bond overlap with the developing p orbital. Stabilization may also occur from conjugation with adjacent carbon-carbon bonds (σ- or $\pi\sigma$-conjugation).

In neighboring group participation, the brilliant studies of Winstein and co-workers in the 1940s established that donor groups (G) β to the reaction center could greatly enhance the rates of solvolytic reactions (eq 3).[19] Both π- and σ-($\pi\sigma$-) participation have been observed.

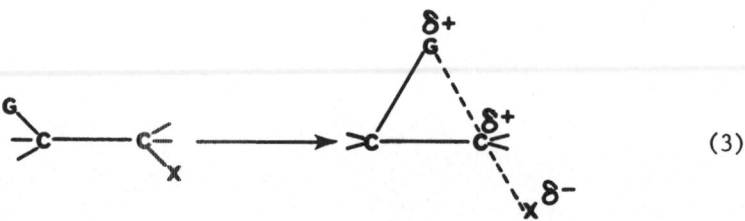

$$\tag{3}$$

It is a basic principle that the more stable a developing cationic center, the less demand that center will make on neighboring groups for additional stabilization through participation (or conjugation).[20] In benzylic derivatives the electron demand at the developing cationic center can be increased by varying the substituent on the aryl group. Thus, the mechanistic probe of increasing the electron demand should be ideal for studying neighboring group effects. Indeed, since Gassman and

Fentiman first applied this technique in the examination of the 7-norbornyl system,[21] numerous systems have been evaluated for both π- and σ-(πσ-) conjugation and participation.

The criteria for demonstrating neighboring group stabilization using this probe are a ρ^+ different from that of a model system and/or a break in the log k-σ^+ plot.

π-CONJUGATION

The effects of allylic and benzylic groups in enhancing rates of solvolytic reactions are well documented.[18] The π systems stabilize the adjacent developing cationic center by donating electrons. Since the positive charge is delocalized into the double bond, allylic and benzylic cations are substantially more stable than analogous saturated carbonium ions (4).

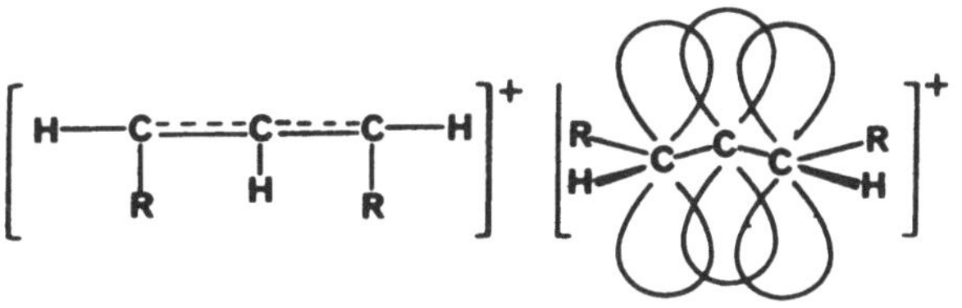

$$ \tag{4} $$

The technique of increasing the electron demand was used to elucidate the effects of π-conjugation in the solvolysis of allylic and benzylic systems. This probe has been applied to cyclohexyl (17) and cyclohex-2-enyl p-nitrobenzoates (18) to estimate their relative electronic contributions.[22] The rate of solvolysis of cyclohex-2-enyl derivatives increases relative to that of the corresponding cyclohexyl derivatives. The data not only reveal the large rate-accelerating effect of the allylic double bond, but also reveal clearly that π-conjugation involving the allylic double bond increases with increasing electron demand at the cationic center. A major increase in the electronic con-

SUBSTITUENT IN Ar	17 REL RATE	18
p– H	1.0	1.52×10^5
p– CF$_3$	1.0	1.9×10^6
m,m'–(CF$_3$)$_2$	1.0	2.3×10^7

tributions from the cyclohex-2-enyl system, as compared to the cyclohexyl system, is likewise revealed by comparing the value of ρ^+ with that for the cyclohexyl system. The data for 17 provided a value of ρ^+ of −4.60 whereas 18 gave a value of −2.52; $\Delta\rho^+$ of 2.08. Hence the stabilization provided by the double bond is a linear function of the electron demand of the incipient carbonium ion center.

Similar effects were observed in the solvolysis of 1-cyclohexyl-1-aryl-1-ethyl (19) and 1-(Δ^1-cyclohexenyl)-1-aryl-1-ethyl p-nitrobenzoates (20).[23] The data for 19 gave a ρ^+ value of −4.71 and 20 a value of −2.35; $\Delta\rho^+$ of 2.36.

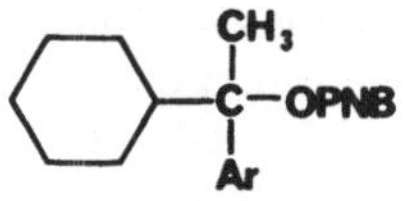

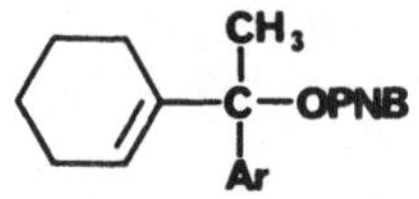

SUBSTITUENT IN Ar	REL RATE	
	19	20
p–CH$_3$O	1.0	83.4
p–H	1.0	6460
p–CF$_3$	1.0	178,200
m,m′–(CF$_3$)$_2$	1.0	1,592,000

The effect in benzylic systems was evaluated in 2-aryl-3-methyl-2-butyl (21) and 1-aryl-1-phenyl-1-ethyl p-nitrobenzoates (22).[24] These results reveal that benzylic systems are not as

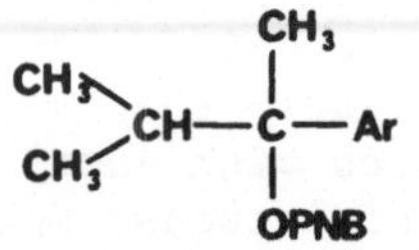

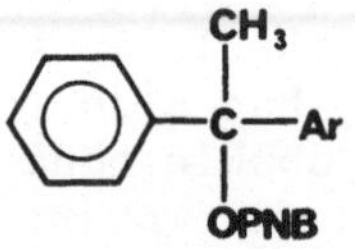

SUBSTITUENT IN Ar	REL RATE	
	21	22
p–CH$_3$O	1.0	4.4
p–H	1.0	84
p–CF$_3$	1.0	670
m,m′–(CF$_3$)$_2$	1.0	3030

effective in stabilizing an adjacent cationic center as are the
allylic double bonds in 18 and 20. This is also exemplified by
comparing the ρ^+ values. The data for 21 gave a ρ^+ value of -4.76
whereas 22 gave one of -3.23; $\Delta\rho^+$ of 1.53.

Clearly the probe of increasing the electron demand can be
used to detect the large stabilization provided by π-conjugation
in allylic and benzylic systems.

π-PARTICIPATION

Systems homologous to allylic and benzylic structures, with
a carbon atom interposed between the cationic center and the π
electron group, are designated homoallylic and homobenzylic.[25]
During solvolysis, a homoallylic or homobenzylic group can stabilize
a developing cationic center by donating π electrons through a 1,3
interaction (π-participation) (23).

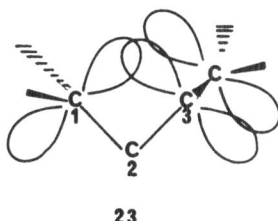

23

7-Norbornenyl. Certain homoallylic double bonds can have an
enormous effect on rates of solvolysis. For example, the rate
factor of 10^{11} observed in anti-7-norbornenyl tosylate (25) (vs.
24) is attributed to π-participation.[26] This system is of historical interest in the development of the technique of increasing

24   25

REL RATE   1.0 10^{11}

the electron demand. In their original study, Gassman and
Fentiman examined the effect of stabilizing groups on the enormous
π-participation present in 7-anti-norbornenyl derivatives.[21] For
the solvolysis of 7-aryl-7-norbornyl (26) and 7-aryl-7-anti-
norbornenyl p-nitrobenzoates (27) in 70 percent dioxane, the rate

enhancement decreases from the value of 10^{11}, observed in the parent secondary derivatives, with the introduction of stabilizing groups at the 7 position and effectively vanishes with the p-anisyl group. The ρ^+ values provide a convenient measure of the

SUBSTITUENT IN Ar	REL RATE	
p–CH₃O	1.0	3.4
p–H	1.0	41.4
p–CF₃	1.0	34,000
m,m′–(CF₃)₂	1.0	255,000

effect.   Thus, ρ^+ for <u>26</u> is –5.27 and it is –2.30 for <u>27</u>; $\Delta\rho^+$ is 2.97.

   Hence neighboring group participation diminishes as the incipient cationic center is stabilized by substitution on the aryl group.

   Cyclopentenyl. Steric requirements are very important in π-participation. For example, Bartlett and Rice have examined the solvolysis of 4-bromocyclopentene (<u>28</u>).[27] They noted that

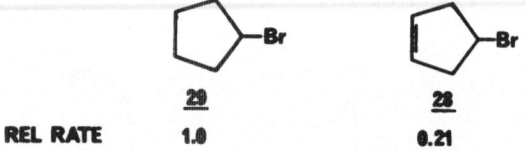

	<u>29</u>	<u>28</u>
REL RATE	1.0	0.21

this system reacted slower than cyclopentyl bromide (<u>29</u>) and concluded that the amount of orbital overlap between C_1 and C_3 must be negligible, and that the structure of the unsaturated bromide is not bent sufficiently toward the form of 7-norbornenyl to produce any significant overlap. Calculations showed that the strain energy involved in reaching a suitable conformation is greater than the stabilization afforded by the orbital overlap.[27] These results are in line with the conclusion that the degree of

puckering in the cyclopentene ring and, hence, the distance between the π electron cloud of the double bond and the developing
cationic center must be crucial in determining the overall effect
of anchimeric assistance.[28] The solvolysis of 1-arylcyclopent-3-
en-1-yl (30) and 1-arylcyclopentyl p-nitrobenzoates (31) reveal
the absence of any significant amount of 1,3 interaction between
the π electrons and the cationic center in 30.[29] The 1-aryl-
cyclopent-3-en-1-yl derivatives undergo solvolysis at a rate about

	31	**30**
SUBSTITUENT IN Ar	**REL RATE**	
p–CH$_3$O	1.0	0.45
p–H	1.0	0.31
p–CF$_3$	1.0	0.25
m,m'–(CF$_3$)$_2$	1.0	0.31

three times slower than their saturated derivatives. The small
rate retardation is in the direction anticipated for the inductive
effect of the double bond on the rates of solvolysis, with no
evidence for a trend which could be attributed to π-participation.
The ρ^+ values for 30 and 31 (-3.92 and -3.82, respectively),
suggest the absence of π-participation. The small $\Delta\rho^+$ of -0.10
is in the direction opposite that anticipated for participation.

$\underline{\Delta^2\text{-Cyclohexenylcarbinyl}}$. Similar effects are observed in
the solvolysis of 1-(Δ^2-cyclohexenyl)-1-aryl-1-ethyl p-nitrobenzoates (32). Where the double bond is homoallylic, no participation is observed with the first three derivatives (p-CH$_3$O, p-H,
p-CF$_3$).[23] In fact, these derivatives solvolyze 4-5 times slower
than the corresponding saturated derivatives, presumably due to
the inductive rate-retarding effect of the double bond.[27,29]
The m,m'-(CF$_3$)$_2$ derivative of 32 shows a modest rate enhancement,
the rate of solvolysis being quite comparable with that of the
saturated derivative. The log k-σ^+ plot reveals a definite break,
indicating the inception of participation of some kind. Such
participation could be π-, similar to that recently postulated by
Lambert and Reatherman for solvolysis of the cyclohexen-4-yl
tosylates in trifluoroacetic acid.[30] Alternatively, the participation may involve the transfer of the allylic hydrogen atom to

SUBSTITUENT IN Ar	REL RATE	
	19	**32**
p—CH$_3$O	1.0	0.27
p—H	1.0	0.22
p—CF$_3$	1.0	0.21
m,m'—(CF$_3$)$_2$	1.0	1.02

the developing cationic center, producing a stabilized allylic cation (5). Unfortunately, the solvolytic products proved to be

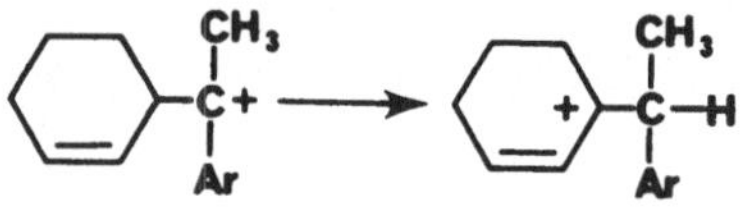

$$(5)$$

complex and did not permit conclusions to be drawn as to these two possible interpretations.

The double bond in the 1-(Δ^3-cyclohexenyl)-1-aryl-1-ethyl system should be even more favorable for π-participation, if such participation is involved. Accordingly, we synthesized 1-(Δ^3-cyclohexenyl)-1-(3,5-bistrifluoromethylphenyl)-1-ethyl p-nitrobenzoate 33 and studied its solvolysis. However, this derivative did not show any rate enhancement over the saturated derivative 34.

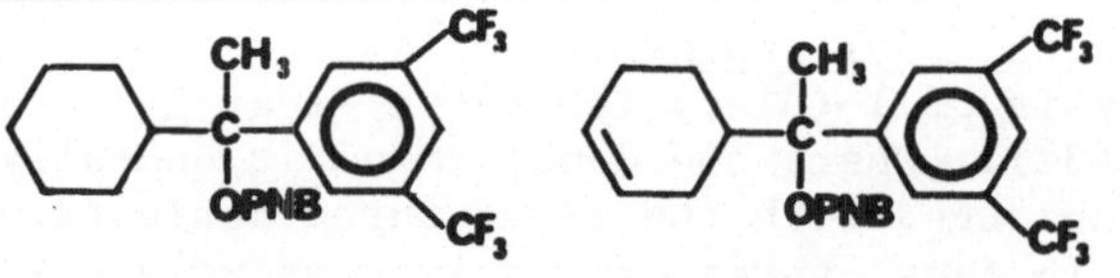

REL RATE	1.0	0.45

Thus, it was concluded that 1-(Δ^3-cyclohexenyl)-1-aryl-1-ethyl derivatives solvolyze without significant participation. It appears probable, therefore, that the enhanced rate for the m,m'-(CF$_3$)$_2$ derivative of 32 is the result of a rearrangement to the

related allylic cation (eq 5). The ρ^+ value for 19 is -4.71 and
-4.83 for 32. The small negative $\Delta\rho^+$ (-0.12) is not indicative
of neighboring group stabilization.

2-Norbornenyl. In estimating neighboring group effects, it
is customary to compare the rate with a related rate not involving
participation.[31,32] For the assessment of π-participation in
homoallylic systems, it is often useful to compare the rate of
solvolysis of the homoallylic compound with the rate of solvolysis
of its saturated analog.[23,26,29] However, this procedure is not
always followed. For example, in the 2-norbornenyl system (35,
36) the observed exo:endo ratio of 7000 for the brosylates was

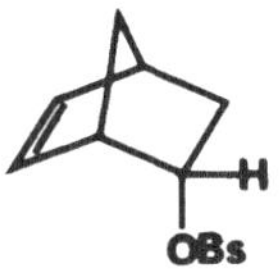

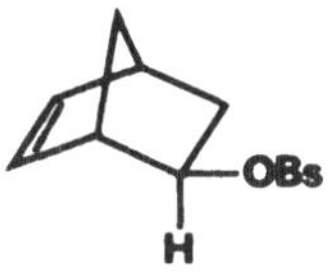

	35	36
REL RATE	1.0	7000

attributed to π-participation in the exo isomer.[33] However,
steric effects may also make major contributions to high exo:endo
rate ratios.[34] Thus the original conclusion as to the magnitude
of π-participation in this system, based solely upon a high exo:endo
rate ratio, would appear to be questionable. Accordingly, the
probe of increasing the electron demand was applied to 2-norbornenyl
system.

In the 2-aryl-2-norbornenyl system (37, 38) the exo:endo

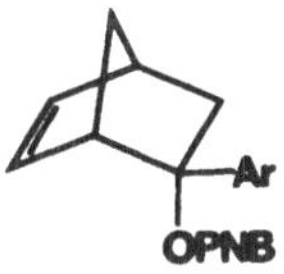

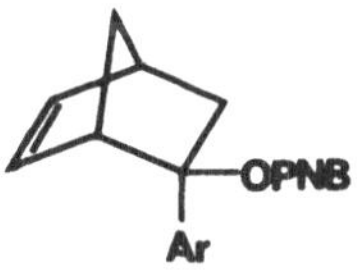

	37	38
SUBSTITUENT IN Ar	REL RATE	
p-CH$_3$O	1.0	312
p-H	1.0	202
p-CF$_3$	1.0	283
m,m'-(CF$_3$)$_2$	1.0	447

rate ratio remains essentially constant when the substituent is varied from p-CH$_3$O to p-CF$_3$.[35] If π-participation is a significant factor in the exo:endo rate ratio, then one should observe increasing neighboring double bond participation, accompanied by increasing exo:endo rate ratios as the electron demand of the carbonium ion center is increased. However, with the more deactivating m,m'-(CF$_3$)$_2$ the exo:endo rate ratio exhibits only a modest increase to 447.[36] Thus it appears that as the electron demand of the cationic center is further increased in the m,m'-(CF$_3$)$_2$ derivatives, π-participation finally becomes a contributing factor in the rate determining step.

An alternative means of examining the effect of increasing electron demand is the change in ρ^+ for the reaction. The 2-aryl-endo-norbornenyl derivatives exhibit a ρ^+ value of -4.17. For the exo isomers a break in the log k-σ^+ plot occurs between p-trifluoromethyl and m,m'-bis(trifluoromethyl)phenyl substituted compounds. This break in the log k-σ^+ plot indicates a change in the mechanistic aspects of solvolysis of the exo isomers. Thus as the cationic center is made more electron demanding by substituents less capable of stabilizing a positive charge than a p-trifluoromethylphenyl group, participation by the π electrons of the 5,6-double bond becomes significant. For substituents such as p-anisyl, phenyl, p-trifluoromethylphenyl, which are more effective than the m,m'-(CF$_3$)$_2$ derivative in stabilizing the carbonium ion center, participation by the π electrons is not needed and hence does not occur. Indeed, these substituents yield a ρ^+ value of -4.21, which is comparable with the ρ^+ value exhibited by the endo isomers. The predominant solvolysis product of 2-aryl-2-norbornenyl derivatives with p-CH$_3$O, p-H, and p-CF$_3$ substituents is the corresponding 2-aryl-exo-norbornenol (39). This is in accord with the indicated absence of significant interaction of the double bond in the ionization step. On the other hand, in the case of the m,m'-(CF$_3$)$_2$ derivative, the predominant product is that arising from a rearrangement involving the 5,6 π electrons, 1-[m,m'-bis(trifluoromethyl)phenyl]-3-nortricyclanol (40) (eq 6). This result is in agreement with the rate data. Thus, the involvement of the π electrons of the double bond becomes more important as the electron demand of the incipient carbonium ion is increased by the m,m'-bis(trifluoromethyl)phenyl group.

Consequently, some other factor, such as steric hindrance to ionization must be responsible for the high exo:endo rate ratios.[9] If we assume that the exo:endo rate ratio of approximately 350 in these tertiary derivatives, attributed to steric factors, can be carried over to the parent 2-norbornenyl system, we are left with a factor of 20 attributable to π-participation. One can argue that the steric factor of approximately 350, observed for the tertiary derivatives, should not be carried over to the secondary. Possibly, some new, smaller factor should be used. Unfortunately, it is not now possible to estimate that factor.

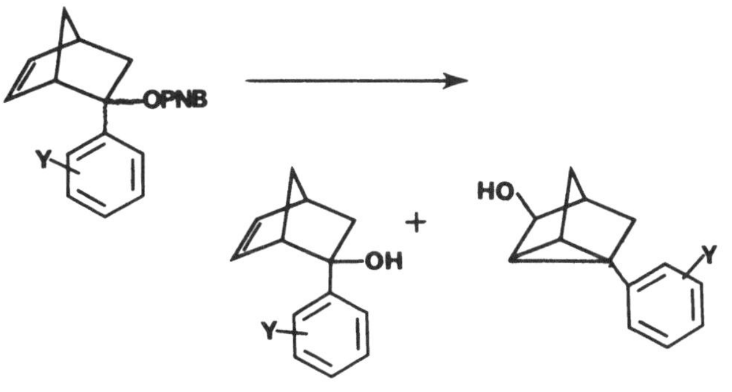

Y	**39**	**40**	(6)
p–CH₃O	> 99 %	—	
p–H	95 %	5 %	
p–CF₃	88 %	12 %	
m,m′–(CF₃)₂	23 %	77 %	

An alternative approach permits one to circumvent this argument. The inductive effect of the double bond in the endo isomer underline{35} is 44. In the absence of π-participation one would expect the

| **REL RATE** | 1.0 | 1.0/44.0 |

same rate-retarding effect of the double bond in the exo isomer underline{36}. However, the observed effect is 1.96. Therefore, π-participation must be increasing the rate of the exo isomer by a factor of approximately 22. Consequently, this approach yields the same conclusion: the exo:underline{endo} rate factor in 2-norbornenyl must be made up of a factor of approximately 20 due to π-participation and a factor of 350 attributable to steric hindrance to ionization.[36,37] This is exemplified in equation 7 which shows end-on views, for the endo chloride (for simplicity), of the initial material and of the tight ion-pair postulated to be the first intermediate in the solvolysis. It is evident that the olefinic moiety provides a U-shaped cavity which can hinder ionization and solvation of the departing anion.[38,39].

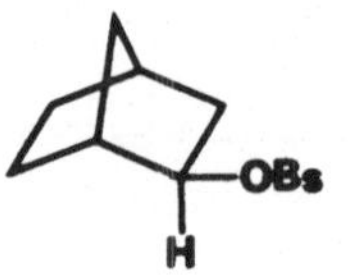

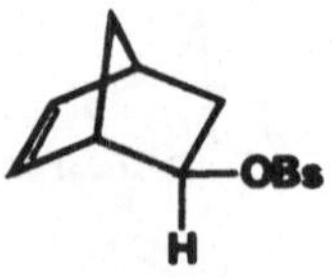

REL RATE   1.0   1.0/1.96

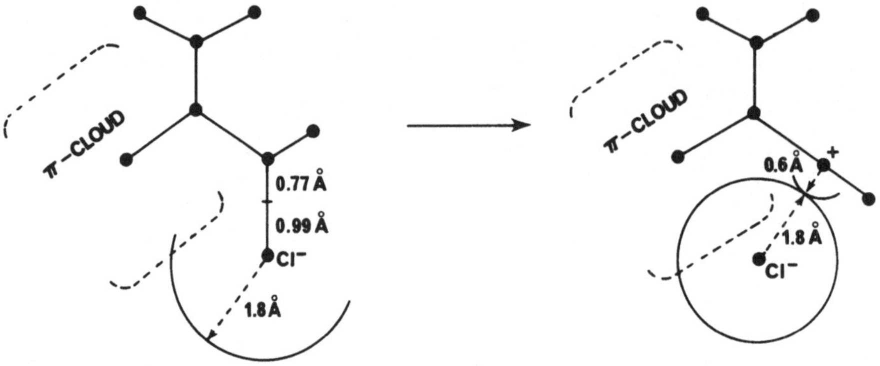

(7)

5-Methyl-2-norbornenyl. In a homoallylic cation the electron
deficiency should reside on C_1 as well as C_4 (eq 8). Groups on

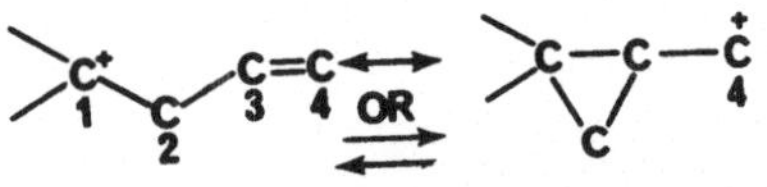

(8)

C_4 with the ability to stabilize carbonium ions should facilitate
the delocalization of positive charge into the homoallylic double
bond. Thus, the introduction of a methyl group onto C_4 greatly
increases the ability of the double bond to participate. Indeed,
in 1958, Sneen demonstrated that a methyl group on the homoallylic
double bond (C_6) in cholesteryl tosylate increases the rate of
solvolysis by a factor of 75.[40] Other workers have also noted
this effect of a methyl group.[41,42]

The effect of a 5-methyl substituent on 2-norbornenyl was
studied.[43] In the 2-aryl-5-methyl-2-norbornenyl system (41 and
42) there is increasing participation with increasing electron
demand at the cationic center. The effect of the 5-methyl sub-

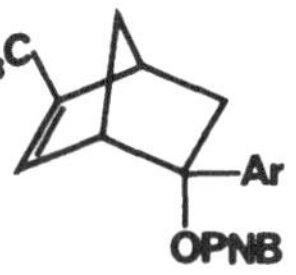

SUBSTITUENT IN Ar		REL RATE
p–CH$_3$O	1.0	354
p–H	1.0	1260
p–CF$_3$	1.0	6700

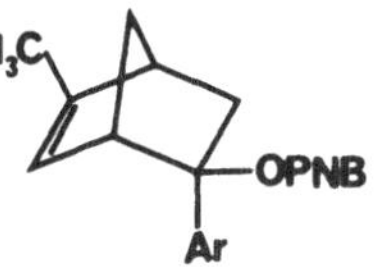

stituent is also indicated by the ρ^+ values.[44] The 2-aryl-5-methyl-endo-norbornenyl p-nitrobenzoates yield a ρ^+ value of -4.19 and the exo isomers yield one of -3.28. The ρ^+ values for the 2-aryl-exo- and -endo-norbornenyl derivatives are -4.21 and -4.17, respectively. The change in the ρ^+ value ($\Delta\rho^+$ = 0.93) for the exo isomers is in the direction anticipated for π-participation. The essential absence of any effect on ρ^+ in the endo isomer is again consistent with the proposed interpretation.

The products produced in the solvolysis of 41 and 42 also correspond to the effect anticipated for increasing participation of the double bond with increasing electron demand of the 2-substituent (eq 9). Only the p-CH$_3$O compound yields the exo alcohol (43) as the predominant product. The other derivatives involving

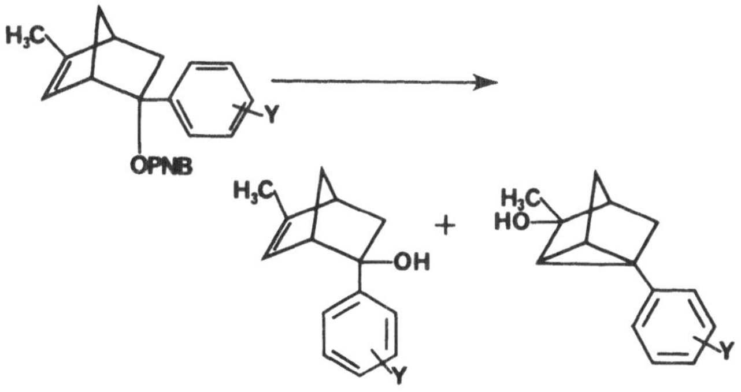

(9)

Y	43	44
p–CH$_3$O	75%	25%
p–H	18%	82%
p–CF$_3$	<2%	>98%

greater electron demand yield chiefly the rearranged product,
1-aryl-3-methyl-3-nortricyclanol (44), resulting from a homoallylic
rearrangement.

    Benzonorbornenyl. The benzonorbornenyl system is an excep-
tionally versatile one for structural studies. It has been subjected to extensive study by various groups.[45,46]

    The importance of π-participation in this versatile system
can be examined by introducing appropriate substituents into the
aromatic ring. Thus it is possible to introduce both activating
and deactivating substituents into the benzo moiety and then
observe the effect on the exo:endo rate ratio at the 2 position.[47,48] Another possibility is to introduce stabilizing substituents at the 2 position[39] and to observe the effect of
decreased electron demand at the reaction center on the rate
ratios.

    Thus, the exo:endo rate ratio in the solvolysis of the
2-aryl-benzonorbornen-2-yl p-nitrobenzoates (45 and 46) remains
sensibly constant at approximately 3000 as the electron demand of
the 2-aryl substituents is varied over a wide range in reactivity.
This observation indicates the absence of π-participation
as a significant factor in the high exo:endo rate ratio (approximately 3000) observed in these 2-arylbenzonorbornen-2-yl
derivatives. The endo derivatives yield a ρ^+ value of −4.51 and
the exo series yield one of −4.50, $\Delta\rho^+$ of 0.01. The data again
fail to reveal any differential electronic effect operating preferentially in the exo position.

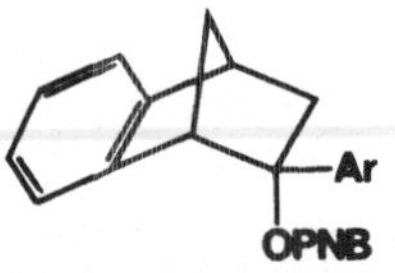

SUBSTITUENT IN Ar	45 REL RATE	46 REL RATE
p−CH$_3$O	1.0	3300
p−H	1.0	2900
p− CF$_3$	1.0	2700
m,m′−(CF$_3$)$_2$	1.0	2800

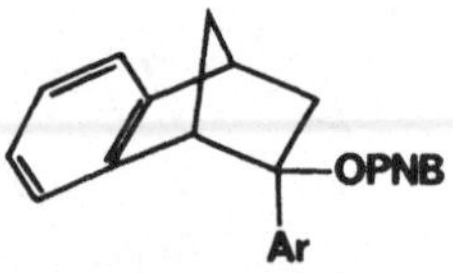

This difference in the exo:endo rate ratios can be nicely accounted for on steric grounds. The steric requirements of π electrons are considerable. In equation 10 is shown an end-on view, of the endo chlorides (for simplicity), of the initial material and the tight ion-pair postulated to be the first intermediate in the solvolysis. It is evident that the aromatic ring provides a more extreme U-shaped cavity, which can hinder ionization and solvation of the departing anion. The secondary benzonorbornenyl tosylate has an exo:endo ratio of 15,000.[45,49] Assuming a factor of 3000 is due to steric hindrance to ionization of the endo isomer, π-participation would contribute a factor of approximately 5 to the exo:endo rate ratio.

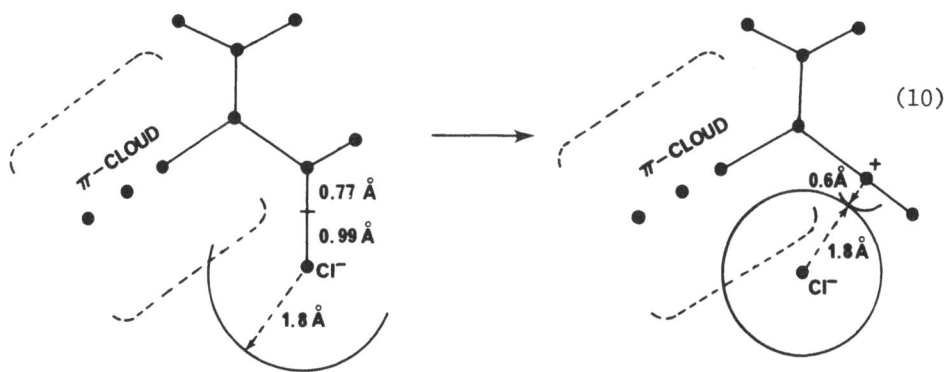

(10)

6-Methoxybenzonorbornenyl. The observation that a 6-methoxy group enhances the exo:endo rate ratio by a factor of 55 in the secondary benzo-2-norbornenyl derivative[48-51] is consistent with the incursion of π-participation arising from the presence of the activating 6-methoxy substituent.

SUBSTITUENT IN Ar	REL RATE	
	47	**48**
p-CH$_3$O	1.0	7000
p-H	1.0	14,500
p-CF$_3$	1.0	25,000
m,m-(CF$_3$)$_2$	1.0	34,000

The exo:endo rate ratio in the solvolysis of the 2-aryl-6-methoxybenzonorbornen-2-yl p-nitrobenzoates (47 and 48) increases with increasing electron demand of the 2-aryl substituents. This observation supports the presence of π-participation in the solvolysis of the 2-aryl-6-methoxy-exo-benzonorbornenyl derivatives.[52] The data for 47 provides a ρ^+ of −4.05 whereas 48 gives a value of −3.72. The $\Delta\rho^+$ of 0.33 is indicative of the presence of a small amount of participation.

σ- AND πσ-CONJUGATION

Highly strained sigma bonds can stabilize an adjacent cationic center. The effect of neighboring cyclopropyl and cyclobutyl groups has been investigated.[53] It is generally considered that such cationic species are stabilized during solvolysis by the overlap of the developing p orbital with the orbitals of the strained bonds (see 49).

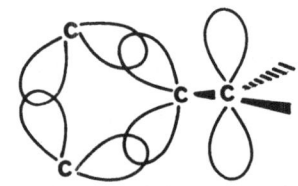

49

Cyclopropylcarbinyl. The fascinating behavior of cyclopropylcarbinyl derivatives in solvolytic processes and other carbonium ion transformations was explored in detail by J. D. Roberts and co-workers.[53,54] The cyclopropyl group is remarkably more effective than other alkyl groups in stabilizing carbonium ions. The mechanistic probe of increasing electron demand was applied to this system.

With increasing electron demand at the cationic center the rate of solvolysis of the cyclopropyl derivative (50), relative to that of a model system (21), increases enormously.[55] The introduction of methyl groups into the cyclopropyl ring (51) increases the rate of solvolysis because of the more effective charge delocalization.[56] On the other hand, chlorine substituents have the opposite effect. The rates of solvolysis of 52 are greatly decreased.[56] Clearly, the two chlorine atoms in 52 are quite effective in reducing charge delocalization into the cyclopropyl ring. The ρ^+ values further reveal the effect of substituents. This is exemplified by varying R in 53 from isopropyl (21), cyclopropyl (50), 2,2-dimethylcyclopropyl (51), to

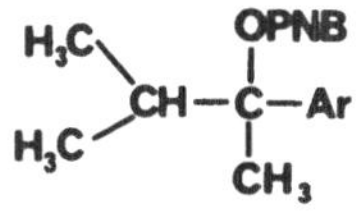

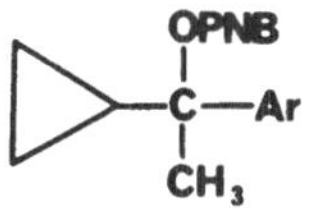

SUBSTITUENT IN Ar	REL RATE	
	21	50
p−CH₃O	1.0	505
p−H	1.0	25,300
p−CF₃	1.0	285,000
m,m′−(CF₃)₂	1.0	1,210,000

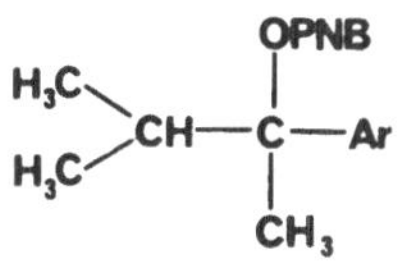

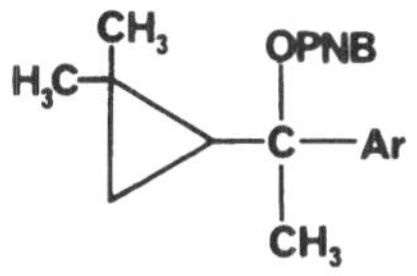

SUBSTITUENT IN Ar	REL RATE	
	21	51
p−CH₃O	1.0	1,700
p−H	1.0	314,000
p−CF₃	1.0	10,900,000
m,m′−(CF₃)₂	1.0	83,000,000

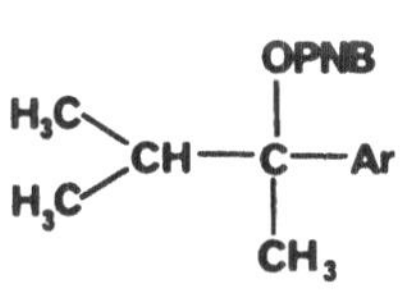

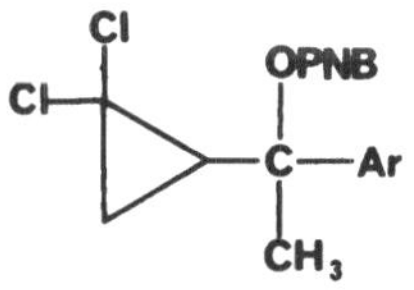

SUBSTITUENT IN Ar	REL RATE	
	21	52
p−CH₃O	1.0	3.6
p−H	1.0	1.9
p−CF₃	1.0	2.0

2,2-dichlorocyclopropyl (52). The large $\Delta\rho^+$ for 50 and 51 reveal a huge stabilization by these cyclopropyl groups. It is noteworthy that ρ^+ for 50 is lower than the ρ^+ value of -3.23 for the benzyl system (22, i.e., 53 where R is phenyl). Thus, it appears that the cyclopropyl group in 50 stabilizes the cation more than a phenyl group would. Similar observations, that a cyclopropyl group stabilizes an adjacent cation more than a phenyl group, have been reported in the literature.[57-59]

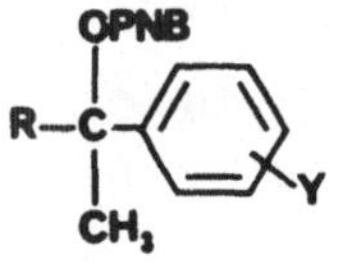

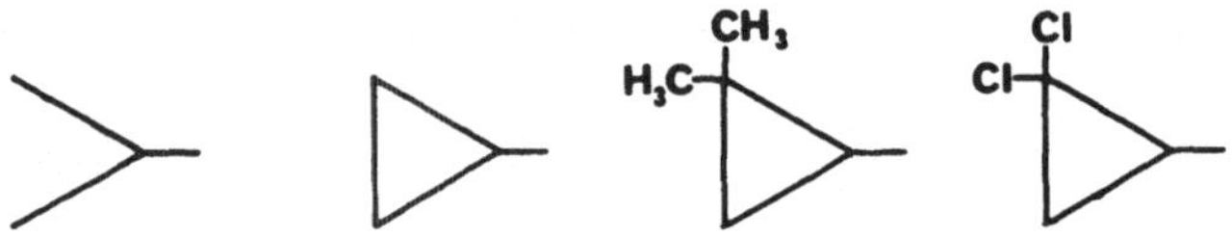

R	21	50	51	52
ρ^+	-4.76	-2.78	-2.06	-4.99
$\Delta\rho^+$	—	1.98	2.70	-0.23

Nortricyclyl. In investigations of the effect of structure on reactivity, the selection of a suitable model structure is crucial. The problem is clearly revealed by the 3-nortricyclyl system, which has been compared both with 2-norbornyl and 7-norbornyl, with opposite conclusions drawn. For example, a comparison of the relative rates for 54 and 55 led to the conclusion that the ion from 55, containing the cyclopropylcarbinyl moiety, does not exhibit stabilization, presumably because of steric difficulties inhibiting stabilization.[54] On the other hand,

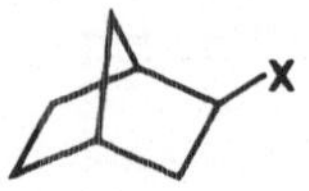

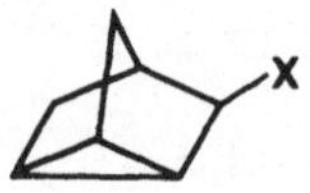

	54	55
REL RATE	1.0	0.07

comparison of 56 and 57 reveals a major contribution of the cyclopropylcarbinyl moiety.[60] The fact is that 55 and 57 are identical.
Yet opposite conclusions were reached as to the effectiveness of
the cyclopropane ring in stabilizing the carbonium ion center
through the particular selection of the model system.

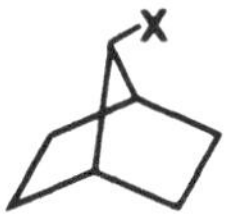

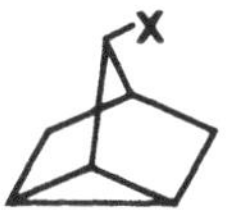

	56	57
REL RATE	1.0	$> 10^8$

The effect of increasing electron demand at the carbonium
ion center offered a method for reducing such ambiguities.[60]
Irrespective of whether we compare 3-aryl-3-nortricyclyl (58) with
7-aryl-7-norbornyl (26) or with 2-aryl-2-norbornyl (59), the ρ^+
values clearly reveal resonance contributions from the cyclopropylcarbinyl moiety. Consequently, this approach circumvents the

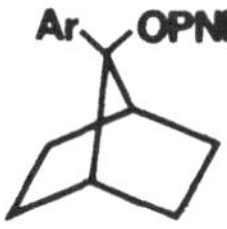

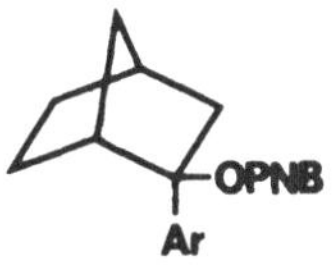

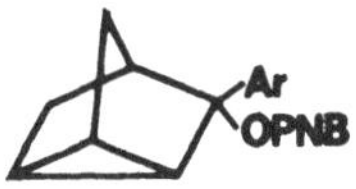

	26	59	58
ρ^+	-5.27	-3.83	-3.27

ambiguities involved in a simple comparison of reaction rates.[62]

From a consideration of the molecular geometries, the 7-
norbornyl system would appear to be a better model for comparison with 3-nortricyclyl. Bond-angle strain should be an important
factor in the rates of solvolysis. Thus, nortricyclene has a
C_2-C_3-C_4 bond angle of 97°,[63] in good agreement with the 96° value
for the C_1-C_7-C_4 angle of norbornane,[64] but considerably smaller
than the 104° value for the C_1-C_2-C_3 angle of norboranane.[64]
Comparison of the rates of solvolysis at 25° for the 3-aryl-3-
nortricyclyl p-nitrobenzoates (58) with those for the corresponding 7-aryl-7-norbornyl derivatives (26) reveals major rate
enhancements with increasing electron demand.

At one time, the high reactivity of cyclopropylcarbinyl

SUBSTITUENT IN Ar	REL RATE	
	26	**58**
p—CH₃O	1.0	7000
p—H	1.0	25,000
p—CF₃	1.0	3,010,000
m,m'—(CF₃)₂	1.0	10,400,000

derivatives in carbonium ion reactions was attributed to the formation of a stabilized species involving a σ bridge from the carbonium carbon and one or both of the far carbon atoms of the ring.[53,54] This would require the intermediate to exist in the parallel conformation to permit the p orbital of the carbonium carbon to be directed toward the ring (60). On the other hand,

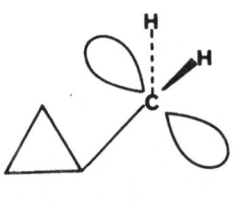

60

the rigid geometry of the 3-nortricyclyl cation forces it to exist in the bisected arrangement. It was believed that this would greatly diminish the stabilization provided by such inter-action of the carbonium carbon with the cyclopropane ring.[54] However, it should be noted that $\Delta\rho^+$ for 21 and 50 (1.98) is very similar to $\Delta\rho^+$ for 26 and 58 (2.00).[65] This indicates that the electronic contributions from the cyclopropane moiety are very similar in the two systems, in spite of the vast difference in geometries and rigidities.[66] These results confirm the con-clusion that the stabilization of cyclopropylcarbinyl is not the result of σ-bridging.[65,67]

p-Cyclopropylphenylcarbinyl. The tool of increasing electron demand was applied to the rates of solvolysis of the 1-(p-cyclo-propylphenyl)-1-arylethyl (61) and 1-(p-isopropylphenyl)-1-aryl-ethyl chlorides (62).[68]

The diarylmethylcarbonium ion is a highly stable one. Con-sequently, the cationic center should make little demand on the p-isopropyl and p-cyclopropyl substituent for further stabilization. The rates of solvolysis of the tertiary chlorides were determined in 97.5 percent acetone-water. The results reveal that, with increasing electron demand at the cationic center, the rate of solvolysis of the cyclopropyl derivative (61) increases moderately when compared to the isopropyl derivative (62). The ρ^+ values of

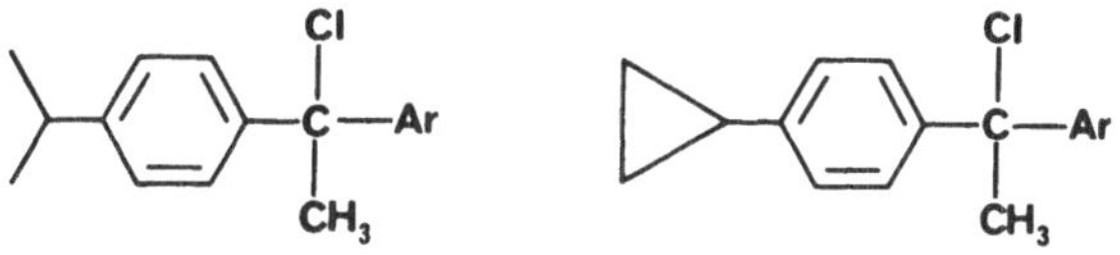

SUBSTITUENT IN Ar		REL RATE
p-CH$_3$O	1.0	1.1
p-H	1.0	2.8
p-CF$_3$	1.0	8.5
m,m'-(CF$_3$)$_2$	1.0	13.5

62 and 61 are -2.91 and -2.24, respectively. Their $\Delta\rho^+$ value of 0.67 is indicative of a small amount of stabilization in 61.

Cyclobutylcarbinyl. The cyclobutylcarbinyl system offers a way to demonstrate that the approach of varying the electron demand can be used to detect small amounts of stabilization by strained σ bonds. For example, methylcyclobutylcarbinyl brosylate (63) undergoes acetolysis 86 times faster than a model system, 64.[69] This rate enhancement is attributable to stabilization of the developing carbonium ion by the strained σ bonds in the adjacent cyclobutyl group. It was observed that with increasing electron

	64	63
REL RATE	1.0	85.5

demand at the cationic center the rate of solvolysis of the aryl-methylcyclobutylcarbinyl derivatives (65) increases slightly, as compared to that of the arylmethylisopropylcarbinyl derivatives (21).[70] The model system, 21, has a ρ^+ of -4.65, while 65 has a

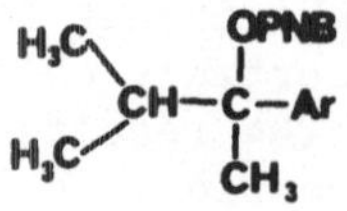

SUBSTITUENT IN Ar	REL RATE	
	21	**65**
p–CH$_3$O	1.0	1.0
p–H	1.0	5.7
p–CF$_3$	1.0	9.0
m,m'–(CF$_3$)$_2$	1.0	17.6

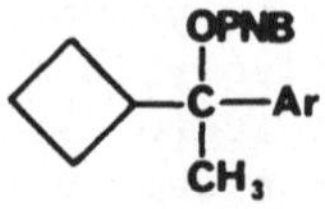

ρ^+ of –3.94. The change in ρ^+ ($\Delta\rho^+$ = 0.71) is in the direction anticipated for a small amount of neighboring group stabilization. Clearly the mechanistic probe of increasing the electron demand can detect both large and small amounts of neighboring group stabilization by strained carbon–carbon bonds, as found in the cyclopropylcarbinyl and cyclobutylcarbinyl systems.

The products of solvolysis of 65 were determined in buffered aqueous acetone (equation 11). The predominant products are those with no skeleton rearrangement (66, 67, and 68). The derivatives with increased electron demand [p–CF$_3$ and m,m'–(CF$_3$)$_2$] give a small amount of rearranged product, 69. These results

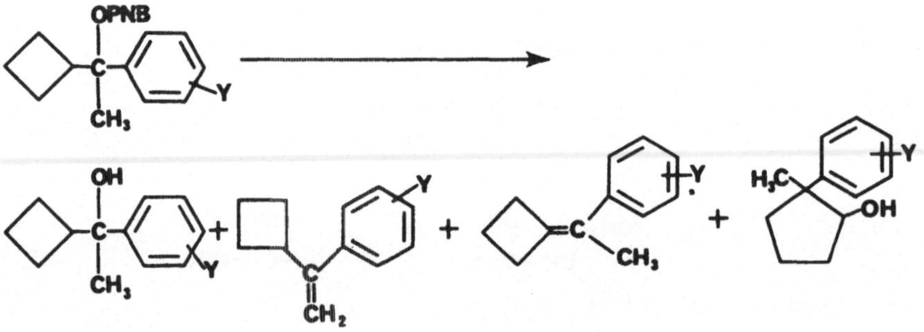

	66	**67**	**68**	**69**
p–CH$_3$O	75%	10%	15%	0%
p–H	63%	23%	14%	0%
p–CF$_3$	50%	30%	15%	5%
m,m'–(CF$_3$)$_2$	37%	40%	12%	11%

are consistent with the kinetic arguments. The fact that the
phenyl derivative reacts with a rate enhancement of almost six
but gives no rearranged product suggests that the formation of
<u>69</u> in the <u>p</u>-trifluoromethylphenyl and <u>m,m</u>'-bis(trifluoromethyl)-
phenyl derivatives occurs after the rate-determining step.

πσ-PARTICIPATION

Remote strained sigma bonds can also stabilize a cationic
center through participation. For example, Coates and Kirkpatrick
have noted that the solvolysis of 9-pentacyclo[$4.3.0.0^{2,4}.0^{3,8}.0^{5,7}$]-
nonyl <u>p</u>-nitrobenzoate (<u>70</u>) is 10^{10} to 10^{12} times faster than
7-norbornyl (<u>71</u>).[71] The rates of solvolysis of 9-aryl-9-penta-

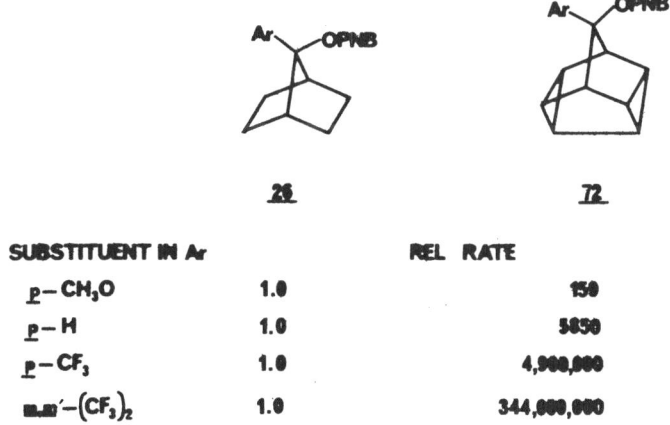

	<u>71</u>	<u>70</u>
REL RATE	**1.0**	$10^{10}-10^{12}$

cyclo[$4.3.0.0^{2,4}.0^{3,8}.0^{5,7}$]nonyl <u>p</u>-nitrobenzoates (<u>72</u>) reveal
major πσ-participation with increasing electron demand at the
cationic center.[72] This major amount of πσ-participation is

	<u>26</u>	<u>72</u>

SUBSTITUENT IN Ar		**REL RATE**
<u>p</u>-CH$_3$O	1.0	150
<u>p</u>-H	1.0	5850
<u>p</u>-CF$_3$	1.0	4,900,000
<u>m,m</u>'-(CF$_3$)$_2$	1.0	344,000,000

also indicated by the ρ^+ values; <u>26</u> has a value of −5.27 and <u>72</u>
a value of −2.05 (Δρ^+ of 3.22).

σ-PARTICIPATION/BRIDGING/ETC.

Several bicylic derivatives have been reported to react with unusually fast rates of solvolysis. It has been suggested that these fast rates were due to neighboring group stabilization by carbon-carbon sigma bonds. This stabilization has been termed σ-participation, σ-bridging, hyperconjugation, and vertical stabilization, but the significance of such stabilization has been questioned. The mechanistic probe of increasing electron demand has been applied to some of these systems.

2-Norbornyl. The high exo:endo rate ratio exhibited in the acetolysis of the 2-norbornyl tosylates (73, 74) has long intrigued chemists.[73] It was originally proposed that the faster

OTs

73   74

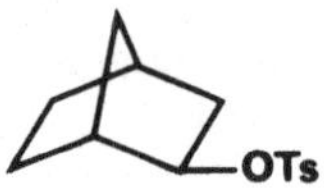

REL RATE   1.0   280

rate of the exo isomer (75) was the result of σ-participation by the 1,6-bonding pair, leading to a stabilized σ-bridged cation (76).[74,75] Such participation is stereoelectronically

⁻OTs

75    76

unfavorable in the endo isomer. (77).

OTs

77

Gassman and co-workers[21] elegantly demonstrated the remarkable stabilizing effect of the p-anisyl group, which causes neighboring group participation to vanish. They observed that the rate enhancement of 10^{11} due to π-participation in the secondary 7-norbornenyl derivative was essentially leveled with the introduction of the p-anisyl group at the 7 position. Hence, a carbonium ion stabilized by a p-anisyl group should make relatively little demand on neighboring groups for further stabilization. However, solvolysis of 2-p-anisylnorbornyl derivatives reveals an exo:endo rate ratio of 284.[76] This compares with a value of 280 for the secondary tosylates.

The gap between the highly stabilized 2-anisyl-2-norbornyl and the corresponding secondary system was bridged by use of the mechanistic probe of increasing electron demand.[77] As the aromatic ring is deactivated (thereby increasing the electron demand at the carbonium ion center) we should observe increasing σ-participation and, provided such σ-participation is a significant factor, increasing exo:endo rate ratios. The data reveal, however, an essentially constant exo:endo rate ratio in the 2-aryl-2-norbornyl systems (78 and 79). Clearly σ-participation

78		79

SUBSTITUENT IN Ar		REL RATE
p- CH₃O	1.0	284
p- CH₃	1.0	232
p- H	1.0	127
p- CF₃	1.0	187
m,m'-(CF₃)₂	1.0	176

cannot be a significant factor in the high exo:endo rate ratios in these derivatives. The exo derivatives had a ρ^+ of −3.82 and the endo derivatives one of −3.72. The small difference ($\Delta\rho^+$ of −0.10) is in the direction opposite that anticipated for participation.

A great majority of the reactions of the norbornyl system exhibit high exo:endo ratios. These high ratios are presumably the result of decreased rates of reaction in the sterically hindered endo portion of the U-shaped norbornane structure. Consequently, it appeared appropriate to consider that the high

exo:endo rate ratio in the solvolysis of the tertiary 2-aryl-2-
norbornyl and the parent secondary 2-norbornyl derivatives may
actually be the result of a normal exo rate combined with a very
slow endo rate.[78] Consider the endo-norbornyl chloride structure.
In the ionization process, the chlorine substituent would be
expected to move along a curved path away from the carbon atom
at the 2-position, maintaining the chlorine substituent perpendi-
cular to the face of the developing carbonium ion so as to retain
maximum overlap of the orbitals undergoing separation. In this
way the system should pass through the transition state to the
first intermediate, the idealized ion pair shown in equation 12.

(11)

Clearly, there would be a major steric overlap of the chlorine
substituent with the endo-6-hydrogen. Moreover, the group undergoing ionization should be strongly solvated by the medium, yet
the U-shaped structure obviously makes solvation of the developing anion difficult. An alternative model for ionization has
been suggested.[73] In this model the departing group would move
initially along the direction of the C-Cl bond, leading to the
first intermediate, the idealized ion pair shown in 12. This path
does not avoid the steric difficulty, although in this model it
is transferred largely to the hydrogen atom or other group at the
2-position.

     The large steric interactions of both models will presumably
cause some other path, providing decreased steric interactions
at the cost of poorer overlap, to be selected as a compromise.
Such a compromise would still result in an increase in the energy
of the transition state as compared to that for a derivative
without this particular structural feature.

     (2-Norbornyl)carbinyl. The high exo:endo rate ratios in
the solvolysis of 2-norbornyl derivatives could be the result of
an electronic contribution (hyperconjugation and vertical
stabilization) which stabilizes the exo transition state more than
the endo.[78] This stereoelectronic approach differs from the
older nonclassical ion proposal in that σ-participation and

σ-bridging would not be involved in the transition state and major distortion of the structure is not essential for the operation of the electronic contribution facilitating ionization of the exo isomer.[79] For example, Jensen and Smart observed that the benzoylation of 2-phenylnorbornanes is somewhat faster for the exo isomer (81) than for the endo (80).[80] They suggested that the

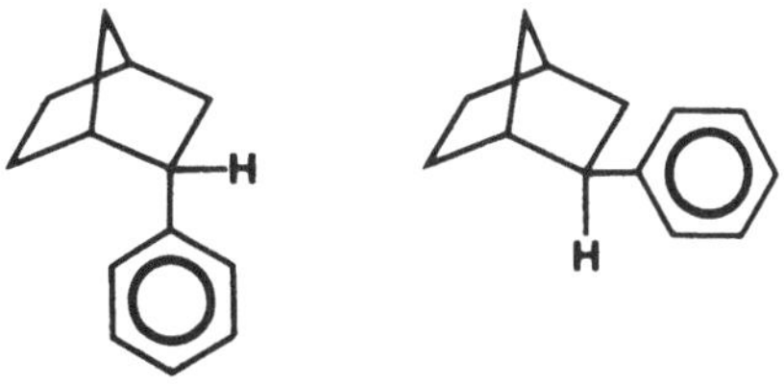

REL RATE   **80**   **81**
                 1.00   1.57

1,6-bonding pair is in ideal position to stabilize the electron deficiency in the σ-complex of the benzoylation reaction, without σ-bridging being involved.[80] The interpretation is similar to that later advanced by Traylor and co-workers.[79] According to this stereoelectronic interpretation, "vertical stabilization" or conjugation involving the 1,6-bonding pair is more favorable in the exo derivative than in the endo. No major distortion of the structure, such as is involved in the σ-bridging interpretation, is required.

The relative abilities of various groups to contribute to electron deficient centers was tested in the tert-cumyl system and a small difference between exo- and endo-norbornyl (82, 83) was observed.[81] The effects are all very small. It has been

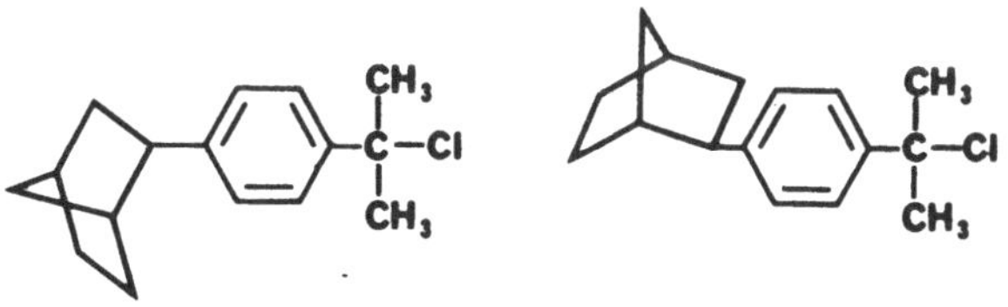

REL RATE   **82**   **83**
                 1.00   1.15

argued that the amount of positive charge delocalized to the para position of the tert-cumyl system is relatively small. Such a small deficiency can make but a small demand on the alkyl group for electronic contributions.

In the case of other groups, the observed effect is significant. Thus, the cyclopropyl group in the <u>tert</u>-cumyl system (<u>86</u>) is unambiguously better than a simple alkyl group in providing electronic stabilization.[82] By placing the developing

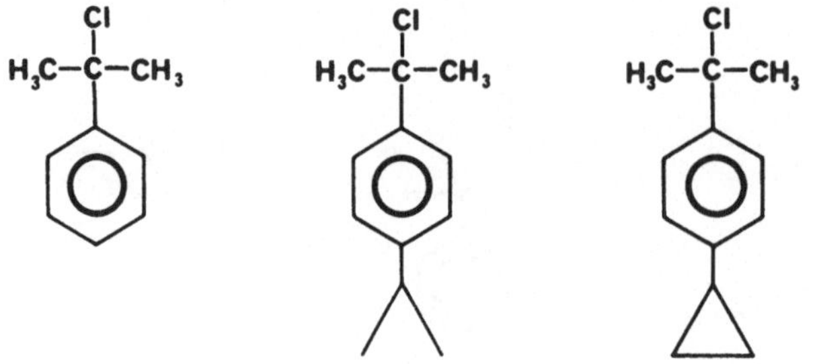

	84	**85**	**86**
REL RATE	1.0	18.0	154

charge α to the group, the electronic demand and the observed effect should be much larger. For example, cyclopropyldimethyl-carbinyl <u>p</u>-nitrobenzoate (<u>87</u>) solvolyzes enormously faster than the corresponding isopropyl derivative (<u>88</u>).[57] It appeared that

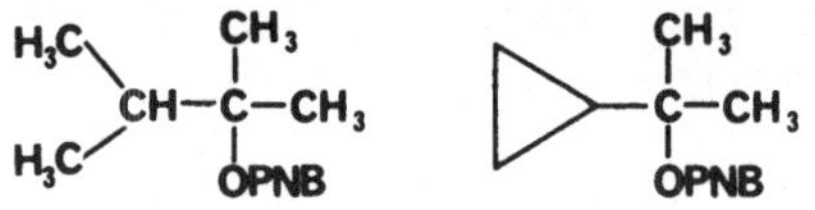

	88	**87**
REL RATE	1.0	170,000

application of the mechanistic probe of increasing electron demand to the aryl(2-norbornyl)methylcarbinyl <u>p</u>-nitrobenzoates (<u>89</u>, <u>90</u>) should provide a reasonable test for these stereoelectronic proposals.[83] Clearly the rates of solvolysis fail to support the proposal that there must be significant differences in the electronic contributions of the two isomeric groups. The <u>exo</u> derivatives yield a value of ρ^+ of -4.44, almost identical with the value of ρ^+ of -4.47 observed for the <u>endo</u> derivatives. It is concluded that the application of the technique of increasing electron demand to these systems fails to reveal any significant electronic factor in the <u>exo</u> isomers (not also present in the corresponding <u>endo</u> derivatives) facilitating their ionization.

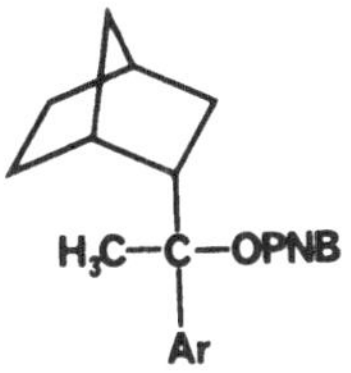

SUBSTITUENT IN Ar	REL RATE	
	89	**90**
$\underline{p}$ – CH_3O	1.0	0.16
$\underline{p}$ – H	1.0	0.33
$\underline{p}$ – CF_3	1.0	0.22
$\underline{m,m'}$ – $(CF_3)_2$	1.0	0.19

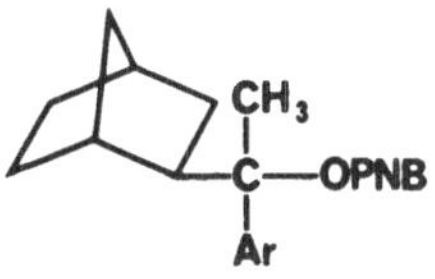

Camphenilyl. The solvolysis of camphenilyl derivatives possess historical implications for the development of σ-partici-pation and the nonclassical ion theory. The exo:endo rate ratios in the 2-aryl-2-camphenilyl p-nitrobenzoates (91, 92) are much higher than those for 2-aryl-2-norbornyl derivatives, but there is no observable increase in exo:endo rate ratios as the electron-withdrawing substituents are introduced into the aromatic ring, over the range examined.[84]   The actual rate constants reveal

SUBSTITUENT IN Ar	REL RATE	
	91	**92**
$\underline{p}$ –CH_3O	1.0	44,000
$\underline{p}$ –H	1.0	49,000
$\underline{p}$ –CF_3	1.0	24,000

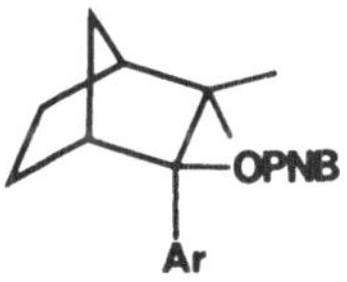

that the 2-aryl-exo-camphenilyl derivatives (92) undergo solvoly-sis at a rate quite comparable with the corresponding 2-exo-norbornyl derivatives (79). The greatly increased exo:endo rate ratios in this system arise as a consequence of a sharp decrease in rate of 2-aryl-endo-camphenilyl derivatives (91). Consequently, as discussed earlier, the slow rate of endo isomers (91) could be accounted for in terms of more major steric difficulties in the solvation of the incipient anion and its departure than are present in the parent 2-norbornyl system. The value of ρ^+ in the solvolysis of 92 is -3.65. This compares with a ρ^+ value of -3.47 for 91. The similar values of ρ^+ confirm the conclusion that σ-participation is not an appreciable factor in the solvolysis of these derivatives.

CONCLUSIONS

The mechanistic probe of increasing electron demand is a versatile tool for the physical organic chemist and can be used to elucidate structure-property relationships. The ρ^+ values provide a convenient measure of reactivity change as a function of structural change. Thus, an increase in strain energy results in a more negative ρ^+ value. On the other hand, an increase in steric bulk results in a less negative value.

In the study of neighboring group effects, this method has found great utility. The available data reveal that ρ^+ values are very sensitive to neighboring group stabilization. Such stabilization can be classified by the $\Delta\rho^+$ found by comparison with a suitable model system. For example, the large amounts of stabilization found in cyclopropylcarbinyl, allylic, and benzylic systems, the π-participation found in 7-norbornenyl and the σ-participation found in 9-pentacyclo[4.3.0.0^{2,4}.0^{3,8}.0^{5,7}]nonyl have $\Delta\rho^+$ between 1.4 and 3.22. Systems with small levels of stabilization, such as the 1-(p-cyclopropylphenyl)-1-arylethyl, cyclobutylcarbinyl, 6-methoxybenzonorbornenyl, and the 5-methyl-2-norbornenyl systems are characterized by $\Delta\rho^+$ of 0.3-0.9. Systems which undergo solvolysis with no significant neighboring group stabilization as in tertiary benzonorbornenyl, 2-norbornyl, 2-norbornenyl, and Δ^3-cyclopentenyl systems have no difference in ρ^+ values ($\Delta\rho^+$, 0.01 to -0.08).

Finally, the mechanistic probe of increasing the electron demand has been used in the study of cations in strong acids and extended to correlate secondary derivatives with their analogous tertiary benzylic derivatives.[85,86] Unfortunately, space limitations do not permit discussion of these applications here.

REFERENCES AND NOTES

1.   (a) P. R. Wells, "Linear Free Energy Relationships," Academic Press, New York, N.Y., 1968; (b) J. E. Leffler and E. Grunwald, "Rates and Equilibria of Organic Reactions," Wiley, New York, N.Y., 1963; (c) M. Charton, CHEMTECH, 502 (Aug 1974); 245 (April 1975); (d) C. D. Ritchie and W. F. Sage, *Prog. Phys. Org. Chem.*, 2, 323 (1964).

2.   (a) C. D. Johnson, "The Hammett Equation," Cambridge University Press, London, 1973; (b) H. H. Jaffé, *Chem. Rev.*, 53, 191 (1953).

3.   (a) L. M. Stock and H. C. Brown, *Adv. Phys. Org. Chem.*, 1, 35 (1963); (b) H. C. Brown and L. M. Stock, *J. Am. Chem. Soc.*, 84, 3298 (1962); (c) Y. Okamato and H. C. Brown, *J. Org. Chem.*, 22, 485 (1957).

4.   H. C. Brown, M. Ravindranathan, E. N. Peters, C. Gundu Rao, and M. M. Rho, *J. Am. Chem. Soc.*, 99, 5373 (1977).

5.   H. Tanida and H. Matsumura, *J. Am. Chem. Soc.*, 95, 1586 (1973).

6.   Calculated at 25°; See E. N. Peters, *J. Org. Chem.*, 42, 1419 (1977).

7.   H. C. Brown and M. Borkowski, *J. Am. Chem. Soc.*, 74, 1896 (1952).

8.   H. C. Brown and G. Ham, *J. Am. Chem. Soc.*, 78, 2735 (1956).

9.   H. C. Brown and K. Ichikawa, *Tetrahedron*, 1, 221 (1957).

10. H. C. Brown and R. S. Fletcher, and R. B. Johannesen, *J. Am. Chem. Soc.*, 73, 212 (1951).

11. J. D. Roberts and V. C. Chambers, *J. Am. Chem. Soc.*, 73, 3176 (1951).

12. J. D. Roberts and V. C. Chambers, *J. Am. Chem. Soc.*, 73, 5034 (1951).

13. S. Winstein, B. K. Morse, E. Grunwald, H. W. Jones, J. Corse, D. Trifan, and H. Marshall, *J. Am. Chem. Soc.*, 74, 1127 (1952).

14. C. H. DePuy, L. G. Schnack, and J. W. Hausser, *J. Am. Chem. Soc.*, 88, 3344 (1966).

15. H. C. Brown, C. Gundu Rao, and M. Ravindranathan, *J. Am. Chem. Soc.*, 99, 7663 (1977).

16. G. S. Hammond, *J. Am. Chem. Soc.*, 77, 334 (1955).

17. H. Tanida and T. Tsushima, *J. Am. Chem. Soc.*, 92, 3397 (1970).

18. For example see A. Streitwieser, Jr., "Solvolytic Displacement Reactions," McGraw-Hill, New York, 1962.

19. S. Winstein, *Chimica Teorica*, VIII, 239 (1965).

20. S. Winstein, B. K. Morse, E. Grunwald, K. C. Schreiber and J. Corse, *J. Am. Chem. Soc.*, 74, 1113 (1952).

21. P. G. Gassman and A. F. Fentiman, Jr., *J. Am. Chem. Soc.*, 92, 2549 (1970).

22. H. C. Brown, M. Ravindranathan, and M. M. Rho, *J. Am. Chem. Soc.*, 98, 4216 (1976).

23. H. C. Brown, M. Ravindranathan, and C. Gundu Rao, *J. Am. Chem. Soc.*, 100, 1218 (1978).

24. H. C. Brown, M. Ravindranathan, and E. N. Peters, *J. Org. Chem.*, 42, 1073 (1977).

25. M. Simonetta and S. Winstein, *J. Am. Chem. Soc.*, 76, 18 (1954).

26. S. Winstein, M. Shatavsky, C. Norton, and R. B. Woodward, *J. Am. Chem. Soc.*, 74, 1113 (1952).

27.   P. D. Bartlett and M. R. Rice, *J. Org. Chem.*, 28, 3351 (1963).

28.   B. A. Hess, Jr., *J. Am. Chem. Soc.*, 93, 1000 (1971).

29.   E. N. Peters and H. C. Brown, *J. Am. Chem. Soc.*, 97, 7454 (1975).

30.   J. B. Lambert and S. I. Featherman, *J. Am. Chem. Soc.*, 99, 1542 (1977).

31.   S. Winstein, and E. Grunwald, *J. Am. Chem. Soc.*, 70, 828 (1948).

32.   S. Winstein, C. R. Lindegren, H. Marshall, and L. L. Ingraham, *J. Am. Chem. Soc.*, 75, 147 (1953).

33.   S. Winstein, H. Walborsky, and K. C. Schreiber, *J. Am. Chem. Soc.*, 72, 5795 (1950).

34.   H. C. Brown, "Boranes in Organic Chemistry," Cornell University Press, Ithaca, N.Y., 1973.

35.   E. N. Peters and H. C. Brown, *J. Am. Chem. Soc.*, 95, 2398 (1973).

36.   H. C. Brown and E. N. Peters, *J. Am. Chem. Soc.*, 97, 7442 (1975).

37.   E. N. Peters and H. C. Brown, *J. Am. Chem. Soc.*, 94, 5899 (1972).

38.   E. N. Peters and H. C. Brown, *J. Am. Chem. Soc.*, 94, 7920 (1972).

39.   H. C. Brown, S. Ikegami, K.-T. Liu, and G. L. Tritle, *J. Am. Chem. Soc.*, 98, 2531 (1976).

40.   R. A. Sneen, *J. Am. Chem. Soc.*, 80, 3982 (1958).

41.   H. C. Brown, E. N. Peters, and M. Ravindranathan, *J. Am. Chem. Soc.*, 97, 2900 (1975).

42.   (a) P. G. Gassman and D. S. Patton, *J. Am. Chem. Soc.*, 91, 2160 (1969); (b) P. D. Bartlett and G. D. Sargent, *J. Am. Chem. Soc.*, 87, 1297 (1965).

43.   H. C. Brown, M. Ravindranathan and E. N. Peters, *J. Am. Chem. Soc.*, 96, 7351 (1974).

44.   H. C. Brown, E. N. Peters, and M. Ravindranathan, *J. Am. Chem. Soc.*, 97, 7449 (1975).

45.   P. D. Bartlett and W. P. Giddings, *J. Am. Chem. Soc.*, 82, 1240 (1960).

46.   W. P. Giddings and J. Dirlam, *J. Am. Chem. Soc.*, 85, 3900 (1963).

47.   H. Tanida, T. Tsuji, and S. Teratake, *J. Org. Chem.*, 32, 4121 (1967).

48.   H. Tanida, H. Ishitobi, and T. Irie, *J. Am. Chem. Soc.*, 90, 2688 (1968).

49.   D. V. Braddon, G. A. Wiley, J. Dirlam, and S. Winstein, *J. Am. Chem. Soc.*, 90, 1901 (1968).

50.   H. Tanida, T. Irie, and T. Tsushima, *J. Am. Chem. Soc.*, 92, 3404 (1970).

51.   H. C. Brown and G. L. Tritle, *J. Am. Chem. Soc.*, 90, 2689 (1968).

52.   H. C. Brown and K.-T. Liu, *J. Am. Chem. Soc.*, 91, 5909 (1969).

53. For example, see K. B. Wiberg, B. A. Hess, Jr., and A. J. Ashe, III, in "Carbonium Ions," Vol. III, G. A. Olah and P. V. R. Schleyer, Ed., Wiley, New York (1972) Chapter 26; and H. G. Richey, Jr. *ibid.*, Chapter 25.

54. J. D. Roberts and R. H. Mazur, *J. Am. Chem. Soc.*, 73, 2509, 3542 (1951).

55. E. N. Peters and H. C. Brown, *J. Am. Chem. Soc.*, 95, 2397 (1973).

56. H. C. Brown, E. N. Peters, and M. Ravindranathan, *J. Am. Chem. Soc.*, 99, 505 (1977).

57. H. C. Brown, E. N. Peters, *J. Am. Chem. Soc.*, 95, 2400 (1973).

58. N. C. Deno, H. G. Richey, Jr., J. S. Liu, D. N. Lincoln, and J. O. Turner, *J. Am. Chem. Soc.*, 87, 4533 (1965).

59. H. C. Brown and E. N. Peters, *J. Am. Chem. Soc.*, 99, 1712 (1977).

60. H. G. Richey, Jr., and N. C. Buckley, *J. Am. Chem. Soc.*, 85, 3057 (1963); P. R. Story and S. R. Fahrenholtz, *ibid.*, 86, 527 (1964).

61. H. C. Brown and E. N. Peters, *J. Am. Chem. Soc.*, 97, 1927 (1975).

62. For example 58 (Ar = Ph) undergoes solvolysis at a slower rate (1/21) than 2-phenyl-2-endo-norbornyl OPNB.

63. E. Heilbronner and V. Schomaker, *Helv. Chim. Acta*, 35, 1385 (1952).

64. J. F. Chiang, C. F. Wilcox, Jr., and S. H. Bauer, *J. Am. Chem. Soc.*, 90, 3149 (1968).

65. D. F. Eaton and T. G. Traylor, *J. Am. Chem. Soc.*, 96, 1226 (1974).

66. G. A. Olah and G. Liang, *J. Am. Chem. Soc.*, 95, 3792 (1973).

67. Although the NMR spectra of secondary and tertiary cyclopropylcarbinyl cations have been interpreted in terms of the classical bisected structures, data for the primary cyclopropyl cation have been interpreted in favor of a σ bridged species: G. A. Olah, C. L. Jeuell, D. P. Kelly, and R. D. Porter, *J. Am. Chem. Soc.*, 94, 146 (1972). On the other hand, W. J. Hehre and P. C. Hiberty, *ibid.*, 96, 302 (1974), have reported that their ab initio molecular orbital calculations favor equilibrating classical bisected cyclopropylcarbinyl cations.

68. H. C. Brown and M. Ravindranathan, *J. Am. Chem. Soc.*, 97, 2895 (1975).

69. S. Winstein and N. J. Holness, *J. Am. Chem. Soc.*, 77, 3054 (1955).

70. E. N. Peters, *J. Org. Chem.*, 42, 3015 (1977).

71. R. M. Coates and J. L. Kirkpatrick, *J. Am. Chem. Soc.*, 92, 4883 (1970).

72. H. C. Brown and M. Ravindranathan, *J. Am. Chem. Soc.*, 99, 297 (1977).

73.   P. v.R. Schleyer, M. M. Donaldson, and W. E. Watts, *J. Am. Chem. Soc.*, 87, 375 (1965).

74.   S. Winstein and D. Trifan, *J. Am. Chem. Soc.*, 74, 1147, 1154 (1952).

75.   P. D. Bartlett, "Nonclassical Ions," Benjamin, New York, 1965.

76.   H. C. Brown and K. Takeuchi, *J. Am. Chem. Soc.*, 90, 2691 (1968).

77.   H. C. Brown, M. Ravindranathan, K. Takeuchi, and E. N. Peters, *J. Am. Chem. Soc.*, 97, 2899 (1975).

78.   (a) H. C. Brown, "The Nonclassical Ion Problem," Plenum, New York, N.Y., (1977); (b) H. C. Brown and E. N. Peters, *Proc. Nat. Acad. Sci.*, 71, 132 (1974).

79.   T. G. Traylor, W. Hanstein, H. J. Berwin, N. A. Clinton, and R. S. Brown, *J. Am. Chem. Soc.*, 93, 5715 (1971).

80.   F. R. Jensen and B. E. Smart, *J. Am. Chem. Soc.*, 91, 5686, 5688 (1969).

81.   H. C. Brown, B. G. Gnedin, K. Takeuchi, and E. N. Peters, *J. Am. Chem. Soc.*, 97, 610 (1975).

82.   H. C. Brown and J. D. Cleveland, *J. Org. Chem.*, 41, 1792 (1976).

83.   H. C. Brown and M. Ravindranathan, *J. Am. Chem. Soc.*, 100, 1865 (1978).

84.   H. C. Brown, K. Takeuchi, and M. Ravindranathan, *J. Am. Chem. Soc.*, 99, 2684 (1977).

85.   H. G. Richey, Jr., J. D. Nichols, P. G. Gassman, A. F. Fentiman, Jr., S. Winstein, M. Brookhart, and R. K. Lustgarten, *J. Am. Chem. Soc.*, 92, 3783 (1970); D. G. Farnum and A. D. Wolf, *ibid.*, 96, 5166 (1974); and D. G. Farnum and R. E. Botto, *Tetrahedron Lett.*, 4013 (1975).

86.   E. N. Peters, *J. Am. Chem. Soc.*, 98, 5627 (1976).

CHARGE AND SPIN DELOCALIZATION TO THE TRIFLUOROMETHYL GROUP

Leon M. Stock and Michael R. Wasielewski

Department of Chemistry, The University of Chicago

Chicago, Illinois 60637

INTRODUCTION

Herbert Brown has a long-standing interest in the concepts of hyperconjugation and non-classical behavior as influences in carbonium ion chemistry. It seems appropriate to consider at this symposium the analogous problem of delocalization to saturated substituents such as the trifluoromethyl group in negatively charged ions and in radicals.

The chemical and spectroscopic properties of radicals, radical-anions, and anions have been studied with special interest following the suggestion of Roberts and his students that carbon-fluorine hyperconjugation was responsible for the seemingly enhanced electron-withdrawing properties of this substituent.[1]

The concept of carbon-fluorine hyperconjugation is related to the concepts of carbon-hydrogen and carbon-carbon hyperconjugation. In carbon-hydrogen hyperconjugation, the three hydrogen atom 1s orbitals are combined to form three orthogonal group orbitals. One of these orbitals has the proper symmetry for interaction with the p orbital at the electron-deficient carbon atom.[2,3] Structure I illustrates this model.

$$H_3 \equiv CCH_2^{+}$$

I

The introduction of an electronegative fluorine atom complicates the matter. Hoffmann and Rossi and their associates have proposed the orbital energy diagram shown in Figure 1 for the monofluoromethyl

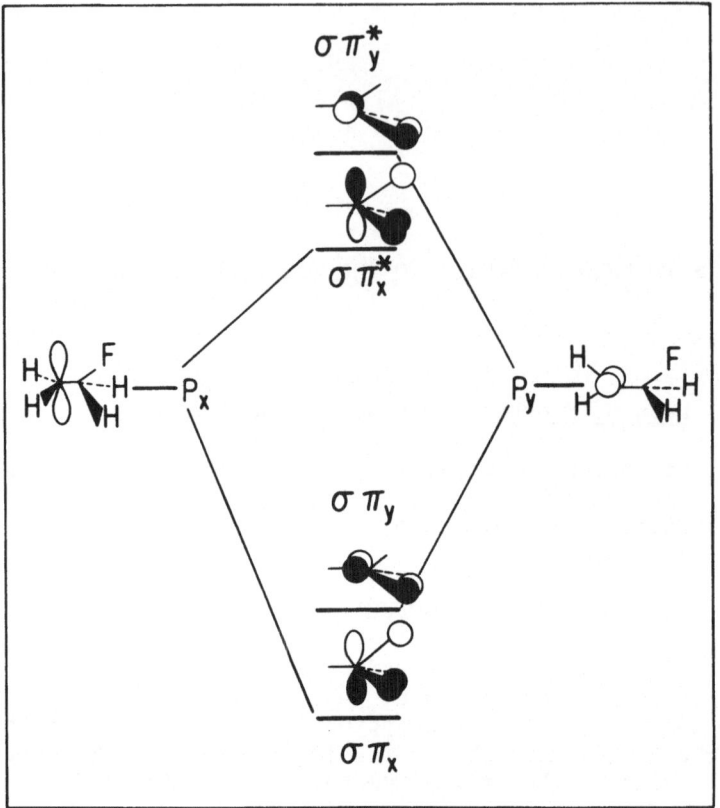

Figure 1

This diagram illustrates the possible orbital interactions between
the group orbitals of the fluoromethyl group and the carbon 2p orbital

group.[4,5] They point out that the orbital energies for $\sigma\pi_y$ and $\sigma\pi_y^*$
are, to a first approximation, unchanged for the monofluoromethyl
group but that $\sigma\pi_x$ and $\sigma\pi_x^*$ are reduced. This analysis is deficient
in the sense that the orbitals of the non-bonding electrons of the
fluorine atom are not included. However, the analysis does introduce
the key idea that the electronegative atom alters the energy levels
such that the interactions between the p orbital of the carbon atom
and the anti-bonding molecular orbitals of the fluoroalkyl group
become more probable.

The corresponding model for the trifluoromethyl group is considerably more complex. The twelve atomic orbitals of the fluorine
atoms combine to produce twelve group orbitals. The E orbitals
shown in Figure 2 have the proper symmetry for interaction with the

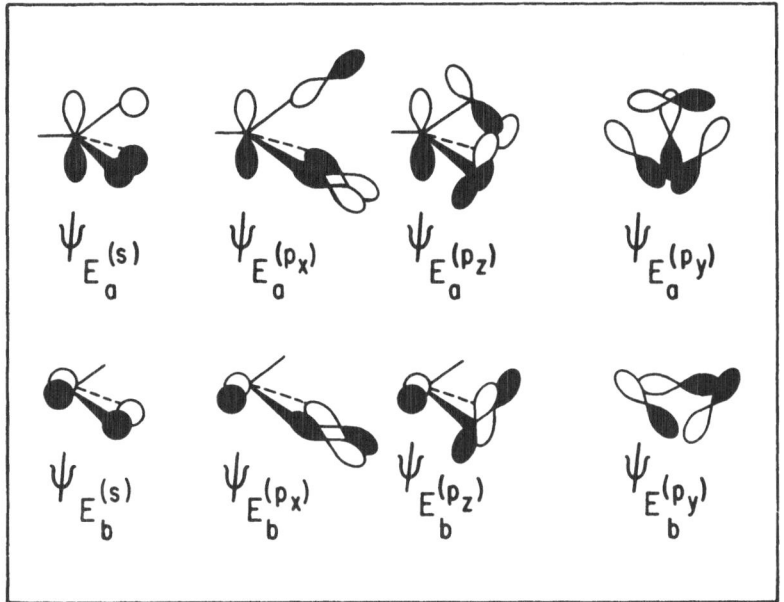

Figure 2

The group orbitals of the trifluoromethyl group

p_x or p_y orbitals of the other carbon atoms. The orbital energies vary enormously. It is especially pertinent that the energy content of the lowest unoccupied π* molecular orbital of the trifluoromethyl group in fluoroform is only modestly greater than that of a carbon 2p atomic orbital.[6],[7] Hence, it is not unreasonable to propose that electron delocalization to an unoccupied, antibonding molecular orbital influences the chemistry and spectroscopy of substances with trifluoromethyl groups.

Several investigators have proposed that 1,3 p-p interactions, II, influence the behavior of molecules with fluoroalkyl groups.

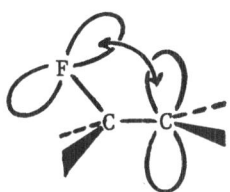

II

Such interactions are formally incorporated into the carbon-fluorine hyperconjugation model. Under certain circumstances, however, this interaction may be treated as a separate phenomenon.

REACTION CHEMISTRY

The situation for reaction chemistry seems reasonably clear. The rate and equilibrium constants for several reactions have been carefully studied. In addition, Liotta and his students have determined the thermodynamic parameters for the dissociation of phenol.[8] They showed that there were no unusual features in the dissociation constants or the enthalpy or entropy of reaction. Thus, the resulting σ_{p-CF_3} and σ_{m-CF_3} values calculated by Holtz[9] may be discussed with confidence.

Table I

Hammett Sigma Constants for the Trifluoromethyl Group

Reaction	Sigma Constants		
	σ_{m-CF_3}	σ_{p-CF_3}	$\sigma_{p-CF_3}/\sigma_{m-CF_3}$
A. Dissociation, Benzoic Acids	0.42	0.53	1.26
B. Dissociation, Anilines	0.47	0.63	1.34
C. Dissociation, Phenols	0.45	0.56	1.25
D. Exchange, Substituted $C_6H_5CH_3$	0.47	0.61	1.29
E. Exchange, Substituted $C_6H_5CH(CF_3)_2$	0.40	0.50	1.20
F. Substitution, Substituted $2-O_2NC_6H_4Cl$	0.44	0.63	1.43

A. In 50% ethanol-water at 25°, ref. 10. B. In water at 25°, ref. 10, 11. C. In water at 25°, ref. 8, 12. D. With lithium cyclohexylamide in cyclohexylamine at 50°, ref. 9, 13. E. With sodium methoxide in methanol-dimethylsulfoxide, ref. 14. F. With sodium thiophenolate in methanol at 35°, ref. 15.

Generally, the σ_p/σ_m ratio is between 1.2 and 1.3 for substituents for which resonance is impossible. Thus, the finding that this ratio ranges from 1.25 to 1.34 for the dissociation reactions suggests that there is nothing unusual about the trifluoromethyl group. The observations for the exchange reactions are similar.

$$CH_3OD + R_FC_6H_4CH(CF_3)_2 \rightarrow CH_3OH + R_FC_6H_4CD(CF_3)_2$$

However, the data for the substitution reaction suggest that the p-trifluoromethyl group may exhibit an exalted substituent effect. Other work by Miller[16] and Brieux[17] also indicates that somewhat larger values of σ_{p-CF_3} (0.63-0.74) are required to accommodate the data for reactions proceeding through very electron-rich transition states leading to benzenanions, III. Such results are compatible with the view that electron delocalization to the trifluoromethyl group is important.

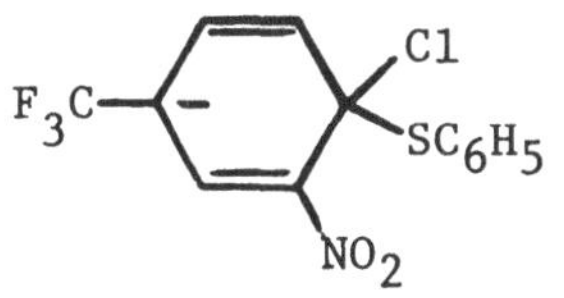

III

The influence of fluorine atoms on the rates of base-catalyzed exchange reactions of aliphatic molecules has been investigated by Andreades[18] and by Streitwieser and Holtz.[19] Unfortunately, many factors, including polar effects and hybridization changes, influence the rates of these exchange reactions. We infer that the information now available for aliphatic molecules is insufficient for a confident decision concerning the role of carbon-fluorine hyperconjugation. Nevertheless, it should be pointed out that the perfluorocyclobutane ylides, IV and V, exhibit remarkable stability[20] and it is tempting to ascribe this stability to the influence of carbon-fluorine hyperconjugation.

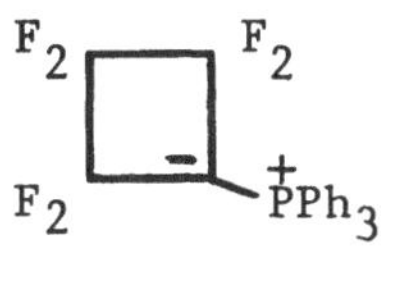

IV 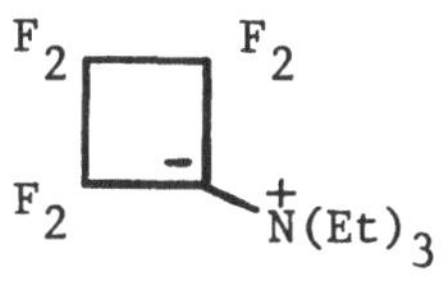   V

PHYSICAL AND SPECTROSCOPIC PROPERTIES

The physical and spectroscopic properties of molecules with trifluoromethyl groups have also been examined. Roberts and his students found that the dipole moment of N,N-dimethyl-4-trifluoro-methylaniline was enhanced.[1] They attributed this result to hyper-conjugation as illustrated by VIC. Holtz and Sheppard pointed out that the moment of N,N-dimethyl-4-perfluoroisopropylaniline was also enhanced, even though the opportunities for carbon-fluorine hyper-conjugation were considerably reduced. Ibbitson and his associates resolved this issue by a study of the *meta* derivatives.[22] They

VIA   VIB   VIC

found that the moments of *meta* derivatives, for example VII, were also enhanced. Their work strongly suggests that a pi inductive interaction rather than carbon-fluorine hyperconjugation is respon-sible for the enhanced moments of both the *meta* and *para* isomers.

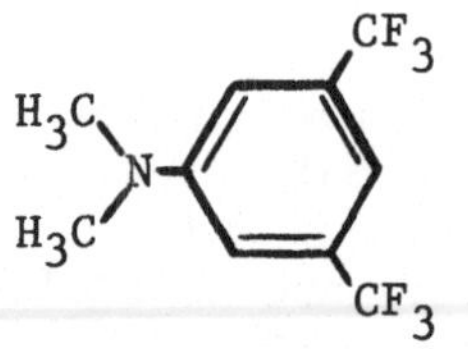

VII

Nuclear magnetic resonance, infrared, and ESCA spectroscopy have all been used to study the effectiveness of the trifluoromethyl group as an electron acceptor in aromatic molecules. This experi-mental work indicates that the trifluoromethyl group is, as expected, a very polar substituent. The work does not indicate that the group engages in hyperconjugation. The conclusions reached through these studies are the same as the conclusions reached through the studies of reaction chemistry. There is no secure evidence for carbon-fluorine hyperconjugation in uncharged molecules.[23]

The investigations of the EPR spectra of radicals and anion
radicals have proved more interesting. In this discussion, we shall
focus on contributions from this laboratory. The results for the
semiquinones, VIII-X, reveal that the coupling constants for the
nuclei in the position β to the aromatic pi orbital are similar.

$$a_\beta^H = 1.8G \qquad a_\beta^C = 0.74G \qquad a_\beta^F = 2.8G$$

VIII	IX	X

The s orbital spin density is 3.6×10^{-3} for hydrogen 1s, 7.3×10^{-4}
for carbon 2s, and 1.6×10^{-4} for fluorine 2s.[24] Other observations for other radicals also indicate that the s orbital spin
density is generally smaller for fluorine atoms than for hydrogen
or carbon atoms.

The angular dependence of the a_β^F constant provides an experimental method for the separation of carbon-fluorine hyperconjugation from a 1,3 p-p interaction. Structure XI illustrates the idea
that carbon-fluorine hyperconjugation decreases to zero when Θ_F, the
dihedral angle between the axis of the carbon 2p orbital and the
carbon-fluorine bond, is 90°. Structure XII illustrates the idea
that the 1,3 p-p interaction remains important when $\Theta_F = 90°$. To
test this point, we prepared fluorotriptycene, XIII, to determine
a_β^F for a structure in which Θ_F was required to be 90°. The a_β^F
value is 0.85G.

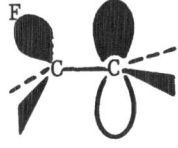

XIA   XIB

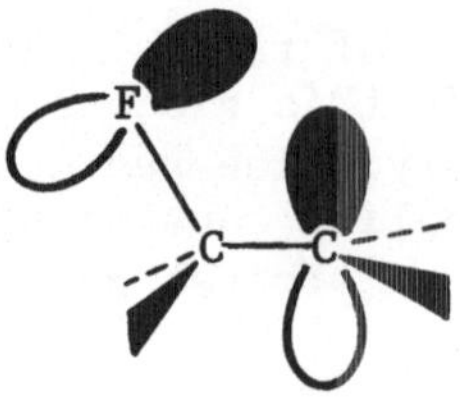

XIIA 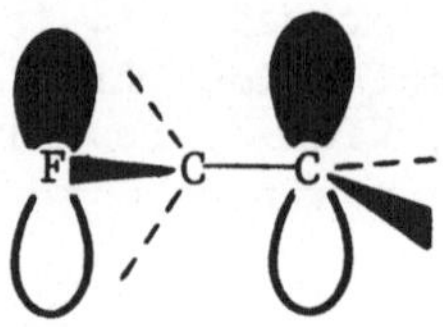   XIIB

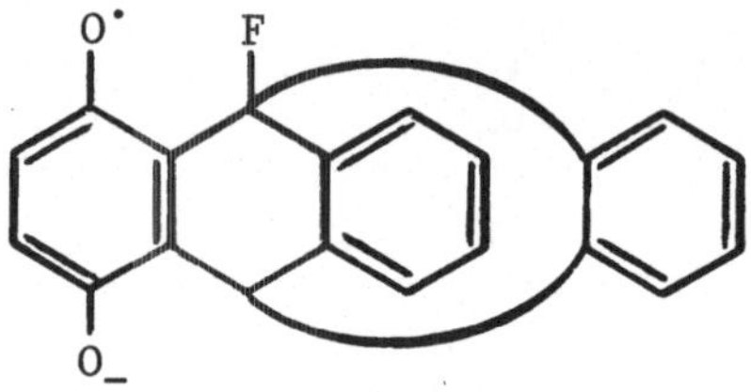

XIII

This constant is only reduced by a factor of three from the result
for X. There are two possible explanations. First, the result may
mean that spin delocalization occurs by a 1,3 p-p interaction or
second, that there is a direct interaction between the oxygen and
fluorine atoms in this molecule.[25] It is clear that direct interactions can be very large, as revealed by the contribution of
Norman and Gilbert, who showed that a^F for the *syn* isomer, XIV, is
large compared to the a^F value for the *anti* isomer, XV.[26] Mallory

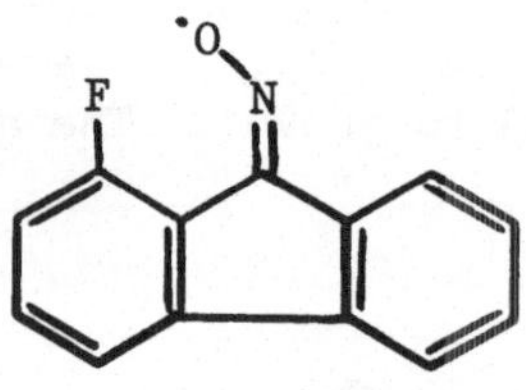

XIV   XV

$a^F = 13.5G$ $a^F = 0G$

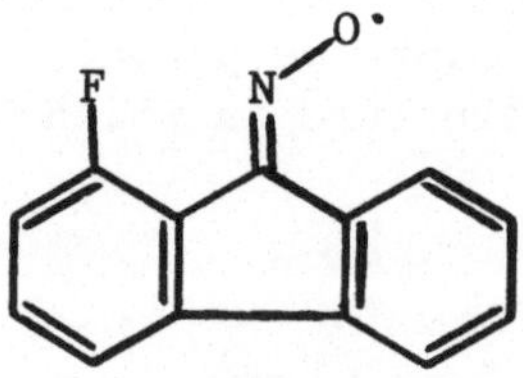

has discussed the relevance of direct interactions of this kind in
considerable detail.[27] Thus, the large a^F_β value for XIII may arise
from a direct interaction, XVI.

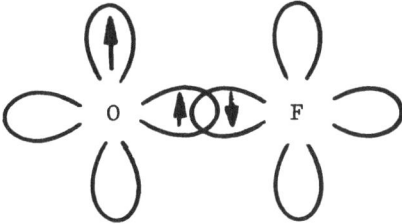

XVI

The next step in the resolution of the problem was the preparation and study of XVII and XVIII.[28] These compounds were selected
for study to evaluate the importance of the direct interaction

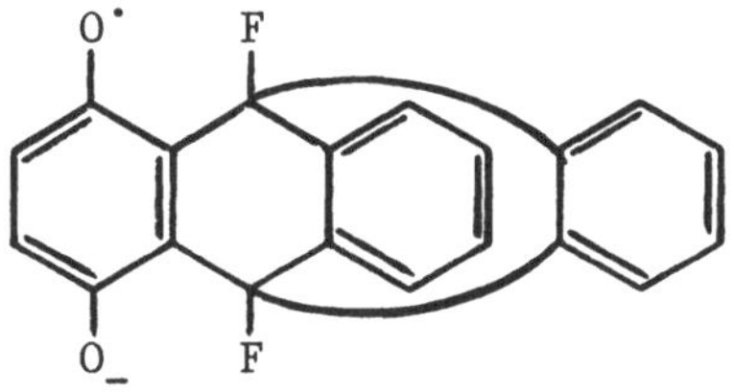

XVII

$a_\beta^F = 0.98G$

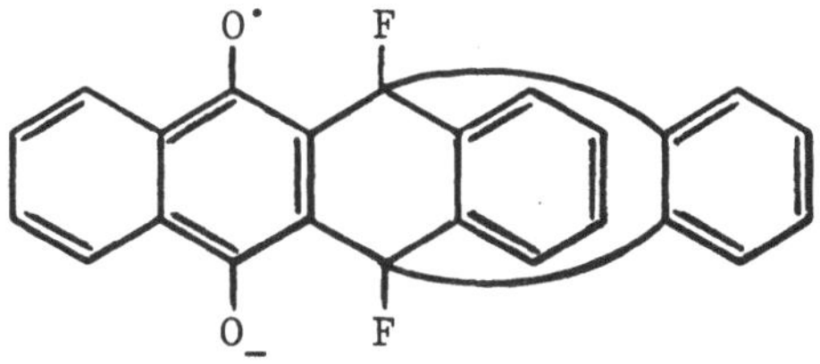

XVIII

$a_\beta^F = 1.09G$

because the pi spin density at the oxygen atoms of benzosemiquinone
and naphthosemiquinone is essentially equal but the pi spin density
at the 2 position is about 40% greater in the naphthalene derivative.[29] The fact that a_β^F is about the same for both radicals,
therefore, is good evidence for the importance of a direct interaction in XIII.

To avoid the complications of direct interactions, we turned
our attention to nitrobenzene anion radicals. The p-nitrobenzyl
fluoride anion radical was prepared electrolytically in acetonitrile
solution. This unstable radical could be observed at temperatures
below 0°. The coupling constants for this radical are compared with
the coupling constants for other nitrobenzene anion radicals in
Table II.

$$O_2NC_6H_4CH_2F + e^- \xrightarrow[R_4NClO_4]{CH_3CN} O_2NC_6H_4CH_2F^{\cdot-}$$

Table II

Coupling Constants for 4-Substituted Nitrobenzene

Anion Radicals at Ambient Temperature

4-Substituent	Coupling Constant, G					Ref.
	a^N	a_2^H	a_3^H	a_β^H	a_β^X	
H	10.3	3.4	1.0	4.0^a	–	30
CH_3	10.7	3.4	1.1	4.0	–	30
$CH_2C(CH_3)_3$	11.0	3.4	1.1	2.1	–	31
$CH_2N(CH_3)_2$	12.4	3.3	1.2	2.6	1.2^b	31
CH_2OH	13.7	3.4	1.2	2.8	–	32
CH_2F	8.8	3.5	0.9	1.7	$27.4^{c,d}$	33
CH_2F, $-35°$	9.7	3.4	1.0	1.8	$25.7^{c,d}$	31
CHF_2	8.2	3.4	1.0	1.0	14.6^c	33
CF_3	8.0	3.2	0.9	–	9.1^c	34
$CH_2\overset{+}{N}(CH_3)_3$, $-35°$	11.2	3.3	1.0	1.5	2.8^b	31

[a]For the 4-hydrogen atom. [b]For the nitrogen atom. [c]For the fluorine atom. [d]Temperature dependent.

The a_β^H values are customarily analyzed on the basis of equation 1.

$$a_\beta^H = \rho_C^\pi (B_0^H + B_2^H <\cos^2\theta_H>) \qquad (1)$$

The quantities B_0^H and B_2^H are empirical constants of about -1 and 50 Gauss, respectively, ρ_C^π is the spin density in the adjacent p orbital, and θ_H is the dihedral angle between the axis of the p orbital and the carbon-hydrogen bond. If the differences in a_β^H are governed by conformational factors alone then the $a_{CH_2X}^H$

should always be greater than 2.[35] Clearly, $a_{CH_2F}^H$ and $a_{CH_2\overset{+}{N}(CH_3)_3}^H$
are much smaller than this prediction. These
results indicate that electronegative groups localize spin and electron density in the substituent group.[31]

The observations for the mono-, di-, and trifluoromethyl derivatives are equally interesting. These results may be analyzed on
the basis of equation 2.

$$a_\beta^F = \rho_C^\pi (B_0^F + B_2^F <\cos^2\theta_F>) \qquad (2)$$

The symbols in equation 2 have the same significance as the symbols
in equation 1. The experimental value for a_β^F for the benzotrifluor-
ide, Table II, suggests B_2^F is about 115G. The result for the mono-
fluoride requires that B_2^F be much larger, about 150G. These findings parallel the findings made by other workers for simple aliphatic
radicals.[36,37] These results also suggest that fluorine atoms
localize spin and electron density on the fluoroalkyl group.

The EPR spectrum of p-nitrobenzyl fluoride anion radical is
temperature dependent, see Figure 3. The nature of the temperature
dependence indicates that there is a four-fold barrier to rotation,
that conformer XIX is more stable than XX by about 2.5 kcal. mole^{-1},
and that the barriers to rotation are about 4.5 and 2.0 kcal. mole^{-1}.
These barriers are clearly much greater than the barriers observed
for radicals with bulky groups such as p-neopentylnitrobenzene
anion radical. These experimental results provide further evidence
for the strength of the interaction between the fluoroalkyl group
and the aromatic pi electron system.

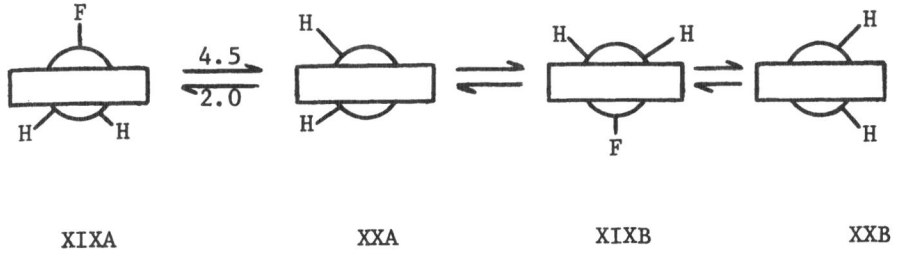

XIXA   XXA   XIXB   XXB

The anion radical of 2-nitro-9,10-difluorotriptycene, XXI, was
studied to test the importance of 1,3 p-p interactions. The a^N and
a^H values correspond closely with the data presented in Table II.
The a_β^F value, 0.45G, is observed for only one of the two fluorine
nuclei and is assigned to the fluorine nucleus adjacent to the *para*

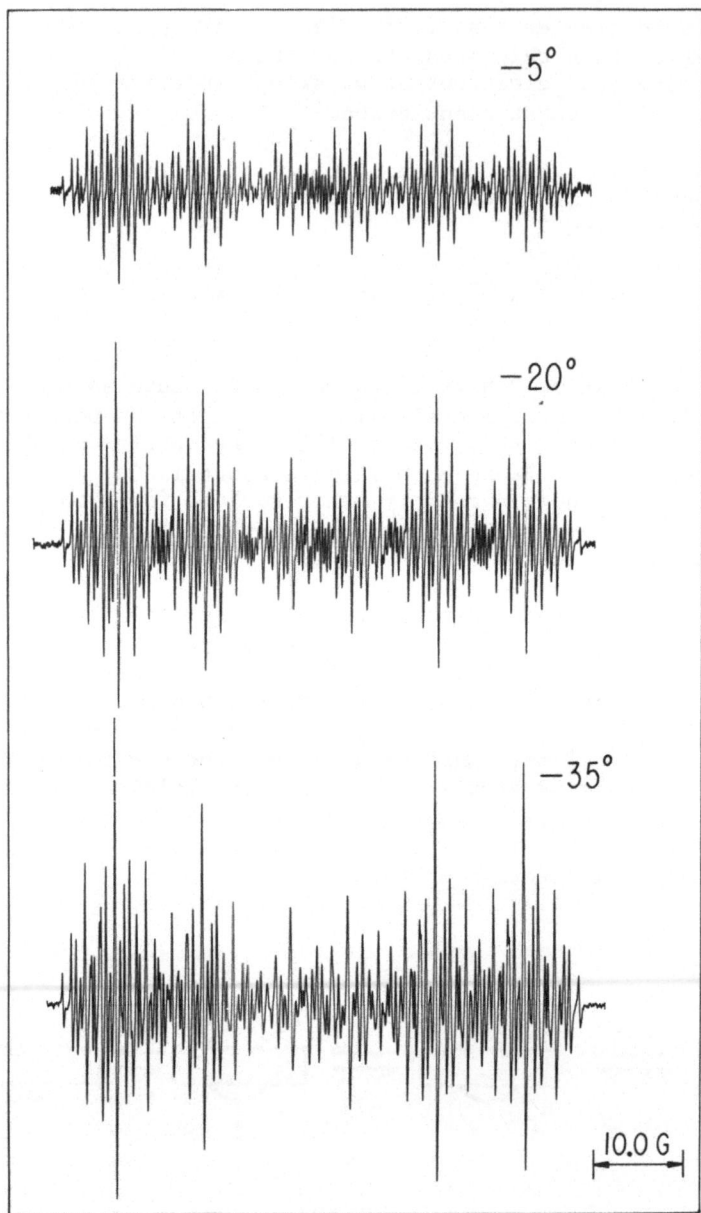

Figure 3

The EPR spectrum of p-nitrobenzyl fluoride anion radical in
acetonitrile

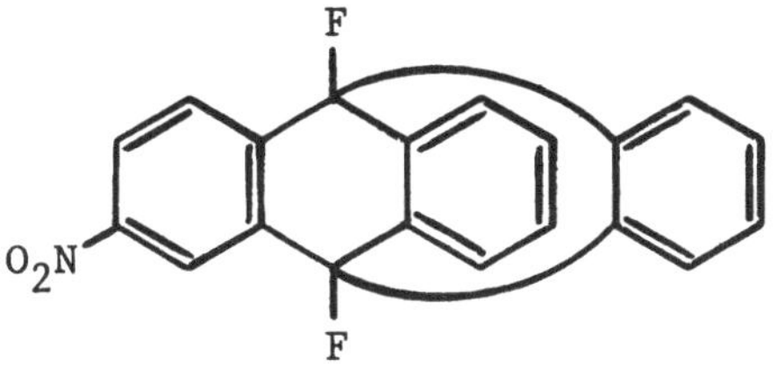

$$a^N = 9.43G$$

$$a_1^H, a^H = 3.78, 3.12G$$

$$a_3^H = 1.06G$$

$$a_\beta^F = 0.45G$$

XXI

position. The difference between a_β^F for the triptycene, 0.45G, and for the p-nitrobenzylfluoride, 25.7G, is remarkably large. This observation provides a clear indication that the carbon–fluorine hyperconjugation model is more suitable for discussion of the coupling interactions of the monofluoroalkyl group than is the 1,3 p-p interaction model. This conclusion is based on the idea that the two models predict quite different dependencies on the dihedral angle, Θ_F. Theory requires that a_β^F approach zero as Θ_F approaches 90° if spin density is delocalized by hyperconjugation. On the other hand, the a_β^F value should remain positive and large if delocalization occurs by a 1,3 p-p interaction because the p-p overlap remains appreciable when Θ_F is 90°.

In another test of this conclusion, we studied the contact chemical shifts of fluorine atoms in paramagnetic nickel complexes of anilines. Eaton, Josey, and Sheppard previously reported the contact shifts of stable bis(phenylaminotroponiminato)nickel(II) complexes.[38] We have examined these shifts in the NMR spectra of nickel acetylacetonate complexes of aniline derivatives. The extent

$$Ni(AcAc)_2 + 2F_3CC_6H_4NH_2 \rightleftharpoons (F_3CC_6H_4NH_2)_2Ni(AcAc)_2$$

of the shift depends on the pi spin density at the *para* carbon atom, XXII. The contact shifts for the fluorine atoms of the three trifluoromethylanilines and the amino-9,10–difluorotriptycene, XXIII, were determined. The value of the fluorine atom contact chemical shift, σ_F, relative to the shift for an *ortho* hydrogen atom, σ_{2-H}, in the same molecule are reported in Table III.

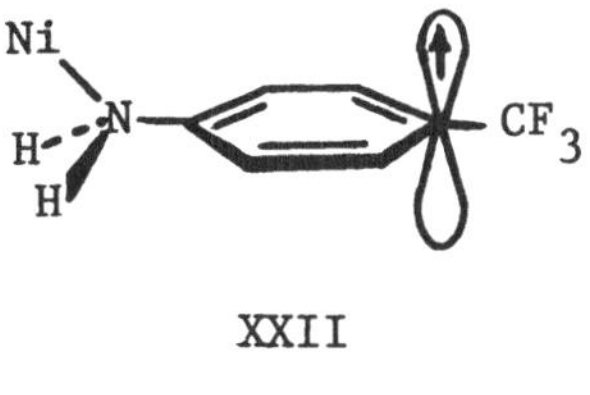

XXII

Table III

Relative Fluorine Atom Contact Chemical Shifts

for Nickel Acetylacetonate Complexes

Compound	Relative Contact Chemical Shift σ_i/σ_{2-H}		
	2-F	3-F	4-F
2-Trifluoromethylaniline	-1.04±0.21		
3-Trifluoromethylaniline		0.43±0.01	
4-Trifluoromethylaniline			-1.63±0.01
XXIII		0.0094±0.007	0.083±0.001

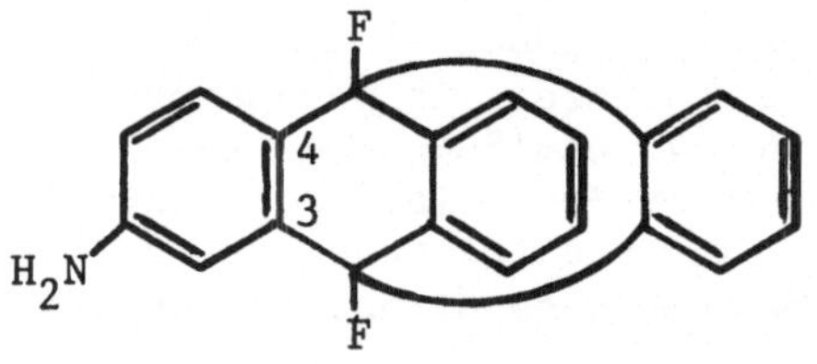

XXIII

The contact shifts for the fluorine atoms and the 2-hydrogen atoms are related by equation 3:

$$\sigma_i/\sigma_{2-H} = (a_i^F/a_2^H)(\gamma_H/\gamma_F) \qquad (3)$$

where γ is the magnetogyric ratio. The shifts for the trifluoro-methylanilines indicate that $a_{2-CF_3}^F$ and $a_{4-CF_3}^F$ are positive and that $a_{3-CF_3}^F$ is negative. These signs depend on the sign of ρ_C^π. Thus, B_2^F is positive in all three cases. This conclusion is in accord with the results obtained by the Eaton group.[38]

The results for the triptycene are particularly informative because the fluorine atoms are constrained to the nodal plane of the pi electron system. The contact shifts of the fluorine atoms in this molecule are very much smaller than the shifts observed for the fluorine atoms in the trifluoromethylanilines. Indeed, the shift observed for the fluorine atom adjacent to the 3 position is so small that the datum cannot be discussed with confidence. The datum for the fluorine atom adjacent to the 4 position quite unambiguously establishes that a_β^F for a fluorine atom in the nodal plane, $\Theta_F = 90°$, is very small and negative. Hence, B_0^F is negative and rather small compared to B_2^F.

$$\frac{a_{\beta-CF_3}^F \quad (<\Theta_F> = 45°)}{a_\beta^F \quad (<\Theta_F> = 90°)} = -20 \qquad (4)$$

The spectroscopic results obtained in this laboratory complement the results of other workers including Bolton, Chachaty, Hudson, Iwasaki, Janzen, Kochi, Krusic, Lontz, McDowell, Polenov, Rogers, Russell, Strom, Terabe, Trapp, Underwood, Whiffen and Wood, and their associates.[23] The data for radicals in solution are usually discussed on the basis of equation 2 with the idea that B_2^F is much larger than B_0^F. However, it is recognized that equation 2 cannot be applied generally, because B_2^F depends on the nature of the radical and on the degree of fluorine substitution. The dependence on the degree of fluorine substitution is illustrated by the major variation in a_β^F for the mono-, di- and tri- fluoromethyl derivatives as already discussed. The dependence on the nature of the radical site is illustrated by the major variation in a_β^F for radicals such as bistrifluoromethylnitroxide and hexafluoroacetone ketyl.[39-41] The B_2^F values which may be inferred from the coupling constants reported

$$CF_3 \overset{\overset{\displaystyle O^\cdot}{|}}{N} CF_3$$

XXIV

$$a_\beta^F = 8.3G$$

$$a_{CF_3}^C = 5.1G$$

$$a_{CO}^C = 23.3G$$

$$CF_3 \overset{\overset{\displaystyle O^\cdot}{|}}{C} CF_3$$

XXV

$$a_\beta^F = 34.9G$$

$$a_{CF_3}^C = 8.0G$$

$$a_{CO}^C = 9.5G$$

for these radicals are 55 and 140G, respectively. Morokuma has proposed the most reasonable explanation. He pointed out that the energy difference between the lowest unoccupied molecular orbital of the trifluoromethyl group and the highest singly occupied molecular orbitals of the nitroxide fragment and the carbonyl fragment differ greatly.[6] Specifically, the energy content of an electron in the carbon-oxygen antibonding molecular orbital is estimated to be 0.10 au compared to -0.46 au for an electron in the nitrogen-oxygen orbital. The energy level of the lowest unoccupied molecular orbital of the trifluoromethyl group is 0.30 au as judged from the analysis of the orbitals of fluoroform.[7] Consequently, hyperconjugation involving spin and charge transfer from the pi orbital of the ketyl to the relatively low-lying antibonding orbital of the trifluoromethyl group is much more favorable for the ketyl than for the nitroxide. This interpretation based on hyperconjugation can account for the three principal experimental findings that the coupling constants depend upon radical structure, the degree of fluorination and the dihedral angle.

CONCLUSION

The dipole moment measurements, the NMR data, and the results of the ESCA work all suggest that electron delocalization is not particularly important in uncharged ground state molecules. In addition, the dissociation constants and rate constants for the reactions of most aliphatic and aromatic compounds can be understood on the basis of the field and pi inductive effects of very polar trifluoromethyl group. On the other hand, the influence of this substituent appears to be enhanced in very electron-rich transition states such as the one leading to the benzenanion in nucleophilic aromatic substitution reactions. The EPR work suggests that spin density is delocalized from the radical center to an antibonding molecular orbital of the trifluoromethyl group by hyperconjugation. The effect is highly dependent upon structure and seems to be most important in anion radicals such as the ketyl and the nitrobenzene anion radical. Thus, we conclude that spin and charge delocalization through carbon-fluorine hyperconjugation can influence the chemical and spectroscopic properties of electron-rich compounds and intermediates.

REFERENCES

1. J. D. Roberts, R. L. Webb, and E. A. McElhill, *J. Am. Chem. Soc.*, 72, 408 (1950).
2. R. S. Mulliken, C. A. Rieke, and W. G. Brown, *J. Am. Chem. Soc.*, 63, 41 (1941).
3. C. A. Coulson, Valence, Oxford, Clarendon Press, 1952, pp 310-314.

4. R. Hoffmann, L. Radom, J. A. Pople, P. von R. Schleyer, W. J. Hehre, and L. Salem, *J. Am. Chem. Soc.*, **94**, 6221 (1972).
5. A. R. Rossi and D. E. Wood, *J. Am. Chem. Soc.*, **98**, 3452 (1976).
6. K. Morokuma, *J. Am. Chem. Soc.*, **91**, 5412 (1969).
7. L. C. Snyder and H. Basch, Molecular Wave Functions and Properties, Wiley, 1972, pp 346-350.
8. C. L. Liotta, D. F. Smith, Jr., H. P. Hopkins, Jr., and K. A. Rhodes, *J. Phys. Chem.*, **76**, 1909 (1972).
9. D. Holtz, *Prog. Phys. Org. Chem.*, **8**, 1 (1971).
10. W. A. Sheppard, *J. Am. Chem. Soc.*, **87**, 2410 (1965).
11. D. D. Perrin, Dissociation Constants of Organic Bases in Aqueous Solution, Butterworths, London, 1965.
12. A. I. Biggs and R. A. Robinson, *J. Chem. Soc.*, 388 (1961).
13. A. Streitwieser, Jr., and H. F. Koch, *J. Am. Chem. Soc.*, **86**, 404 (1964).
14. K. J. Klabunde and D. J. Burton, *J. Am. Chem. Soc.*, **94**, 820 (1972).
15. A. M. Porto, L. Altieri, A. J. Castro, and J. A. Brieux, *J. Chem. Soc.*, *B*, 963 (1966).
16. (a) J. Miller, *Aust. J. Chem.*, **9**, 61 (1956); (b) J. Miller and W. Kai-Yan, *J. Chem. Soc.*, 3492 (1963); (c) K. C. Ho and J. Miller, *Aust. J. Chem.*, **19**, 423 (1966).
17. W. Greizerstein, R. A. Bonelli, and J. A. Brieux, *J. Am. Chem. Soc.*, **84**, 1026 (1962).
18. S. Andreades, *J. Am. Chem. Soc.*, **86**, 2003 (1964).
19. A. Streitwieser, Jr., and D. Holtz, *J. Am. Chem. Soc.*, **89**, 692 (1967).
20. D. J. Burton, R. D. Howells, and P. D. Vander Valk, *J. Am. Chem. Soc.*, **99**, 4830 (1977).
21. W. A. Sheppard and C. M. Sharts, Organic Fluorine Chemistry, Benjamin, New York (1969), Chapter 2.
22. J. D. Hepworth, J. A. Hudson, D. A. Ibbitson, and G. Hallas, *J. Chem. Soc. Perkin II*, 1905 (1972).
23. The data are reviewed by L. M. Stock and M. R. Wasielewski, *Prog. Phys. Org. Chem.*, **14**, 000 (1978).
24. L. M. Stock and J. Suzuki, *J. Am. Chem. Soc.*, **87**, 3909 (1965).
25. D. Kosman and L. M. Stock, *J. Am. Chem. Soc.*, **92**, 409 (1970).
26. R. O. C. Norman and B. C. Gilbert, *J. Phys. Chem.*, **71**, 14 (1967).
27. F. B. Mallory, *J. Am. Chem. Soc.*, **95**, 7747 (1973).
28. Please consult M. R. Wasielewski, Thesis, University of Chicago Library, for a full account of the results discussed in this article.
29. M. Broze, Z. Luz, and B. L. Silver, *J. Chem. Phys.*, **46**, 4891 (1967).
30. L. M. Stock and P. E. Young, *J. Am. Chem. Soc.*, **94**, 7686 (1972).
31. L. M. Stock and M. R. Wasielewski, *J. Am. Chem. Soc.*, **97**, 5620 (1975).
32. P. L. Kolker and W. A. Waters, *J. Chem. Soc.*, 1136 (1964).
33. E. A. Polenov, B. I. Shapiro, and L. M. Yagupol'skii, *Zhur. Strukt. Khim.*, **12**, 163 (1971).

34. E. G. Janzen and J. L. Gerlock, *J. Am. Chem. Soc.*, $\underline{89}$, 4902
     (1967).

35. Under the reasonable assumptions that B_O^H is negligible and that
     ρ_{C4}^π is the same for the nitrobenzene anion radicals, $a_{CH_2X}^H/a_{CH_3}^H$
     is given by $(\cos^2\Theta_H + (\cos^2\Theta_H + 120°))$. The minimum
     value is 0.5.

36. K. S. Chen, P. J. Krusic, P. Meakin, and J. K. Kochi, *J. Phys.
     Chem.*, $\underline{78}$, 2014 (1974).

37. I. Biddles, J. Cooper, A. Hudson, R. A. Jackson, and J. T.
     Wiffen, *Mol. Phys.*, $\underline{25}$, 225 (1973).

38. D. R. Eaton, A. D. Josey, and W. A. Sheppard, *J. Am. Chem. Soc.*,
     $\underline{85}$, 2689 (1963).

39. P. J. Scheidler and J. R. Bolton, *J. Am. Chem. Soc.*, $\underline{88}$, 371
     (1966).

40. E. G. Janzen and J. L. Gerlock, *J. Phys. Chem.*, $\underline{71}$, 4577 (1967).

41. W. R. Knolle and J. R. Bolton, *J. Am. Chem. Soc.*, $\underline{91}$, 5411
     (1969).

MOLECULAR AND CONFORMATIONAL EQUILIBRIA STUDIED BY ELECTRON SPIN RESONANCE SPECTROSCOPY*

Glen A. Russell

Department of Chemistry, Iowa State University

Ames, Iowa 50011

The discovery of aliphatic semidiones by my group in the early 1960's came from several separate investigations.[1] We were one of the first groups of organic chemists to apply electron spin resonance spectroscopy to studies of organic reactions. One of our initial lines of research involved the study of electronic comproportionation reactions, such as that shown in equation 1.[2]

$$\pi + \pi^{2-} \rightleftharpoons 2\pi\bullet^{-} \tag{1}$$

We found that when π was an α-dione and π^{2-} was the dianion of the corresponding α-hydroxy ketone then comproportionation occurred readily, yielding the cyclic semidione ($\underline{1}$) or, in acyclic cases, a mixture of *cis*- and *trans*-semidiones ($\underline{2} \rightleftharpoons \underline{3}$) (ESR spectrum, Figure 1).[3]

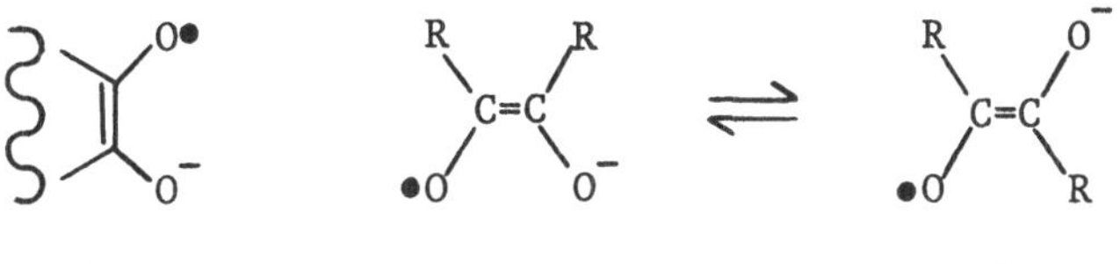

$$\underline{1} \qquad\qquad \underline{2} \qquad\qquad \underline{3}$$

Another line of research that brought us to aliphatic semidiones was a study of the mechanism of autoxidation of carbanions.

*Part 36 in the series "Aliphatic Semidiones".

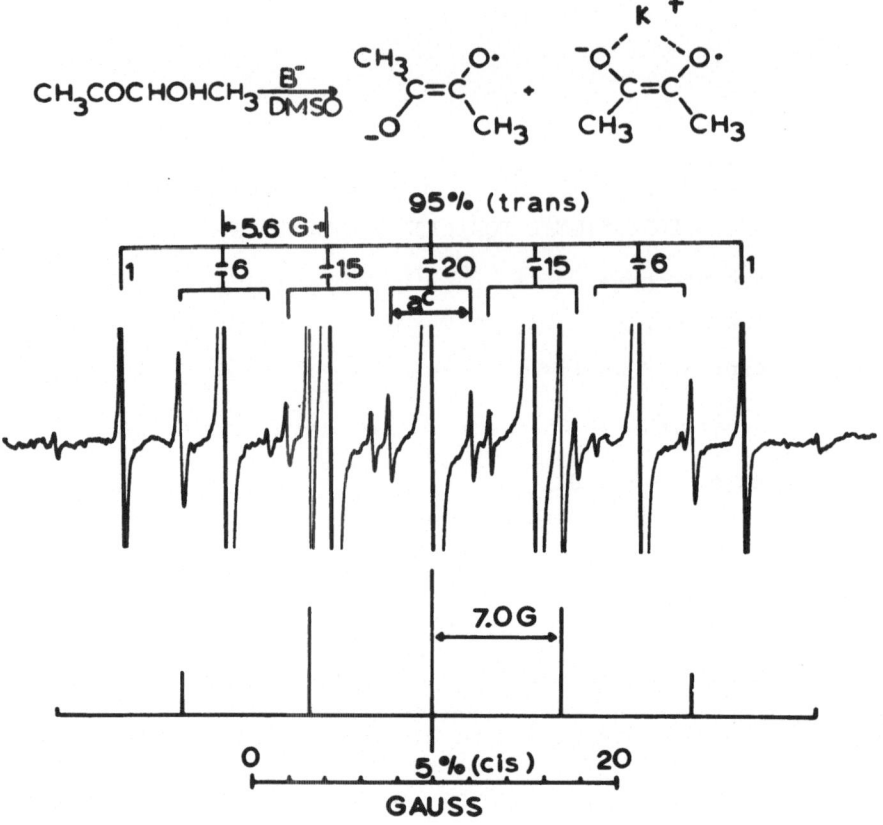

Figure 1. ESR spectrum of a mixture of the *cis* (ion-pair)- and *trans* (free ion)-dimethylsemidiones in Me$_2$SO containing 0.1 $\underline{M}$ K$^+$. The g-value for the *cis* ion-pair (a^H=7.0 gauss) is 2.00483, while the *trans* free ion (a^H=5.6 gauss) has g=2.00497. A^C (natural abundance) is seen for the methyl group of the *trans* semidione (a=4.5). In t-butyl alcohol solution the lines are sharper and $\underline{a}_{CO}^C$ for the *trans* species can be easily resolved. The other values of $\underline{a}^C$ in Me$_2$SO are $\underline{a}_{CO}^C$ (*trans* by isotropic substitution) 0.6 gauss, $\underline{a}_{CH_3}^C$ (*cis*)=5.2 gauss, $\underline{a}_{CO}^C$ (*cis* by isotropic substitution)=1.1 gauss.

We thought that ESR spectroscopy might allow us to detect paramagnetic intermediates in this process, which for many carbanions involves the propagation steps of Scheme I.[4]

$$R\bullet + O_2 \rightarrow ROO\bullet$$

$$ROO\bullet + R:^- \rightarrow ROO^- + R\bullet$$

Scheme I

When we performed the autoxidation of an enolate anion ($-\overset{\overset{O^-}{|}}{C}$=CH$-$) in
the ESR cavity we obtained strong ESR signals for persistent para-
magnetic species which we recognized to be the semidiones.[5] We also
found that α-hydroxy ketones (acyloins), in dimethyl sulfoxide
(Me$_2$SO) containing potassium t-butoxide, would spontaneously form
the semidione without the addition of oxygen.[6] Scheme II illustrates
the reactions most likely involved in the formation of the semidiones in the enolate oxidation as well as in the α-hydroxy ketone
disproportionation. Probably the most general route to aliphatic
semidiones, and the preferred route in view of side reactions, is
the disproportionation of the acyloin.

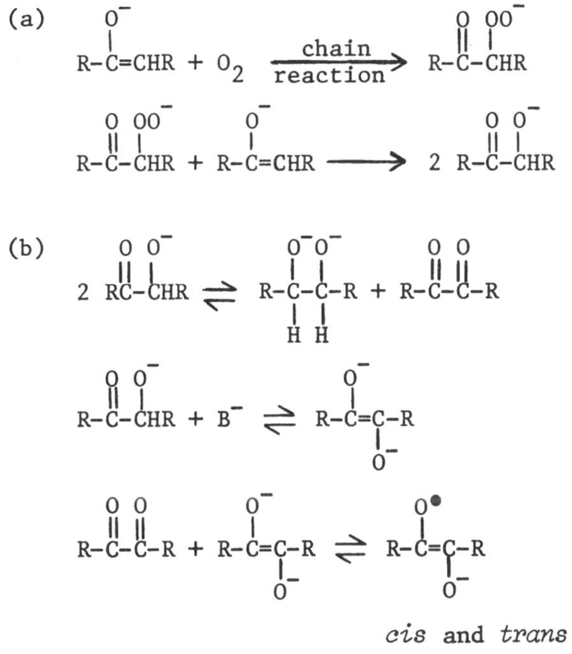

(a)

$$R\text{-}\overset{\overset{O^-}{|}}{C}\text{=CHR} + O_2 \xrightarrow[\text{reaction}]{\text{chain}} R\text{-}\overset{\overset{O}{\|}}{C}\text{-}\overset{\overset{OO^-}{|}}{C}HR$$

$$R\text{-}\overset{\overset{O}{\|}}{C}\text{-}\overset{\overset{OO^-}{|}}{C}HR + R\text{-}\overset{\overset{O^-}{|}}{C}\text{=CHR} \longrightarrow 2\ R\text{-}\overset{\overset{O}{\|}}{C}\text{-}\overset{\overset{O^-}{|}}{C}HR$$

(b)

$$2\ R\overset{\overset{O}{\|}}{C}\text{-}\overset{\overset{O^-}{|}}{C}HR \rightleftharpoons R\text{-}\underset{\underset{H}{|}}{\overset{\overset{O^-}{|}}{C}}\text{-}\underset{\underset{H}{|}}{\overset{\overset{O^-}{|}}{C}}\text{-}R + R\text{-}\overset{\overset{O}{\|}}{C}\text{-}\overset{\overset{O}{\|}}{C}\text{-}R$$

$$R\text{-}\overset{\overset{O}{\|}}{C}\text{-}\overset{\overset{O^-}{|}}{C}HR + B^- \rightleftharpoons R\text{-}\underset{\underset{O^-}{|}}{\overset{\overset{O^-}{|}}{C}}\text{=C-R}$$

$$R\text{-}\overset{\overset{O}{\|}}{C}\text{-}\overset{\overset{O}{\|}}{C}\text{-}R + R\text{-}\underset{\underset{O^-}{|}}{\overset{\overset{O^-}{|}}{C}}\text{=C-R} \rightleftharpoons R\text{-}\underset{\underset{O^-}{|}}{\overset{\overset{O^\bullet}{|}}{C}}\text{=C-R}$$

cis and *trans*

Scheme II

Aliphatic semidiones can be detected as intermediates in acyloin
condensations.[7] A convenient preparation of a semidione is simply
to treat the crude product of an acyloin condensation with Me$_2$SO
(Scheme III) or to perform the acyloin condensation in the ESR cell.

If an enediol dianion is generated in Me$_2$SO by reaction of a
1,2-bis(trimethylsiloxy)alkene with excess dimsylate anion
(CH$_3$SOCH$_2^-$) no detectable amount of semidione will be formed; Me$_2$SO
does not oxidize the dianion to the semidione. One can, however,
generate the semidione from the dianion by adding oxygen or by
neutralizing the excess dimsylate by t-butyl alcohol or trimethyl-

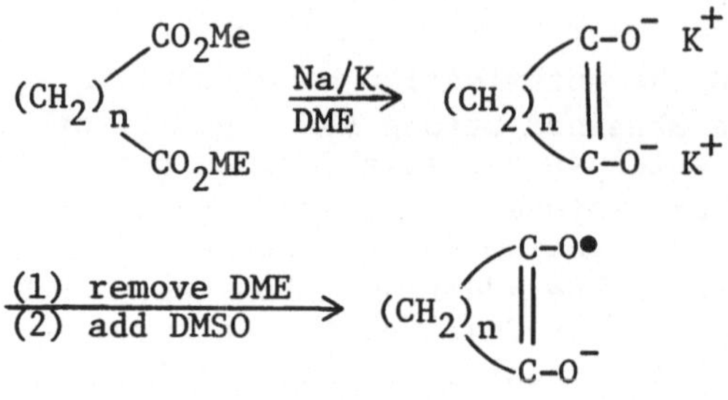

$$(CH_2)_n \overset{\displaystyle CO_2Me}{\underset{\displaystyle CO_2ME}{\Big<}} \quad \xrightarrow[\text{DME}]{\text{Na/K}} \quad (CH_2)_n \overset{\displaystyle C-O^- \; K^+}{\underset{\displaystyle C-O^- \; K^+}{\Big<}} \Big\|$$

$$\xrightarrow[\text{(2) add DMSO}]{\text{(1) remove DME}} \quad (CH_2)_n \overset{\displaystyle C-O\bullet}{\underset{\displaystyle C-O^-}{\Big<}} \Big\|$$

Scheme III

chlorosilane, when the disproportionation reactions shown in Scheme II, part b, occur.

Semidiones can be generated in a variety of solvents including HMPA, DMF, DME, *t*-butyl alcohol and pyridine, as well as in the preferred solvent, Me$_2$SO. In aqueous solution the semidione is short-lived, but it can be detected when generated by continuous techniques such as radiolysis of an aqueous solution of an α-hydroxy ketone or α-dione, or in a flow experiment involving continuous generation of hydroxyl radicals (Scheme IV).[8] At appropriate pH either the radical anion, the neutral radical, or the radical cation can be detected. The neutral semidione radical is quite acidic, having a pKa value of about 4.6.

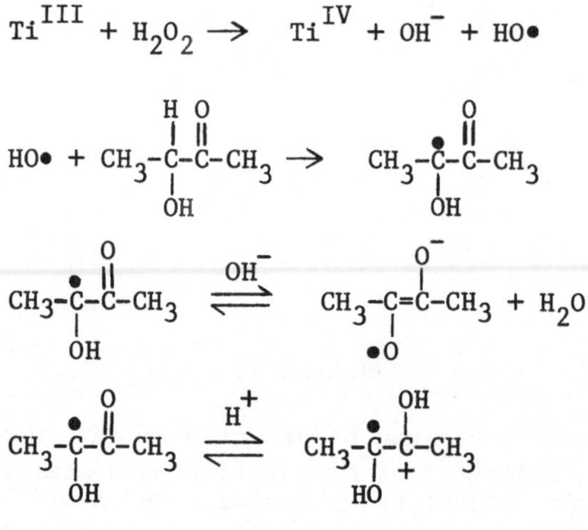

$$Ti^{III} + H_2O_2 \rightarrow Ti^{IV} + OH^- + HO\bullet$$

$$HO\bullet + CH_3-\overset{\overset{\displaystyle H}{|}}{\underset{\underset{\displaystyle OH}{|}}{C}}-\overset{\overset{\displaystyle O}{\|}}{C}-CH_3 \rightarrow CH_3-\overset{\overset{\displaystyle \bullet}{}}{\underset{\underset{\displaystyle OH}{|}}{C}}-\overset{\overset{\displaystyle O}{\|}}{C}-CH_3$$

$$CH_3-\overset{\overset{\displaystyle \bullet}{}}{\underset{\underset{\displaystyle OH}{|}}{C}}-\overset{\overset{\displaystyle O}{\|}}{C}-CH_3 \underset{}{\overset{OH^-}{\rightleftharpoons}} CH_3-\overset{\overset{\displaystyle O^-}{|}}{C}=\overset{}{\underset{\underset{\displaystyle \bullet O}{}}{C}}-CH_3 + H_2O$$

$$CH_3-\overset{\overset{\displaystyle \bullet}{}}{\underset{\underset{\displaystyle OH}{|}}{C}}-\overset{\overset{\displaystyle O}{\|}}{C}-CH_3 \underset{}{\overset{H^+}{\rightleftharpoons}} CH_3-\overset{\overset{\displaystyle OH}{|}}{C}-\overset{}{\underset{\underset{\displaystyle HO}{+}}{C}}-CH_3$$

Scheme IV

Another somewhat specific route to a semidione involves a carbonyl insertion reaction. Alkali metal reductions of strained ring ketones yield , not the ketyl, but the ring-enlarged semidione (eq 2).[9] Apparently the first formed ketyl expels CO or CO$\cdot^-$, which can then add to a second molecule of the ketyl or ketone. Support for this proposal was found in the observation that diisopropylketone, carbon monoxide and potassium form diisopropylsemidione (eq 3).

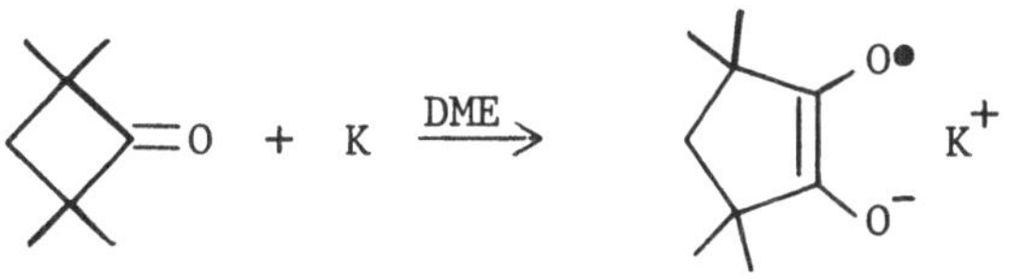

O•
K
DME
K^+
O^-

(2)

O
C
K
CO
DME
O•
C=C
O

(3)

The dimsylate anion ($CH_3SOCH_2^-$) will reduce a variety of α,β-unsaturated systems. Our evidence strongly indicates that the reducing ability of the dimsylate anion is due to Michael addition, followed by one-electron transfer from the resulting resonance-stabilized carbanion to another molecule of the starting material.[10] Thus, the dimsylate anion will reduce α,β-unsaturated ketones or diones, such as:

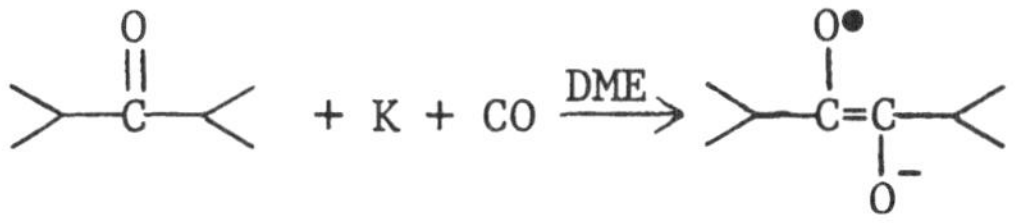

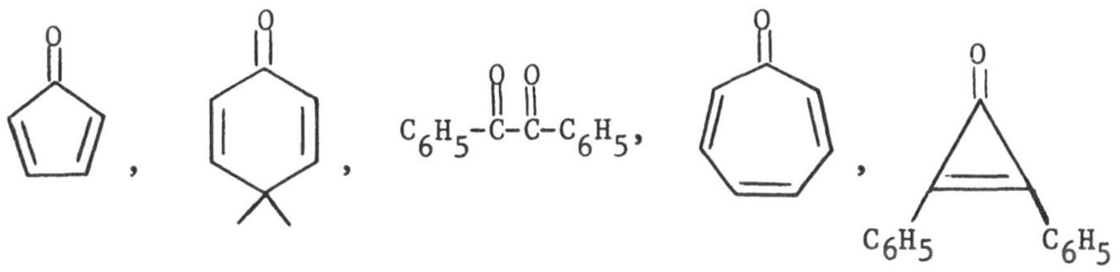

O
O
O O
C_6H_5-C-C-C_6H_5,
O
O
C_6H_5
C_6H_5

to radical anions. In the case of diphenylcyclopropenone, the resulting ketyl is unstable, and there is formed, instead, the semidione of 1,2-diphenylcyclobutene-1,2-semidione (ESR spectrum, Figure 2) (eq 4).[11]

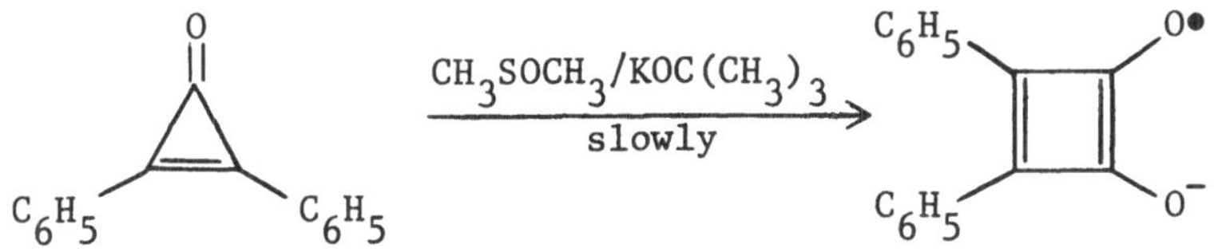

O
C_6H_5
C_6H_5
$CH_3SOCH_3/KOC(CH_3)_3$
slowly
C_6H_5
O•
C_6H_5
O^-

(4)

α-Diones are also reduced by enolate anions to semidiones (eq 5). For example, the enolate anion of propiophenone will reduce bicyclo[2.2.1]heptane-2,3-dione to the semidione.[12] In fact, this

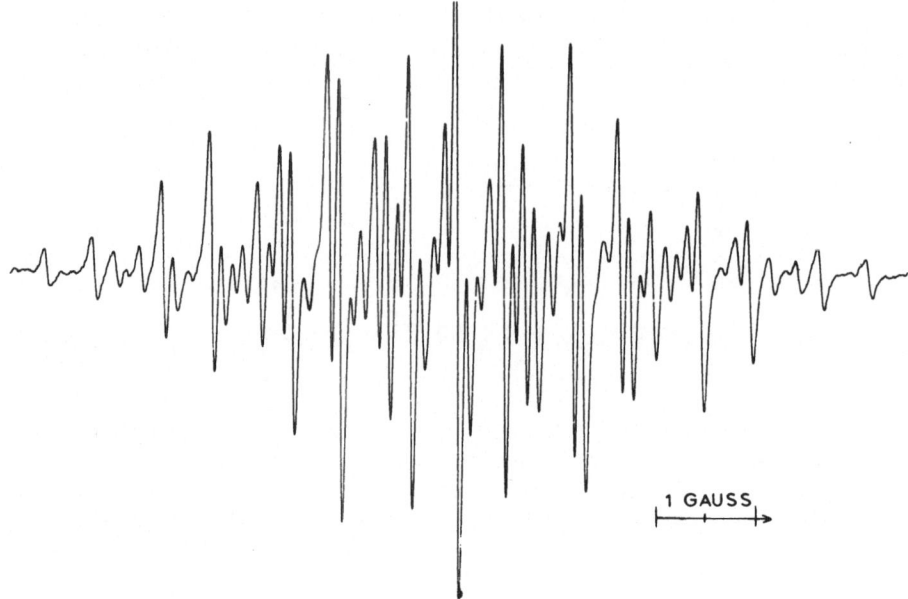

Figure 2. ESR spectrum of 2,3-diphenylcyclobutene-1,2-semidione in Me$_2$SO (K$^+$) at 25°. The aromatic hydrogen hyperfine splitting constants are a_p^H=1.30 (2H), a_o^H=1.20 (4H), a_m^H=0.48 (4H). The semidione was prepared by treatment of either diphenylcyclobutene-1,2-dione or diphenylcyclopropenone with potassium t-butoxide-Me$_2$SO in the absence of oxygen.

$$\begin{array}{c} \underset{\text{R-C-C-R}}{\overset{\text{O O}}{\|\ \|}} + \underset{\text{R'-C=C}}{\overset{\text{O}^-}{|}} \rightarrow \underset{\text{R-C=C-R}}{\overset{\text{O}^\bullet}{|}} + [\text{R'-C-C}] \\ \text{O}^- \qquad\qquad\qquad \text{not persistent} \end{array} \tag{5}$$

represented a third independent line of research which led us to semidiones, since we had also initiated, in the early 1960's, a program to study one-electron transfer processes of the general type shown in equation 6.[13] A subclass of this general reaction consists of systems where R:$^-$ is the conjugate base of π (e.g., p-nitrotoluene).[14] Thus, an α-dione with enolizable α-hydrogen atoms, such as 1,2-cyclohexanedione or 1-phenylpropane-2,3-dione, will form the semidione in basic solution via this pathway when treated with potassium t-butoxide in Me$_2$SO (Scheme V).

$$\text{R:}^- + \pi \rightarrow \pi\bullet^- (\text{persistent}) + \text{R}\bullet (\text{non-persistent}) \tag{6}$$

In view of these reactions it is not surprising that the autoxidation of α-alkoxy ketones in basic solution is also a general route

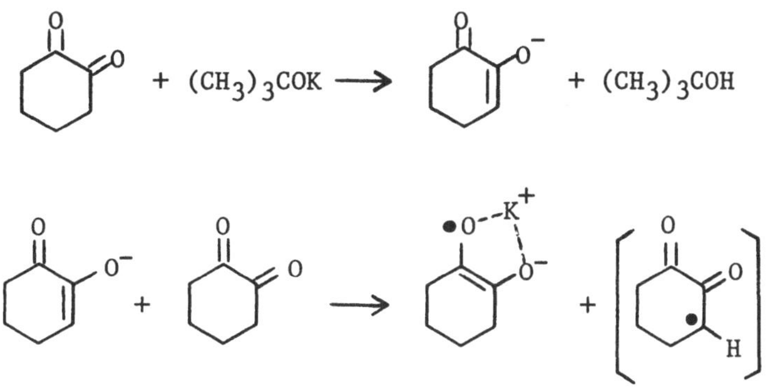

Scheme V

to semidiones (eq 7).[15] However, if one tries to generate biacetyl semidione from the biacetyl and base, self-condensation of the biacetyl is the predominating reaction, and only 2,5-dimethyl-p-benzosemiquinone can be observed by ESR spectroscopy (eq 8).

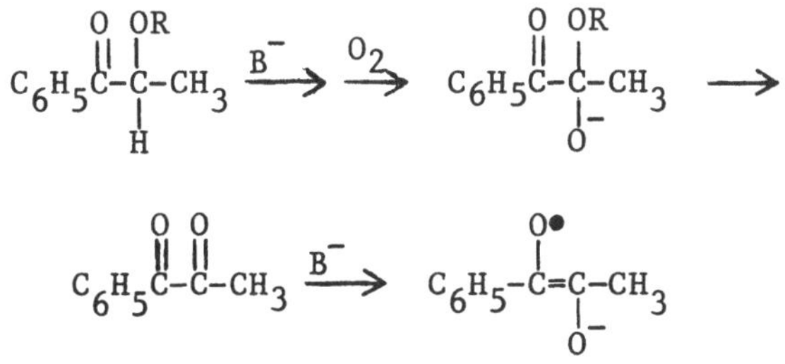

$$C_6H_5\overset{O}{\overset{||}{C}}-\overset{OR}{\overset{|}{\underset{H}{C}}}-CH_3 \xrightarrow{B^-} \xrightarrow{O_2} C_6H_5\overset{O}{\overset{||}{C}}-\overset{OR}{\overset{|}{\underset{O^-}{C}}}-CH_3 \longrightarrow$$

$$C_6H_5\overset{O}{\overset{||}{C}}-\overset{O}{\overset{||}{C}}-CH_3 \xrightarrow{B^-} C_6H_5-\overset{O\bullet}{\overset{|}{C}}=\overset{}{\underset{O^-}{C}}-CH_3 \qquad (7)$$

$$CH_3\overset{O}{\overset{||}{C}}-\overset{O}{\overset{||}{C}}-CH_3 \xrightarrow[Me_2SO]{KOC(CH_3)_3} \qquad (8)$$

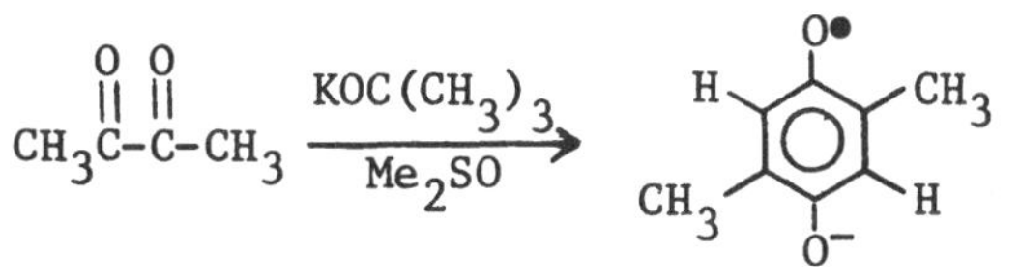

α,β-Unsaturated semidiones are not as persistent as saturated semidiones.[16] In the presence of basic Me$_2$SO an acyclic vinyl semidione such as methylvinylsemidione is converted in a few minutes to pentane-2,3-semidione by the reactions shown in Scheme VI.

Another reaction leading to semidiones – methylation by the dimsylate anion – has been observed with glyoxals. Phenylglyoxal radical anion (best prepared from the α-hydroxy ketone) slowly forms the 1-phenylpropane-2,3-semidione (Scheme VII)[9] in basic Me$_2$SO.

Vinylogs of semidiones are also readily prepared, particularly the 1,4-semidiones.[17] The comproportionation reaction (eq 8) occurs

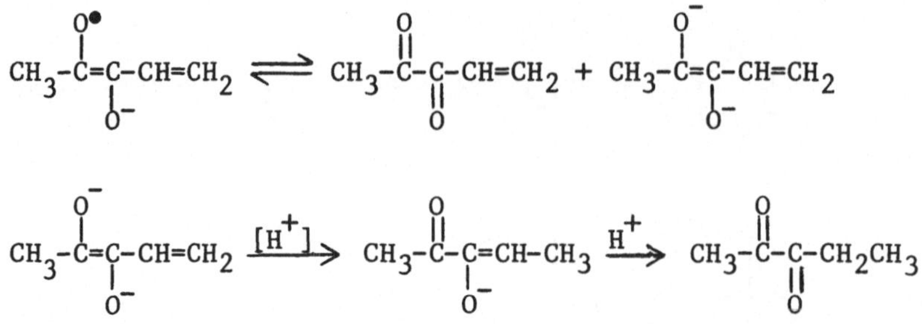

$$CH_3-\overset{\overset{O\bullet}{|}}{\underset{\underset{O^-}{|}}{C}}=C-CH=CH_2 \rightleftharpoons CH_3-\overset{\overset{O}{||}}{C}-\overset{\overset{O}{||}}{C}-CH=CH_2 + CH_3-\overset{\overset{O^-}{|}}{\underset{\underset{O^-}{|}}{C}}=C-CH=CH_2$$

$$CH_3-\overset{\overset{O^-}{|}}{\underset{\underset{O^-}{|}}{C}}=C-CH=CH_2 \overset{[H^+]}{\longrightarrow} CH_3-\overset{\overset{O}{||}}{C}-\overset{}{\underset{\underset{O^-}{|}}{C}}=CH-CH_3 \overset{H^+}{\longrightarrow} CH_3-\overset{\overset{O}{||}}{C}-\overset{\overset{}{}}{\underset{\underset{O}{||}}{C}}-CH_2CH_3$$

Scheme VI

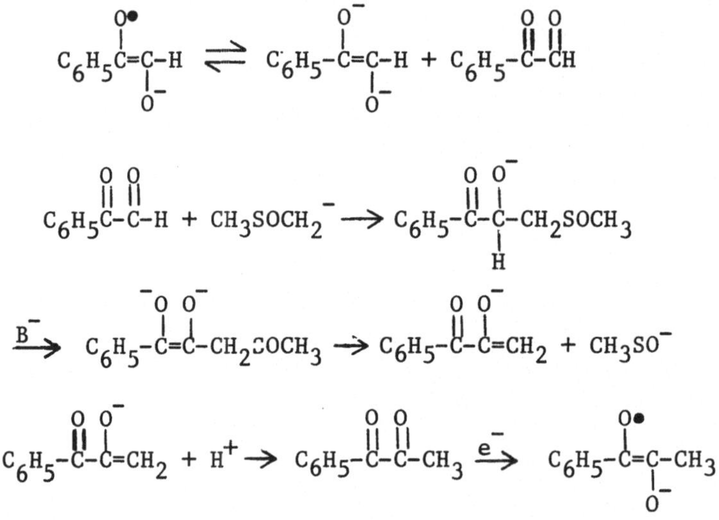

— — — — —

$$C_6H_5\overset{\overset{O\bullet}{|}}{\underset{\underset{O^-}{|}}{C}}=C-H \rightleftharpoons C_6H_5-\overset{\overset{O^-}{|}}{\underset{\underset{O^-}{|}}{C}}=C-H + C_6H_5-\overset{\overset{O}{||}}{C}-\overset{\overset{O}{||}}{C}H$$

$$C_6H_5\overset{\overset{O}{||}}{C}-\overset{\overset{O}{||}}{C}-H + CH_3SOCH_2^- \longrightarrow C_6H_5-\overset{\overset{O}{||}}{C}-\overset{\overset{O^-}{|}}{\underset{\underset{H}{|}}{C}}-CH_2SOCH_3$$

$$\overset{B^-}{\longrightarrow} C_6H_5-\overset{\overset{^-O}{|}}{C}=\overset{\overset{O^-}{|}}{C}-CH_2COCH_3 \longrightarrow C_6H_5-\overset{\overset{O}{||}}{C}-\overset{\overset{O^-}{|}}{C}=CH_2 + CH_3SO^-$$

$$C_6H_5-\overset{\overset{O}{||}}{C}-\overset{\overset{O^-}{|}}{C}=CH_2 + H^+ \longrightarrow C_6H_5-\overset{\overset{O}{||}}{C}-\overset{\overset{O}{||}}{C}-CH_3 \overset{e^-}{\longrightarrow} C_6H_5-\overset{\overset{O\bullet}{|}}{\underset{\underset{O^-}{|}}{C}}=C-CH_3$$

Scheme VII

readily. A particularly easy route to the 1,4-semidione involves
autoxidative dehydrogenation of a 1,4-dione in basic solution.
This conversion undoubtedly follows the route shown in Scheme VIII.

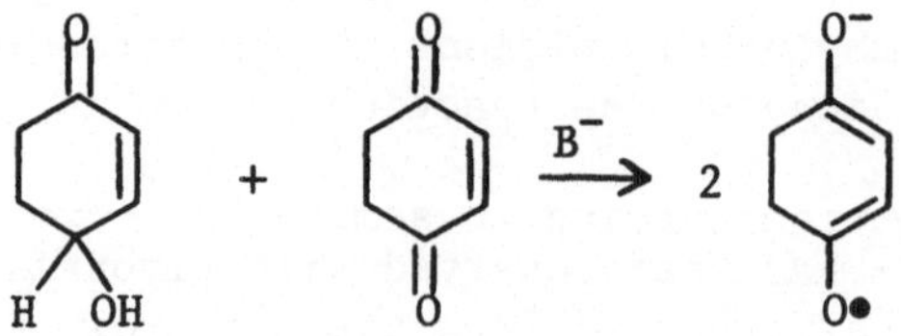

$$(8)$$

Ketyls, semidiones, semitriones, etc., form a series of relatively persistent radical anions. However, the semitriones, semi-

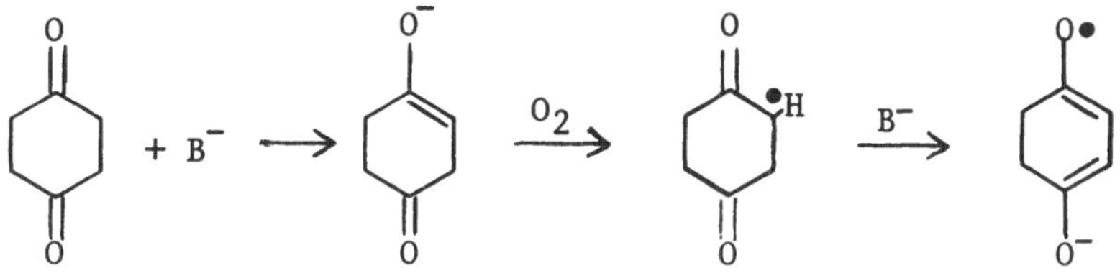

Scheme VIII

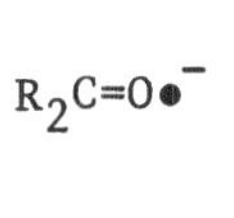

— — — — —

ketyl 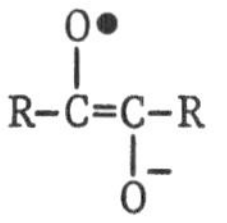   1,2-semidiones 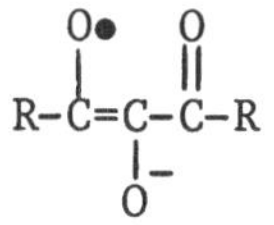   1,2,3-semitrione

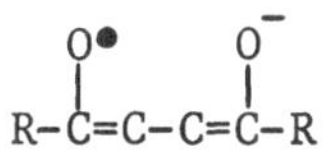

1,4-semidione 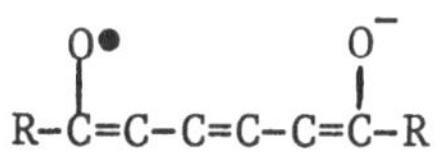   1,6-semidione

tetraones, etc., are difficult to prepare.[18] In basic solution the precursors readily undergo a benzylic acid type rearrangement leading eventually to the semidione (Scheme IX).

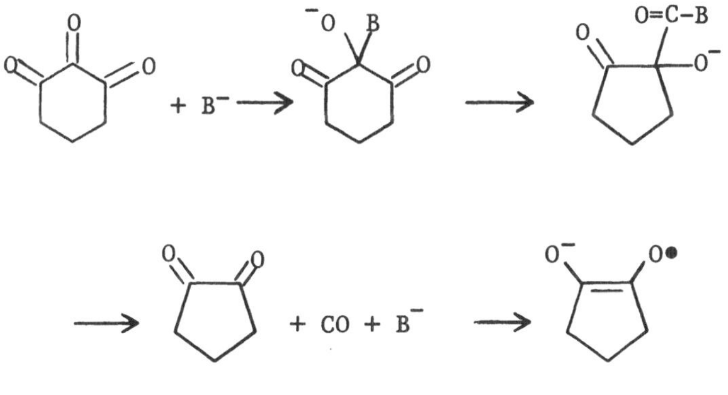

Scheme IX

ION-PAIR FORMATION FOR 1,2-SEMIDIONES

The equilibrium between *cis*- and *trans*-dimethylsemidione ($2 \rightleftarrows 3$) has been investigated as a function of counterion and solvent.[19,20] With cesium as the counterion only the *trans*-semidione can be detected in Me_2SO. It is concluded that for the free ions the thermodynamic *cis/trans* ratio is less than 1/100. On the other hand, with lithium as the counterion only the *cis*-semidione, with well-resolved hyperfine splitting by a single lithium cation, is observed. For sodium, potassium and rubidium as counterions the *cis/trans* ratio is a function of cation concentration and increases from rubidium to potassium to sodium.

The *cis/trans* ratio increases as the polarity of the solvent decreases and ion-pairing becomes more important. Thus, with 0.1 $\underline{M}$ K^+, the *cis/trans* ratio increases from 0.06 in Me_2SO to 0.12 in DMF, 0.33 in HMPA, 4 in triglyme, 10 in pyridine and >20 in *t*-butyl alcohol, DME or THF.

The solutions can be titrated with cryptand complexing agents (eq 9). With sodium or potassium as the counterion, a slight excess

222-cryptand

$$+ \; K^+, \; \text{cryptand} \qquad\qquad (9)$$

of the cryptand over total M^+ in solution gives an ESR spectrum of only the *trans* semidione. With lithium as the counterion, the 222-cryptand has no effect, but the 211-cryptand gives rise to the *trans*-semidione exclusively. The crown ethers are much less effective than the cryptands and, in fact, in Me_2SO solution 18-crown-6 or 15-crown-5 ethers have but a small effect on the *cis/trans* ratios of the sodium or potassium salts of dimethylsemidione. However, the effects observed are again stoichiometric, and the small change in the *cis/trans* ratio occurs sharply when the concentration of the crown ether becomes equivalent to total cation concentration. We conclude that the crown ethers form the crown-complexed *cis*-dimethyl-semidione ion pairs, <u>4</u>. To observe a dialkyl *cis*-semidione without ion-pairing under static conditions one must utilize a cyclic system (<u>1</u>). In a cyclic *cis*-semidione, such as cyclopentane-1,2-semidione, line-broadening of the low and high field peaks is observed, with the line-broadening being most extensive on the high field side. This general phenomenon is observed for many, but not all, cyclic semidiones and is the result of a slow equilibrium between the free ion and the 1:1 ion pair (eq 10).[21]

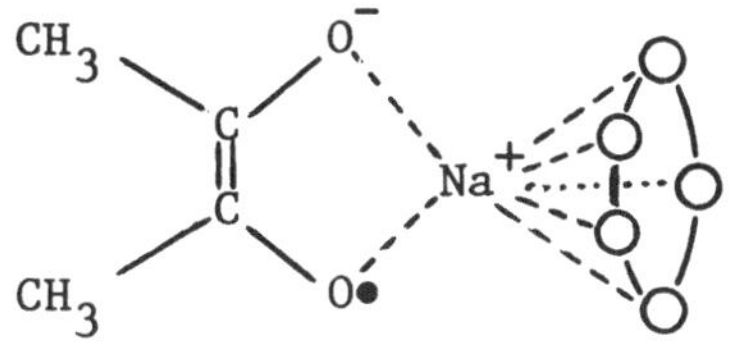

$$\underline{4}$$

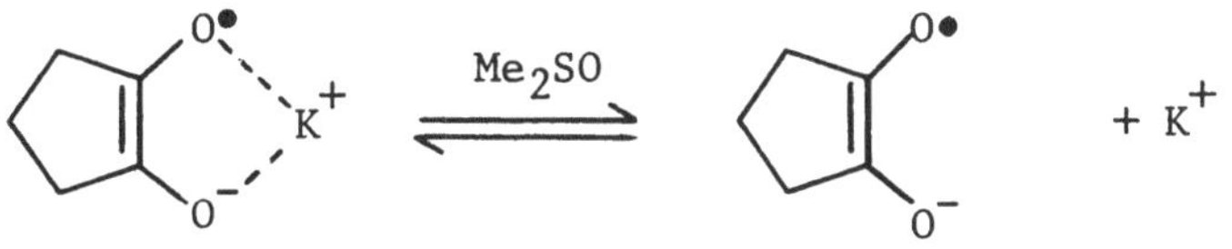

$$\xrightarrow{\text{Me}_2\text{SO}} \qquad + K^+ \qquad (10)$$

Figure 3a shows a typical ESR spectrum of cyclopentane semi-dione in Me_2SO containing 0.1 $\underline{\text{M}}$ total K^+. As the K^+ concentration is increased the line-broadening disappears, giving eventually a spectrum of the ion-pair (Figure 3b) (a^H=13.65 gauss, $\underline{g}$=2.00492). On the other hand, if the K^+ concentration is decreased by dilution, or by addition of 222-cryptand, the spectrum again loses the selective line-broadening and, in the presence of excess cryptand, becomes that of the free radical anion (a^H=12.90 gauss, $\underline{g}$=2.00510) (Figure 3c). From the peak positions and line-broadening at inter-mediate K^+ concentrations, we conclude that for 0.1 $\underline{\text{M}}$ K^+ the two species are present at approximately equal concentrations and that their individual lifetimes are 5×10^{-8} sec in Me_2SO at 25°. We have an amazingly slow interconversion between the free ion and the ion-pair, even with potassium as the counterion. Table I presents data for other ion-pairs of cyclopentane-1,2-semidione.

Table I

Values of $\underline{a}^H$ (gauss) and $\underline{g}$-Value for Cyclopentane-1,2-semidiones

Species	$\underline{a}^H$ (4 equivalent α-hydrogen atoms)	$\underline{g}$-value
Free ion	12.90	2.00510
K^+ ion pair	13.65	2.00492
Na^+ ion pair	13.83 ($\underline{a}^{Na}$=0.52)	2.00489
Li^+ ion pair	14.52 ($\underline{a}^{Li}$=0.55)	2.00472
K^+ 18-crown-6 ion-pair	13.33	2.00500

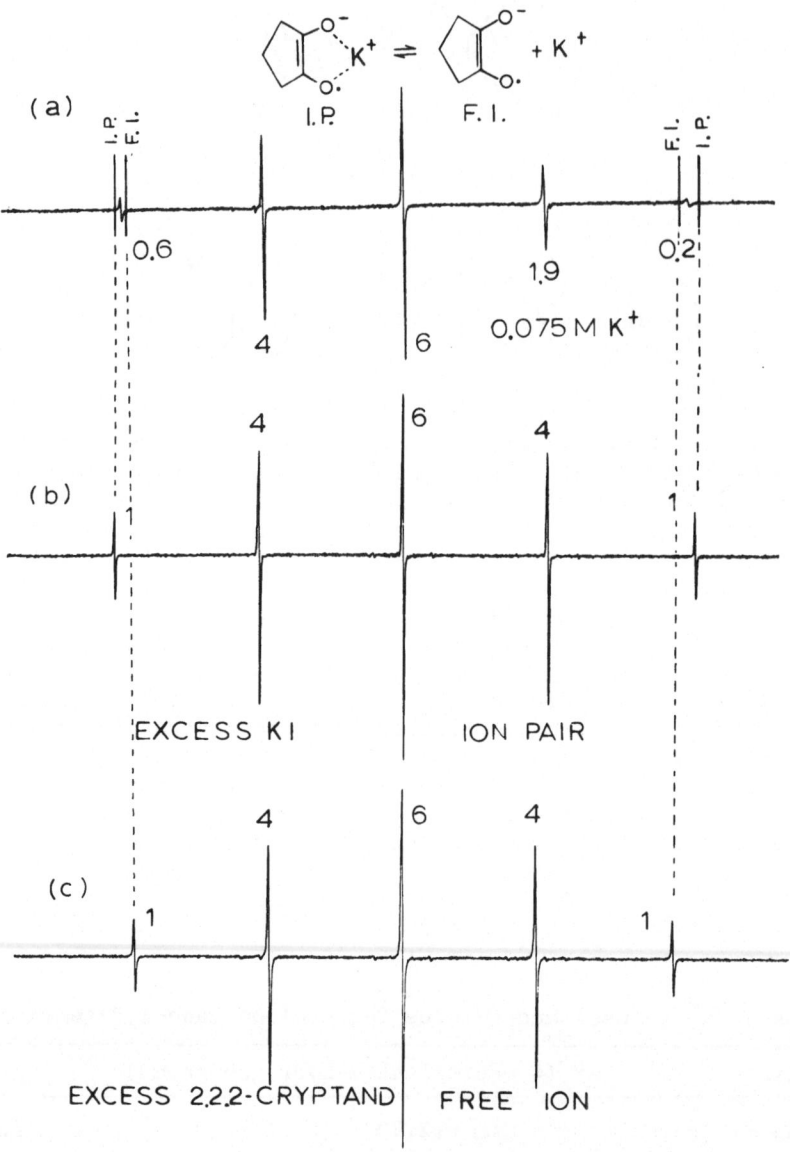

Figure 3. ESR spectra showing ionic equilibria in cyclopentane-1,2-semidione; Me$_2$SO, K$^+$, 25°:   (a) approximate 1:1 mixture of free ions and ion-pair (0.075 $\underline{M}$ K$^+$) showing high field line-broadening in the fast exchange mode; (b) the 1:1 ion-pair observed by adding 0.6 $\underline{M}$ KI to (a); (c) the free ion observed by adding an excess of 2.2.2-cryptand to (a).

The $\underline{g}$-values and $\underline{a}^H$-values of *cis*-semidiones are linearly related. The ion-pair has a low $\underline{g}$-value and a high value of $\underline{a}^H$ because of increased spin density at the carbonyl oxygen atom (resonance structure $\underline{5}$), whereas in the free ion resonance structure $\underline{6}$

$\underline{5}$ $\underline{6}$

is more important, leading to a lower value of $\underline{a}^H$ and a higher $\underline{g}$-value. (The $\underline{g}$-value of an oxygen-centered radical is about 2.015, whereas for a carbon-centered radical, g=2.002.) Cyclobutane- and cyclohexane-1,2-semidiones do not show appreciable line-broadening under the standard ESR conditions of 0.05-0.1 $\underline{M}$ K^+ in Me_2SO. In the case of cyclobutanesemidione the free ion now greatly predominates, whereas for cyclohexanesemidione the ion-pair is the predominant species. In both cases, the ion-pair or free ion can be produced by the appropriate addition of either excess K^+ or 222-cryptand.

In Me_2SO solution the lithium hyperfine splitting in a *cis*-semidione ion-pair is easily resolved ($\underline{a}^{Li}$=0.6 gauss). Hyperfine splitting by sodium is more difficult to observe, in part because of the exchange process shown in equation 11. In the presence of crown ethers, or at very low Na^+ concentrations, this exchange is slowed, and hyperfine splitting by a single sodium ion can be observed ($\underline{a}^{Na}\cong$0.55 gauss). In some cases (e.g., biacetyl radical anion) the sodium hyperfine splitting has an asymmetry suggesting more than one 1:1 ion-pair, possibly an in-plane ($\underline{7}$) and an out-of-plane ($\underline{8}$) complex. Lithium gives a symmetrical hyperfine splitting, presumably from the analog of $\underline{7}$.

$$\tag{11}$$

Perfluorobiacetyl radical anions can be studied easily in ether solvents, since this semidione can be prepared by alkali metal or iodide ion reduction of the dione.[22] Ion-paired *cis*-semidione is formed with Li^+ and Na^+. In addition, there is evidence that an ion-pair is formed in the *trans* species; the trifluoromethyl groups are

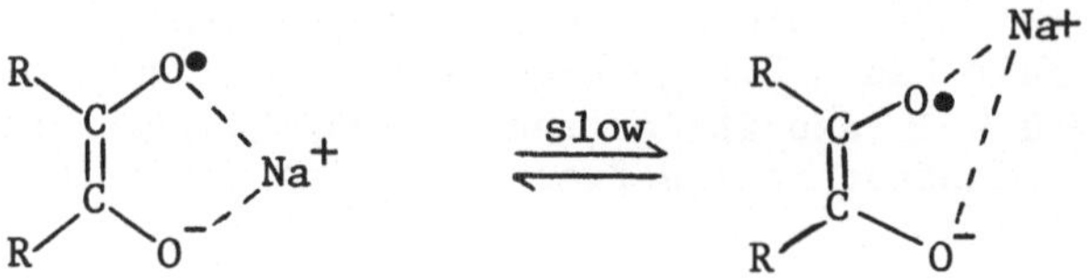

$$\underline{7} \qquad\qquad\qquad \underline{8}$$

magnetically non-equivalent (eq 12). Time-averaging of the trifluoromethyl groups is independent of the concentration of M^+ in solution,
indicating an intramolecular process. The rate of this process
increases as M^+ is changing from Li^+ to Na^+ to K^+ to Rb^+. With Cs^+
there is no evidence for the reaction shown in equation 12. In THF

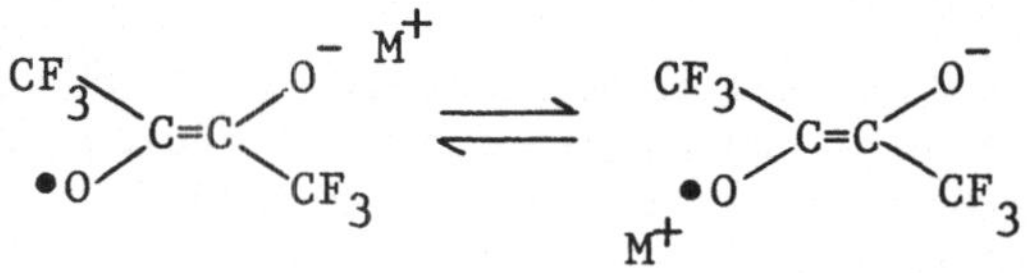

$$\tag{12}$$

solution with mixtures of alkali metal cations (lithium and sodium,
lithium and potassium, or lithium and rubidium) four distinct ion-
paired species can be detected by ESR spectroscopy, with no indication of time-averaging among them at 25° ($\tau > 10^{-5}$ sec). The four
species for Li^+ and Na^+ are $\underline{9} - \underline{12}$. In ethyl ether with Li^+ the

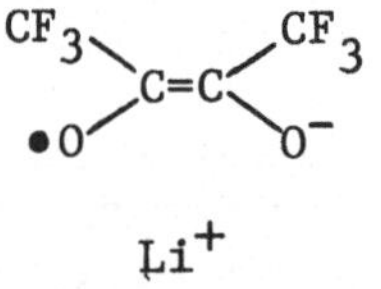

Li^+

$a^F = 11.14(6)$, $a^{Li} = 0.56$ G

$\underline{9}$

Na^+

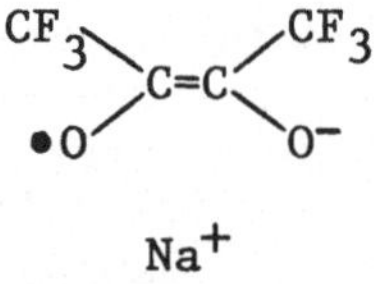

$a^F = 11.36(6)$, $a^{Na} = 0.49$ G

$\underline{10}$

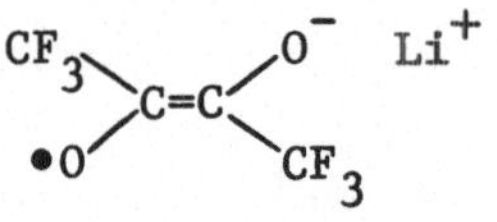

$a^F = 5.25(3)$, $11.35(3)$ G
selective line-broadening
in the slow exchange mode

$\underline{11}$

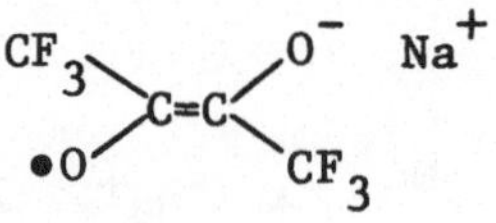

$a^F = 8.62(6)$ G
selective line-broadening
in the fast exchange mode

$\underline{12}$

trans-perfluorobiacetyl semidione exists as a triple ion with hyperfine splitting by two lithium ions (13), but the cis-semidione is
still present as the 1:1 ion-pair.

$$CF_3 \diagdown \qquad O^- \quad Li^+$$
$$C=C$$
$$Li^+ \quad \bullet O \diagup \qquad \diagdown CF_3$$

$\underline{a}^F = 7.9(6)$, $\underline{a}^{Li} = 0.24(2)$ G

<u>13</u>

Potassium hyperfine splitting is not observed for the alkyl semidiones. However, it can be observed even in Me$_2$SO solution for the substituted semidione radical dianions such as <u>14</u> (Figure 4).[23]

$$CH_3 \diagdown \qquad O \bullet$$
$$C=C$$
$$O^- \diagup \qquad \diagdown C=O$$
$$K \overset{+}{\text{---}} O$$

$\underline{a}^H = 5.57(3H)$, $\underline{a}^K = 0.45$ G

<u>14</u>

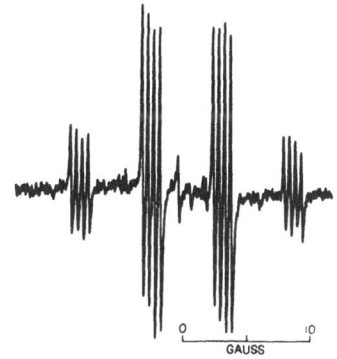

$a^H_{CH_3} = 5.37$ GAUSS

$a^K = 0.45$

Figure 4. Potassium hyperfine splitting observed for <u>14</u> in Me$_2$SO solution.

USE OF THE SEMIDIONE SPIN LABEL TO STUDY CONFORMATIONAL EQUILIBRIA

1,2-Semidiones can be considered to be paramagnetic derivatives of olefins. Similarly, 1,4-semidiones are paramagnetic derivatives of 1,3-dienes. Conformational analysis of these radical anions by ESR spectroscopy is often easier than the corresponding analysis of the parent olefins by NMR spectroscopy. Thus, by ESR spectroscopy $\Delta H^{\ddagger}$ for ring flip of cyclohexane-1,2-semidione (eq 13, R=H) has been measured as 4 kcal/mol, which corresponds to a magnetic coalescence temperature of −85°.[24] This value can be compared with a $\Delta F^{\ddagger}$ of 5.3 kcal/mole observed for ring inversion of *cis*-3,3,4,5,6,6-d_6-cyclohexene in bromotrifluoromethane by PMR (coalescence temperature =−165°).[25] It appears that in this case the semidione is a reasonable spin label for the olefin.

$$(13)$$

$$\underline{15} \qquad\qquad\qquad\qquad\qquad \underline{16}$$

A 4-*t*-butyl group freezes the cyclohexane semidione into conformation $\underline{15}$ up to at least 90°. On the other hand, for a 4-methyl substituent both $\underline{15}$ and $\underline{16}$ are populated, with $\underline{15}$ being preferred by an enthalpy difference of 1.4 kcal/mole ([$\underline{16}$]/[$\underline{15}$]=0.13 at 40°).[24]

Cycloheptane-1,2-semidione is a rigid species ($\tau > 10^{-6}$ sec) up to at least 90°. From an analysis of the ESR hyperfine interactions it can be concluded that the staggered conformation $\underline{17}$ is the only one populated.[26] Cycloheptane semidione shows a rich long-range

$\underline{a}^H$(gauss)	C_3	C_4	C_5
equatorial	0.28	2.05	0.54
axial	6.60	0.28	<0.05

$$\underline{17}$$

hyperfine splitting due to π−σ delocalization. This interaction is maximized when the carbon-hydrogen bond involved forms one leg of a coplanar zigzag arrangement of bonds and the carbonyl carbon

p_z-orbital.[12] Further examples of this long-range interaction are given in structures 18 and 19a,b.

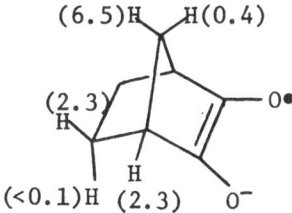

18, a^H in gauss (Ref. 12)

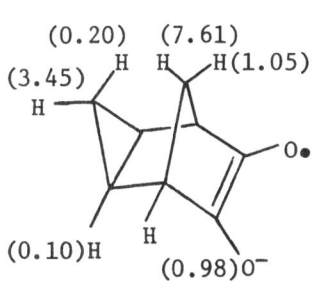

19a, a^H in gauss (Ref. 26)   19b, a^H in gauss (Ref. 26)

The cyclic C_8-1,2-semidione is conformationally labile, while the cyclic C_9-1,2-semidione is in a rigid staggered conformation up to +100°.[1] A reinvestigation of the conformational analysis of the medium-sized cyclic semidiones, free of ion-pairing effects by use of the cryptands, is in progress.

VALENCE ISOMERIZATION OF 1,2-SEMIDIONES

A. Cyclobutane-1,2-Semidione

Cyclobutane-1,2-semidione is not as easy to observe as are the C_5-C_{15} cyclic 1,2-semidiones. However, it can be prepared from the acyloin or 1,2-bis(trimethylsiloxy)cyclobutene by treatment with an approximately stoichiometric amount of potassium t-butoxide or potassium dimsylate in Me_2SO (ESR spectrum, Figure 5). If the siloxyalkene is treated with an excess of dimsylate anion no ESR signal is detected. If, now, oxygen is added, or if the excess dimsylate is neutralized with trimethylchlorosilane or an acid, the spectrum shown in Figure 6 results. Three semidiones are apparent –

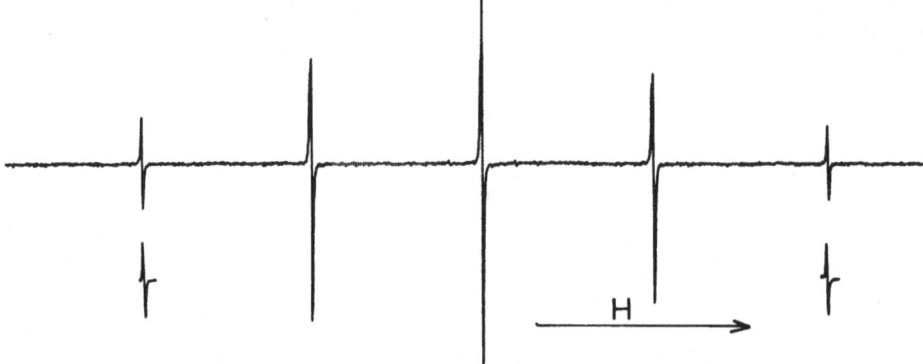

Figure 5. ESR spectrum of cyclobutane-1,2-semidione in Me_2SO (K^+); $\underline{a}^H$=14.0 gauss, $\underline{g}$=2.00508. The high and low field inserts show the peaks scanned within 15 sec of each other.

cyclobutane-1,2-semidione, *trans*-dimethylsemidione, and *trans*-methylethylsemidione, as well as minor amounts of the two *cis* acyclic semidiones. The only explanation we can offer is that a valence isomerization has occurred at the dianion stage, as shown in equation 14. The formation of dimethylsemidione and methylethylsemi-

$$\text{(structure)} \quad + \ 2 \ CH_3SOCH_2^- \ \rightarrow \ \text{(structure)} \ \rightleftharpoons \ \text{(structure)} \qquad (14)$$

dione from the acyclic dianion would occur by the reactions previously described in Scheme VI, which are repeated for the present case in Scheme X. The major effect of addition of oxygen in this system may be the consumption of the excess dimsylate anion.

Treatment of 2,3-bis(trimethylsiloxy)-1,3-butadiene with an excess of dimsylate anion gives a solution without appreciable paramagnetic products. Again, treatment with oxygen, or neutralization of the excess dimsylate anion, yields a mixture of cyclobutane-1,2-semidione, *trans*- and *cis*-biacetyl radical anion and *trans*- and *cis*-methylethylsemidione (see Scheme XI).[27] To observe cyclobutane semidione for an appreciable length of time one must adjust the system so that there is no significant concentration of the enediol dianion present (see Scheme XII). If the basicity of the solution is sufficiently high that the dianion concentration becomes appreciable, valence isomerization of the dianion becomes important, and the dimethyl and methylethylsemidiones are formed as the cyclobutane semidione decays.

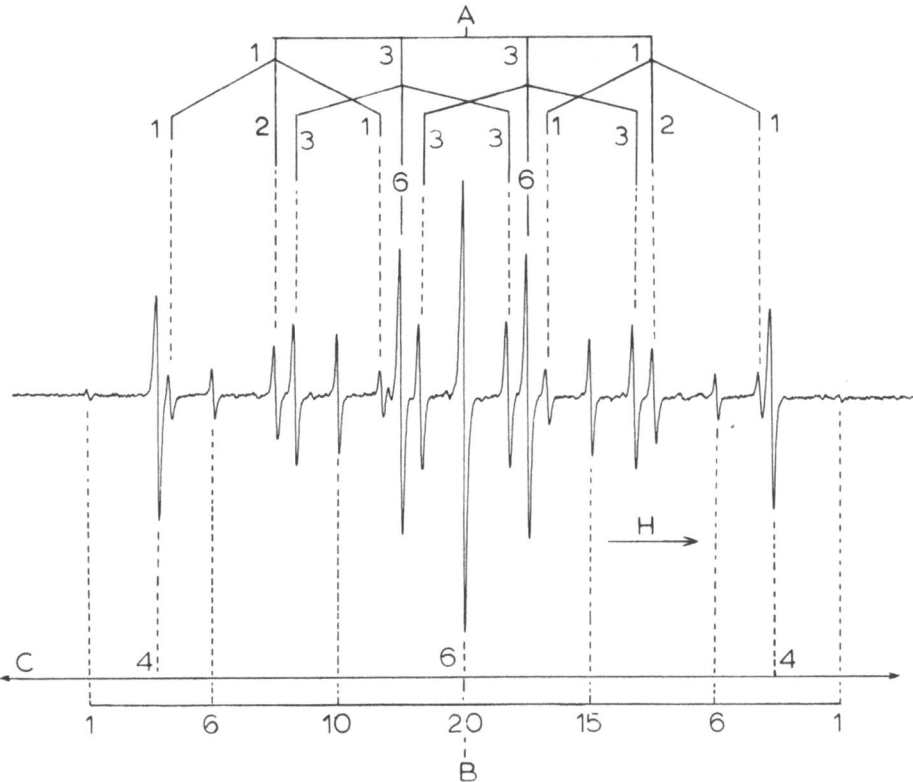

Figure 6. ESR spectrum of a mixture of *trans*-methylethylsemidione
(A), $\underline{a}^H$=5.75 (3H), 4.85 (2H); *trans*-dimethylsemidione (B), $\underline{a}^H$=5.6
(6H); cyclobutane-1,2-semidione (C) (its three central peaks),
$\underline{a}^H$=14.0 (4H) gauss. Minor amounts of the *cis* isomers of A and B can
be seen in the baseline; at higher signal intensities these species
can be identified: *cis*-methylethylsemidione, $\underline{a}^H$=7.05 (3H), 6.10
(2H); *cis*-dimethylsemidione, $\underline{a}^H$=7.0 (6H) gauss.

        It is not certain whether the radical anions participate in
the same valence isomerization as the dianions. If such a reaction
(eq 15) does occur, certainly the equilibrium must lie far on the
side of the cyclobutane-1,2-semidione, since we see no evidence for
the butadiene-2,3-semidione from ESR spectroscopy. Probably the
diene semidione reacts to give the cyclobutane semidione rapidly
and irreversibly. One might expect that the same mixture of radical
anions could be generated from biacetyl in the two stage process
shown in Scheme XIII. However, despite many attempts, the only
paramagnetic product we have ever observed from biacetyl plus base
is the 2,5-dimethyl *p*-benzosemiquinone resulting from condensation.

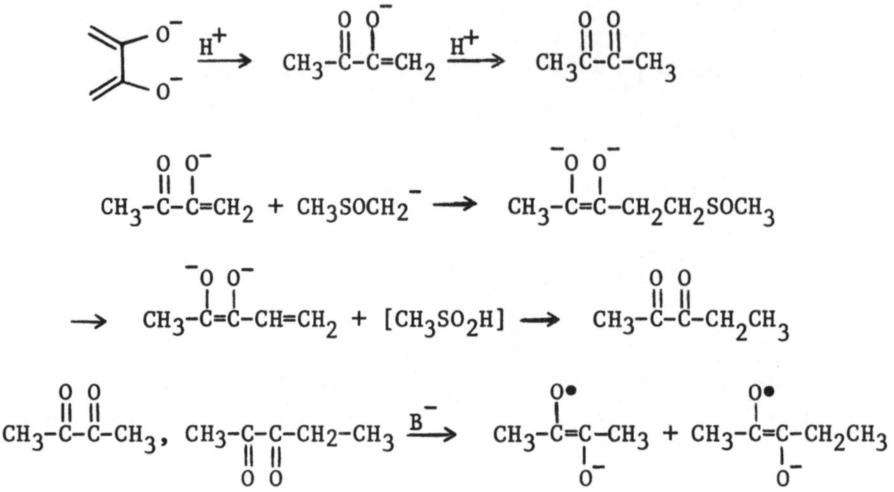

Scheme X

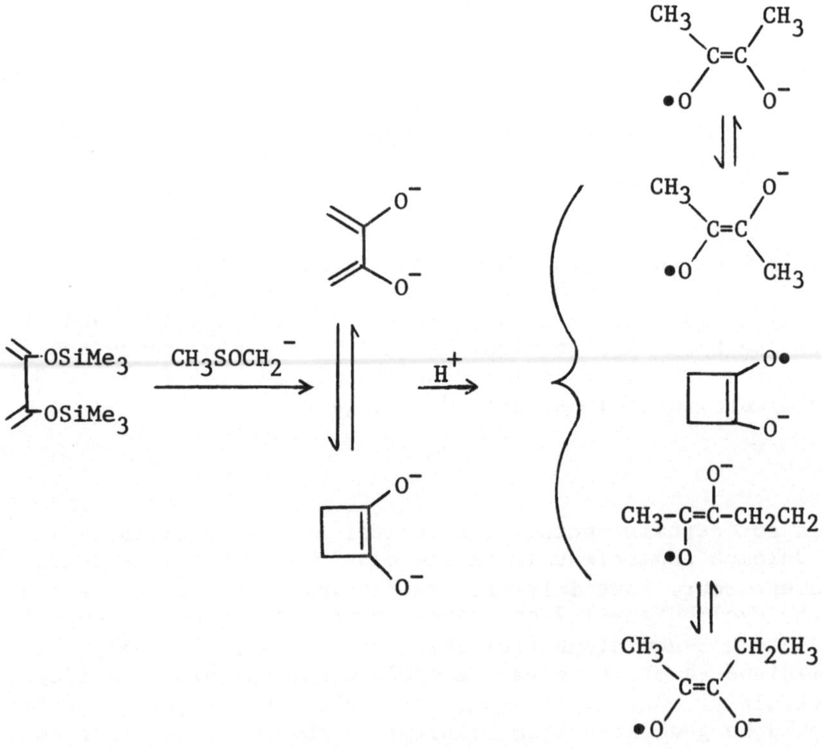

Scheme XI

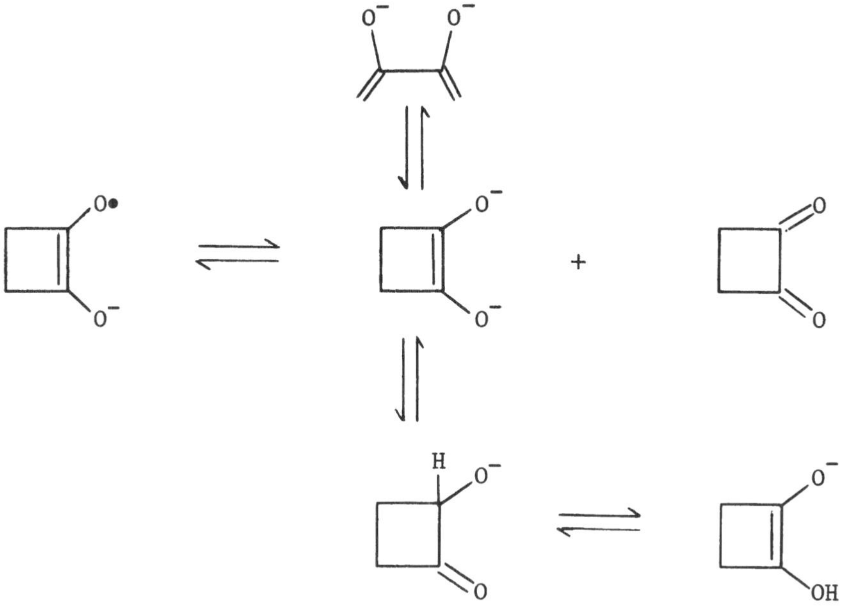

Scheme XII

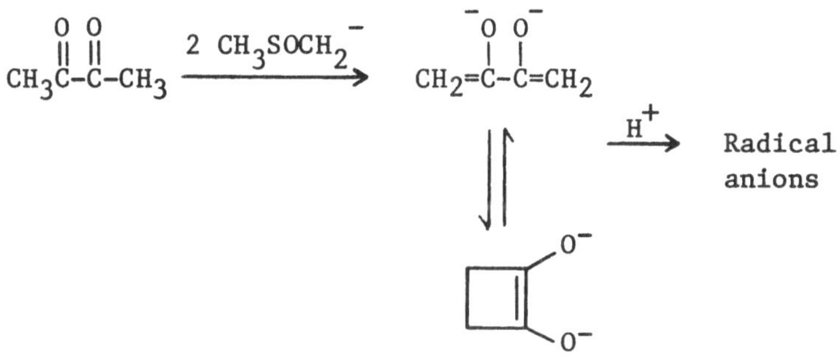

— — — — —

Scheme XIII

— — — — —

(15)

The butadiene-cyclobutene valence isomerization has been found in other systems; the cyclooctatriene derivative shown in equation 16, for example, gives the bicyclic semidione.[28]

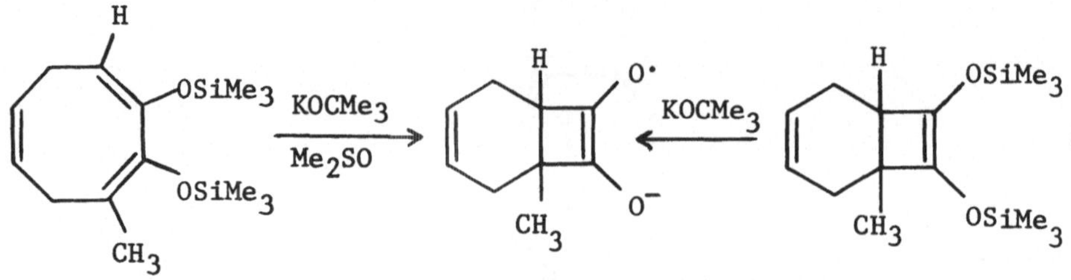

$$\text{(16)}$$

B. Cyclooctatriene 1,2–Semidione

Among the possible valence isomers of cyclooctatriene-1,2-semidione (20 ⇌ 21) are 22 and 23. The conversion of 21 to 22 would be similar to the conversion of butadiene semidione to cyclo-butane-1,2-semidione. However, when 8-(benzoyloxy)cycloocta-2,4,6-trien-1-one was treated with potassium t-butoxide/Me2SO in a flow system, no evidence for 20, 21 or 22 was obtained. Instead, a well-resolved spectrum of a mixture of 24 and 25 was obtained.[29] This is consistent with valence isomerization of 20 ⇌ 21 to 23 (see Scheme XIV).

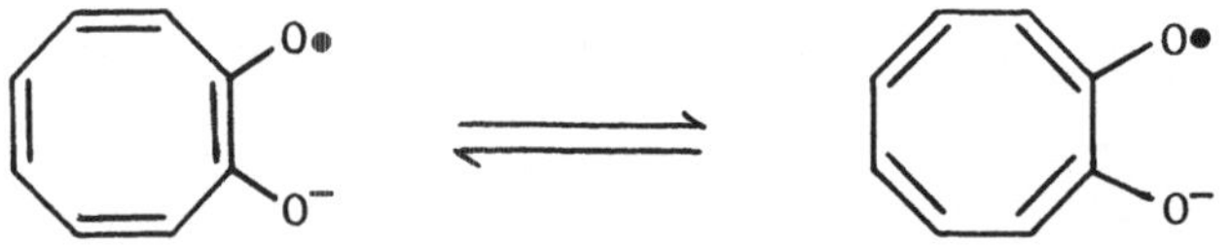

20   21

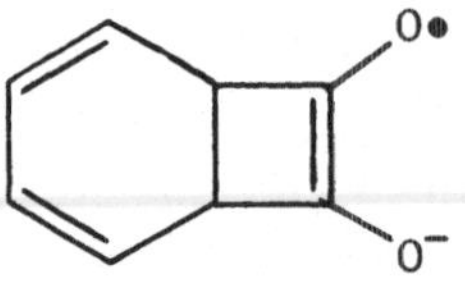

22 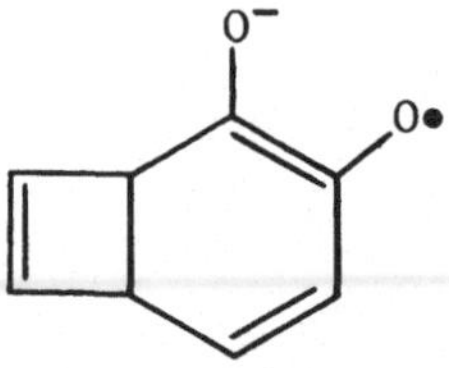   23

The mixture of 24 and 25 is also formed when 2-acetoxy-3-keto-bicyclo[4.2.0]oct-4-ene is treated with base and oxygen in Me2SO (eq 17).[30] The formation of 25 from 24 is not surprising. Numerous other substituted o-quinones or o-semiquinones undergo facile hydrox-ylation at vacant p-positions. For example, the reaction shown in equation 18 occurs with great ease.[31] The process whereby 24 is formed is more surprising. However, this is not an isolated example of this phenomenon. In the corresponding 1,4-semidione 27, or its 2,3-benzo derivatives, we have observed the isomerization of 27 to 28 catalyzed by traces of oxygen (Scheme XV, R=H, CH3).[32]

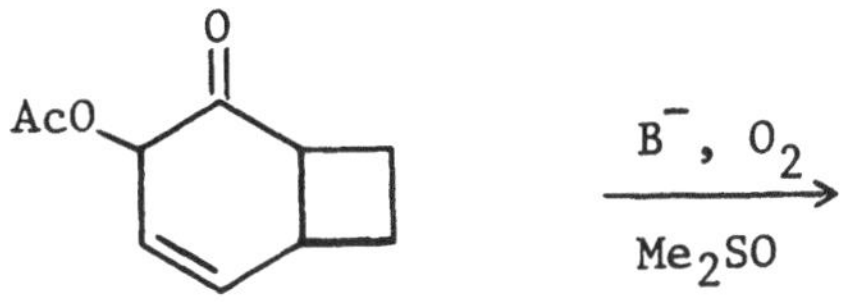

$$\text{(17)}$$

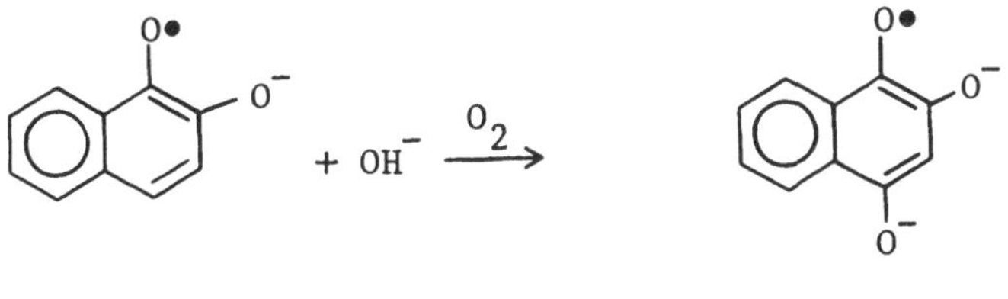

$$\text{(18)}$$

$$\underline{26}$$

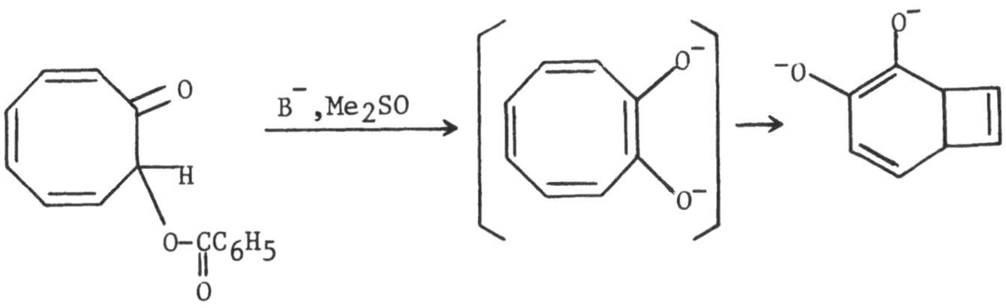

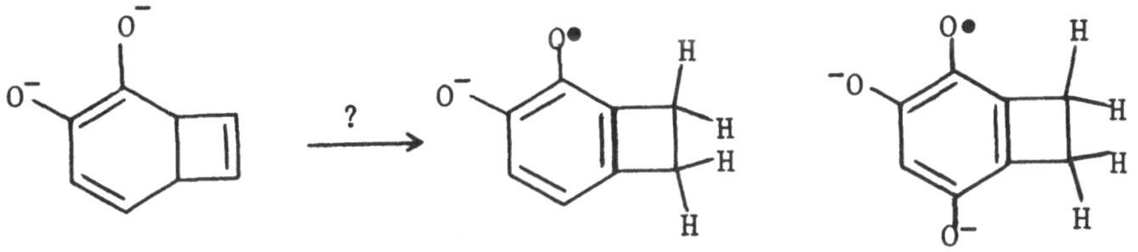

$$\underline{24} \qquad\qquad \underline{25}$$

low g, a^H=5.06,   high g, a^H=1.52,
2.48, 1.37(2),   3.59(2), 0.51(2)
0.51(2) gauss   gauss

Scheme XIV

   Furthermore, when R = CH$_3$ the isomerization of $\underline{27}$ yields $\underline{28}$
with the *cis* geometry.[33] We have suggested that the isomerization
of $\underline{27}$ to $\underline{28}$ is catalyzed by the corresponding cyclobutadiene derivative, $\underline{29}$ (Scheme XVI). We now see how $\underline{24}$ might be formed from a
precursor with the bicyclic ring structure of $\underline{23}$ (the reaction could

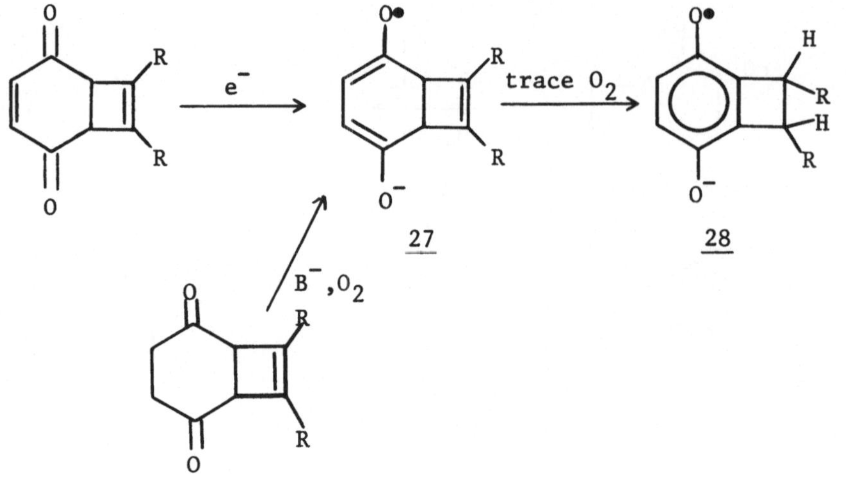

Scheme XV

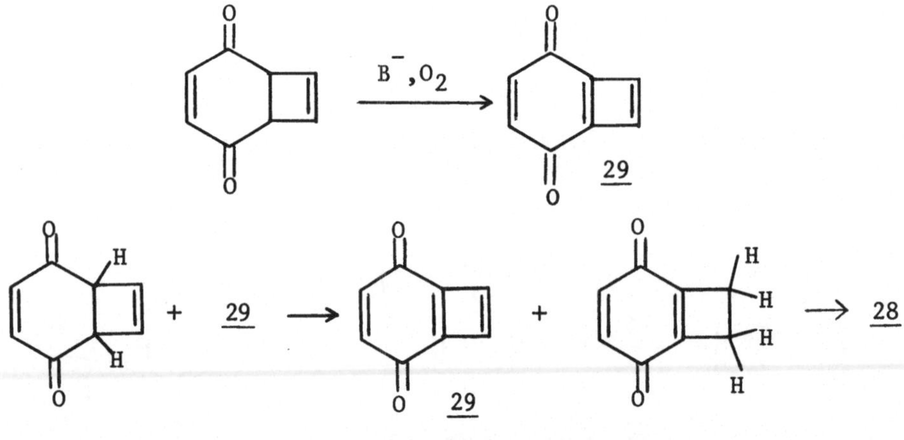

Scheme XVI

occur at the dione, dianion, or radical anion stage, e.g., as shown in equation 19).

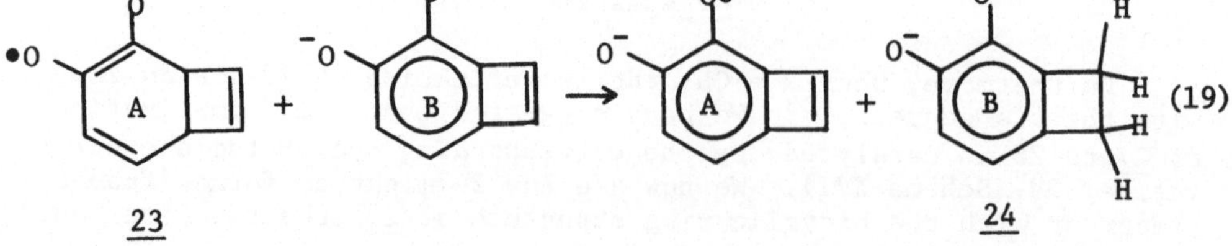

(19)

Another set of reactions requiring the isomerization of $20 \rightleftharpoons$ 21 to 23 involves the formation of 26, a ten-carbon compound, from eight-carbon precursors of $20 \rightleftharpoons 21$. The radical dianion 26 is a very stable semiquinone and is the paramagnetic end-product of many base-catalyzed autoxidation.[31] It is the only paramagnetic product detected upon treatment of the three possible cyclooctatriene-semidione precursors shown in Scheme XVII with base and oxygen in Me$_2$SO.[30] The formation of 26 is easily explained if the o-quinone analogous to 23 is actually involved (Scheme XVIII).

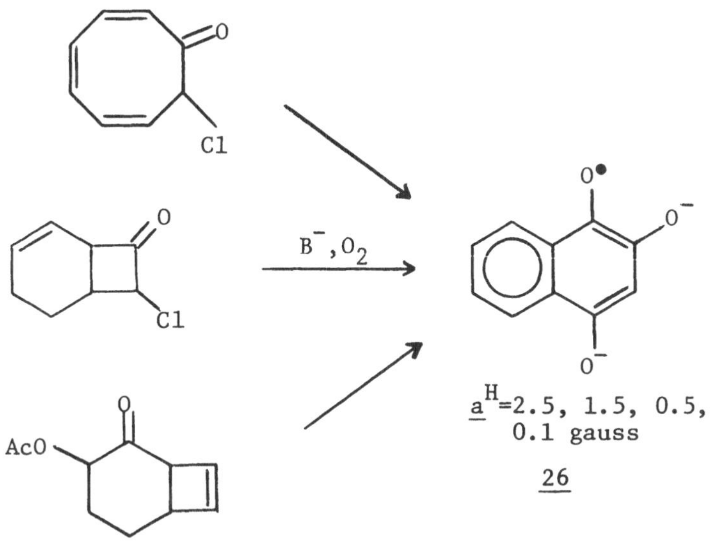

a^H=2.5, 1.5, 0.5, 0.1 gauss

26

Scheme XVII

The results shown in Scheme XVIII require that at some stage 22 be converted to $20 \rightleftharpoons 21$, and that the observed products arise from the further valence isomerization of 21 to 23. From the results obtained to date it seems doubtful that the monocyclic cyclooctatriene-1,2-semidione (20) will ever be observed under static conditions. The parent monocyclic dione, 3,5,7-cyclooctatriene-1,2-dione, is a known compound, but a very unstable one.[34]

We have attempted to synthesize methyl derivatives of this system by two acyloin condensations (eqs 20 and 21). *In situ* reduction of 30 by NaK in DME yielded an ESR spectrum consistent with formation of 31: a^H=0.93 (2H), 0.45 (2H), 0.08 (6H); g=2.00495.[30] Exposure to oxygen produced a new signal, with g=2.00446, a^H=2.6 (2H), 0.45 (6H). These g-value and a^H-values are appropriate for the o-semiquinone 34. We conclude that 34 is considerably more

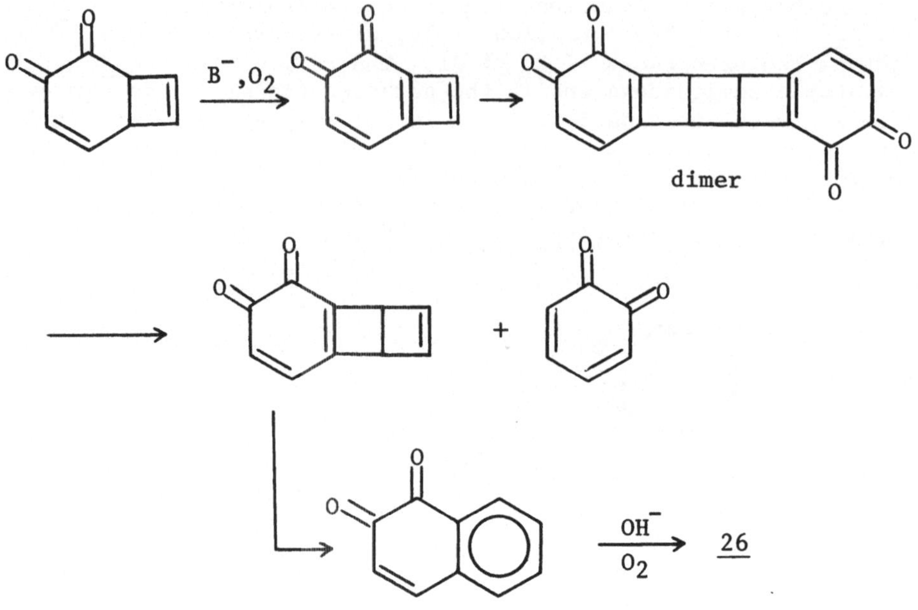

Scheme XVIII

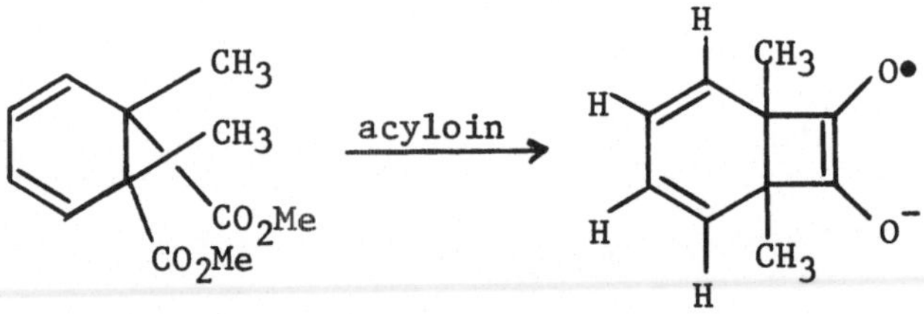

- - - - - -

$$(20)$$

<u>30</u>   <u>31</u>

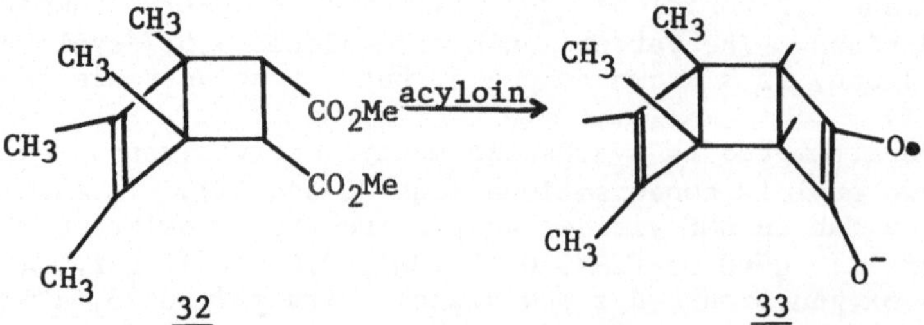

$$(21)$$

<u>32</u>   <u>33</u>

stable than $\underline{29}\cdot^-$ or $\underline{35}$, which we have never detected. The formation of $\underline{34}$ requires a valence isomerization of the type shown in
Scheme XIX.

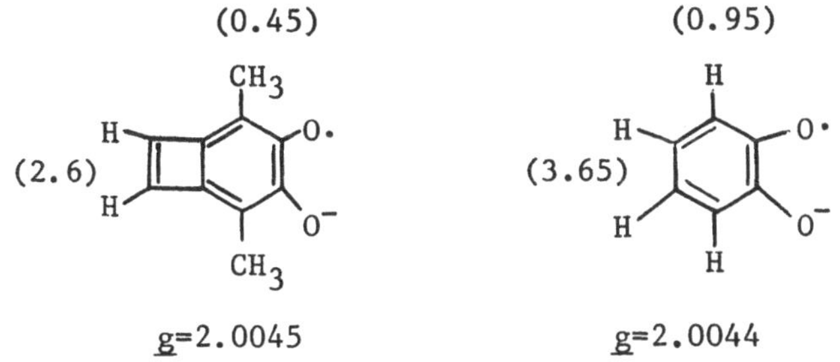

$\underline{34}$

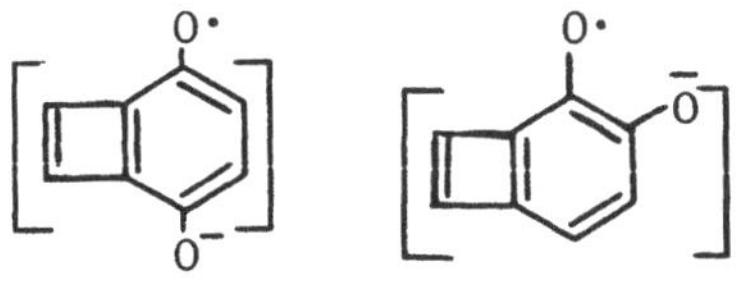

$\underline{29}\cdot^-$ $\underline{35}$

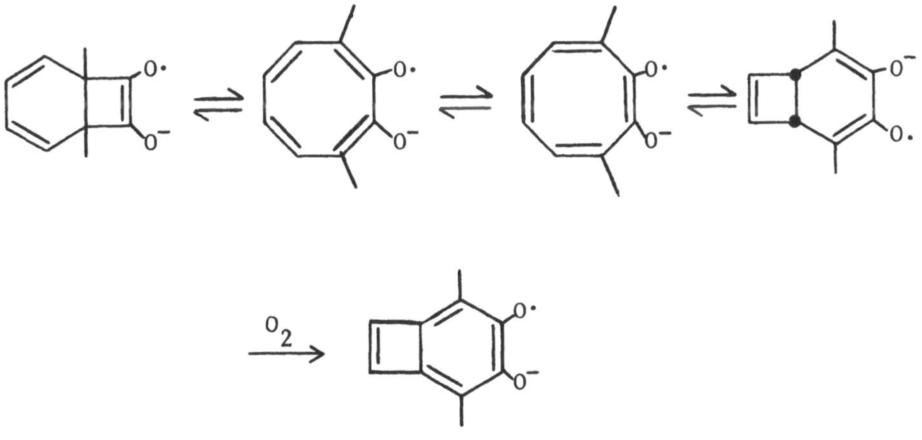

$\underline{34}$

Scheme XIX

The bicyclic diester $\underline{32}$ presented a somewhat different reactivity pattern. Depending upon how the acyloin condensation was

Table II

Semidiones Formed from Acyloin Condensation of 32

Conditions	Semidiones Detected
(1) NaK, Xylene, Me$_3$SiCl, reflux, followed by KOCMe$_3$/Me$_2$SO	A + B + C
(2) Product of (1) chromatographed over florisil	B + C
(3) NaK, Et$_2$O, Me$_3$SiCl, -10° followed by KOCMe$_3$/Me$_2$SO	B + C
(4) K, DME, 10°, in ESR cell	A

performed, different mixtures of three semidiones with the same g-values were obtained.[30] Table II summarizes the results. Semidiones A, B and C had the following values of a^H in gauss: A, 7.2 (2H), 0.48 (6H), 0.24 (6H); B, 13.25 (2H); C, $\overline{12}$.93 (1H), 10.86 (1H), 2.10 (1H). When the three semidiones were present in the same solution it was observed that B and C decayed more rapidly than A, but there was no indication that B or C were converted to A. Semidione B appears to be 33. The values of $\underline{a}^H$ for cyclobutane-1,2-semidiones are typically in the range of 11–14 gauss.[35] Semidione C is obviously an over-reduced species; structure 36 is a possibility. Semidione A (as shown in Figure 7) has a set of splitting constants rather similar to those of 31. Structure 37 is thus suggested. Certainly the magnitude of the $\underline{a}^H$ values observed for A

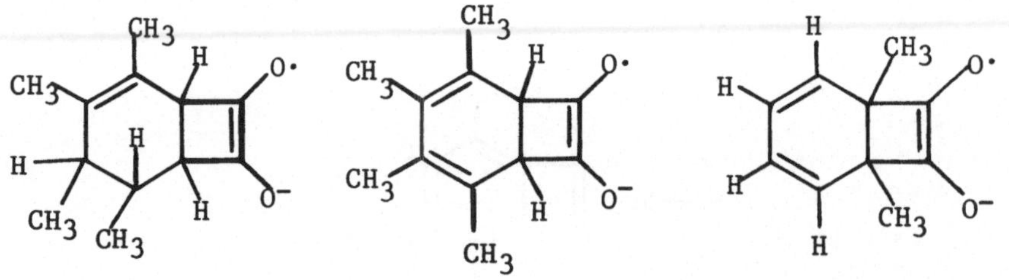

36

37

a^H=0.48 (6H),
$\overline{0}$.24 (6H) G

31

a^H=0.93 (2H),
$\overline{0}$.45 (2H) G

exclude the planar monocyclic cyclooctatriene-1,2-semidione. However, the magnitude of the large triplet splitting in A is a bit perplexing. We have never seen $\underline{a}^H$ in a cyclobutane-1,2-semidione below 10 gauss.[35] Perhaps in 37 (and 31) there is a serious geometric and electronic perturbation due to homoaromaticity, as by a

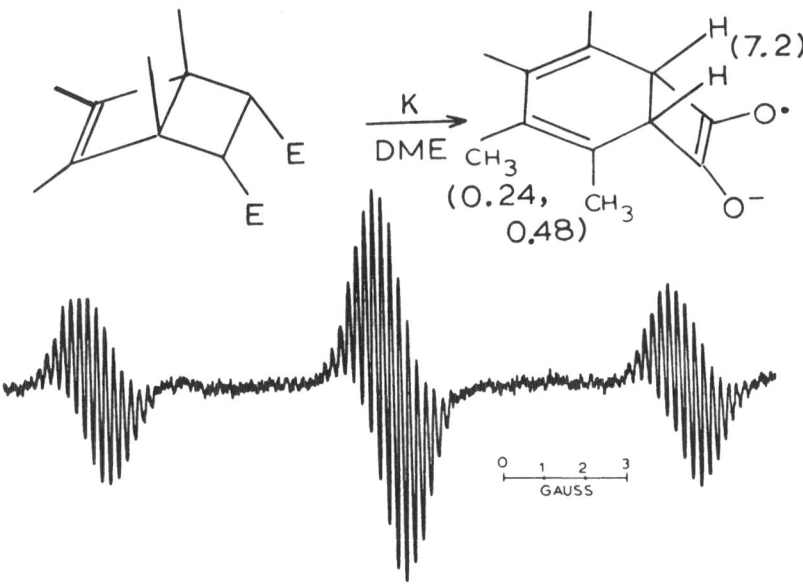

Figure 7. ESR spectrum of 2,3,4,5-tetramethylbicyclo[4.2.0]octa-2,4-diene-7,8-semidione in dimethoxyethane at 25°.

contribution from structure 38. Alternatively, the possibility exists that semidione A might be the non-planar cyclooctatriene-1,2-semidione 39, or perhaps a rapidly time-averaged equilibrium mixture of 39 and 37. Most likely structure 37 or 38 is the correct assignment. The formation of 37 from 32 can be formulated as shown in Scheme XX.

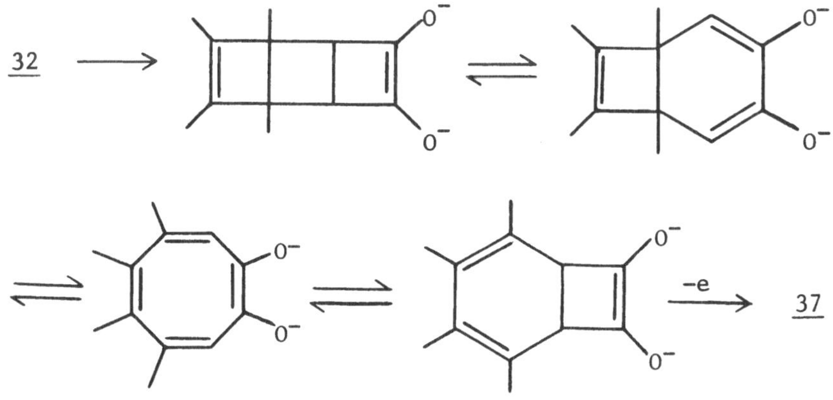

Scheme XX

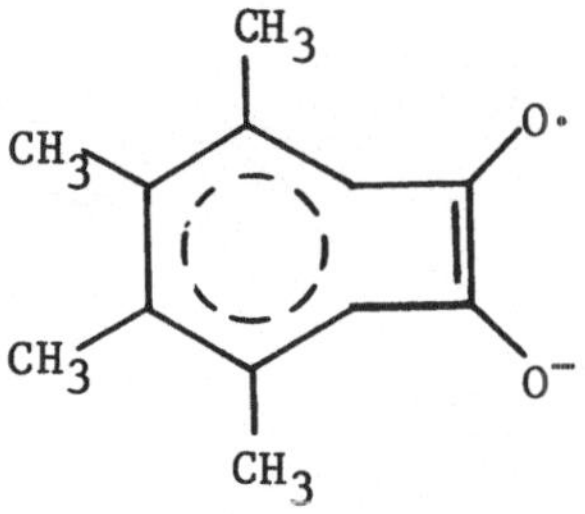

38 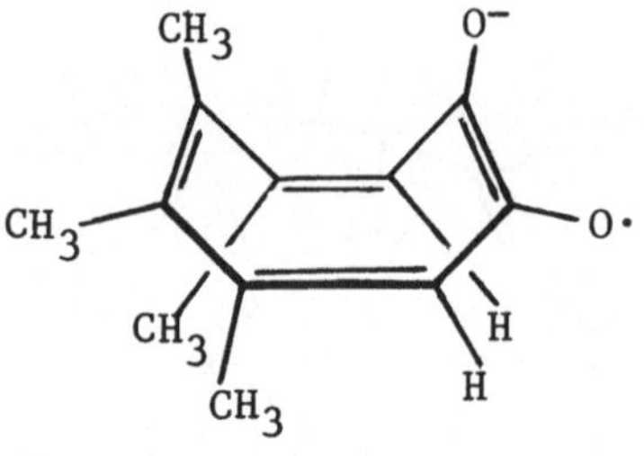   **39**

All of our results in this system indicate that 22 or 23 are
the preferred valence isomers, with alkyl substitution favoring 22
over 23.

C.   Valence Isomers of o-Benzosemiquinone

The two valence isomers of o-benzosemiquinone are 40 and 41.

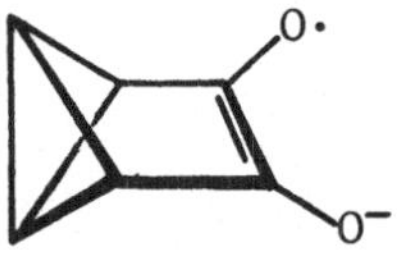

40 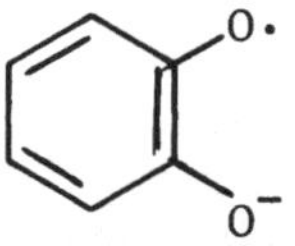 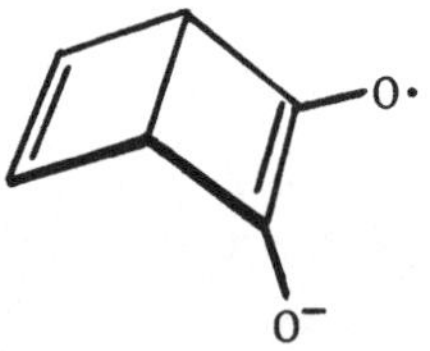   **41**

We have investigated the rearrangement of 40 to o-benzosemiquinone.
Reduction of the dione 42^{36} at -40° by electrolysis produces a
broad ESR singlet which is possibly due to 40⁻. Electrolysis at
25° cleanly produced tetramethyl-o-benzosemiquinone; valence isomerization has occurred at either the semidione or dianion stage.[37]

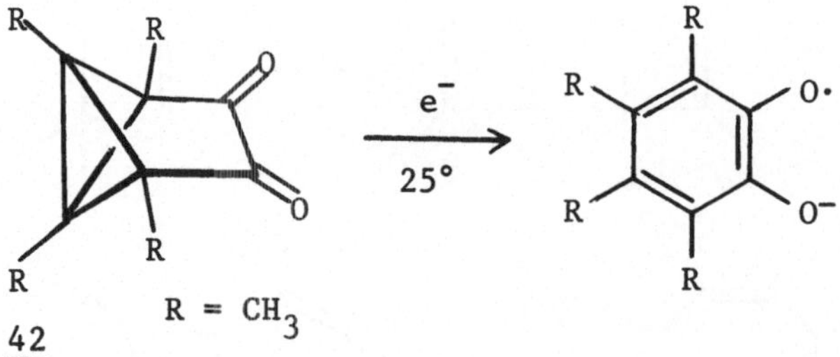

$$ \qquad (22) $$

Semidione 43 has been detected, but it is unusually unstable.[38]
Probably in this case the valence isomerization shown in equation
23 also occurs. The isomerization of 44 (eq 24) is being investigated as another possible route to the cyclooctatriene-1,2-semi-
dione.[39] However, it appears that semidione 44 is stable at 25°.

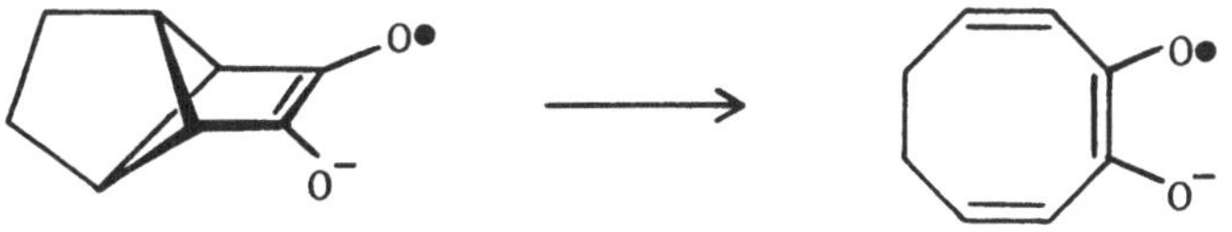

$$(23)$$

$$\underline{43}$$

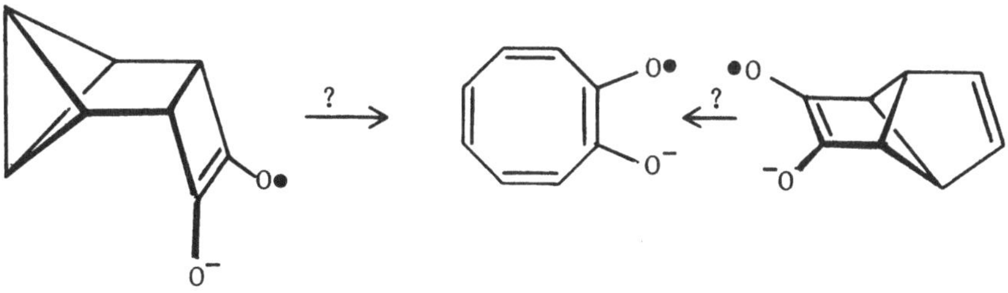

$$\xrightarrow{\text{?}} \quad \xleftarrow{\text{?}}$$

$$(24)$$

$$\underline{44}$$

$\underline{a}^H$=9.31 (2H), 2.81 (1H),
0.20 (1H), 0.10 (2H) gauss
in DMF

D. Cyclopropane Ring Opening in 1,2-Semidiones

Attempts to prepare bicyclo[2.1.0]pentane-2,3-semidione from acyloin condensations have been unsuccessful. Instead, formation of cyclopentane semidione is observed.[40] Again, the dianion can undergo the cyclobutane - butadiene valency isomerization (eq 25).

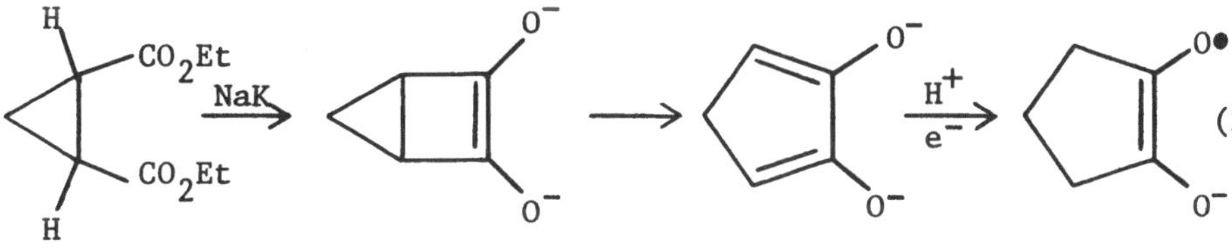

$$\xrightarrow{\text{NaK}} \quad \longrightarrow \quad \xrightarrow[\text{e}^-]{\text{H}^+} \quad (25)$$

Bicyclo[3.1.0]-hexane-2,3-semidiones with a *syn*-6-substituent undergo an isomerization to give the *anti*-6-substituted derivative.[40] We have deduced that this reaction involves the sequence of steps shown in Scheme XXI.

The isomerization shown in Scheme XXI occurs more rapidly the higher the concentration and the stronger the base. This, as well as hydrogen-deuterium exchange at the *exo* C-4 position, implicates the radical dianion. The reaction also requires a steric driving force at 25°. The isomerization is not observed with a *syn*-6-

deuterium substituent and is faster when R = C_2H_5 or CH_2OCH_3 than
with R = CH_3. The reaction goes to completion, presumably because
the isomer with the *anti*-C-6 alkyl group is made thermodynamically
more stable by a decrease in non-bonded interactions.

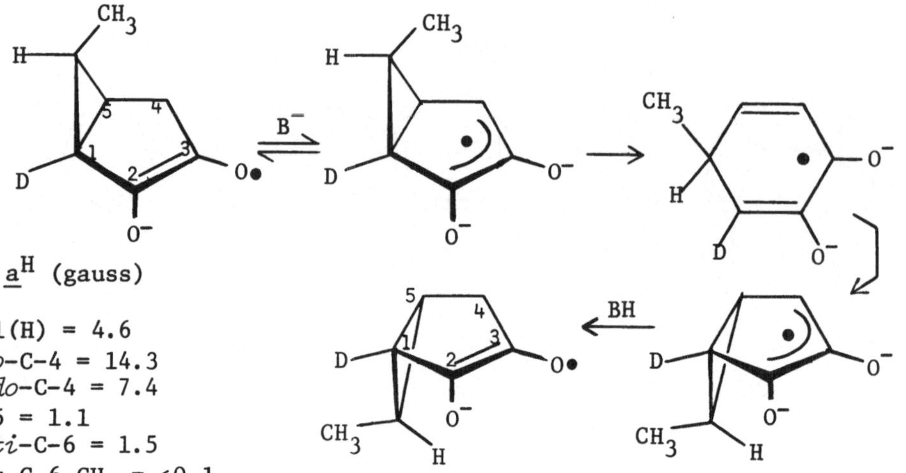

$\underline{a}^H$ (gauss)

C-1(H) = 4.6
exo-C-4 = 14.3
endo-C-4 = 7.4
C-5 = 1.1
anti-C-6 = 1.5
syn-C-6-CH_3 = <0.1

$\underline{a}^H$ (gauss)

C-1(H) = 4.3
exo-C-4 = 14.6
endo-C-4 = 7.6
C-5 = 0.40
syn-C-6 = 0.90
anti-C-6-CH_3 = 0.40

Scheme XXI

Analogous rearrangements have been observed in labeled bicyclo-
[3.1.0]hex-2-en-1-yl radicals.[41] For this allyl radical, not only
can *exo-endo* isomerization at C-6 be detected from product analysis,
but the cyclohexadienyl radicals can actually be observed by ESR
spectroscopy (eq 26), especially in an adamantane matrix.

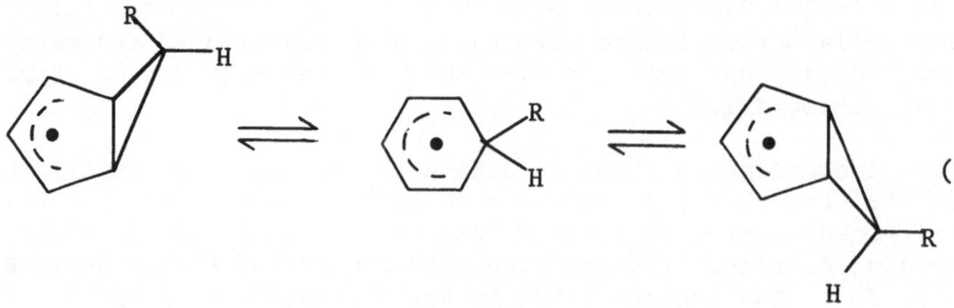

(26)

   The ketone 45 reacts with base and oxygen to form 26 as the
final paramagnetic oxidation product (for spectrum, see Figure 8).[31]
The isomeric ketone 46 also gives the same final product (eq 28).
When the amount of oxygen is carefully controlled in the oxidation
of 45, or when the keto acetate derivative of 46 is treated with
base in the absence of oxygen, semidiones 47 and 48 are formed
(Scheme XXII).[39] We see no interconversion between 47 and 48 at 25°
or in the presence of UV-irradiation. However, molecular oxygen
converts either 47 or 48 cleanly to 26.

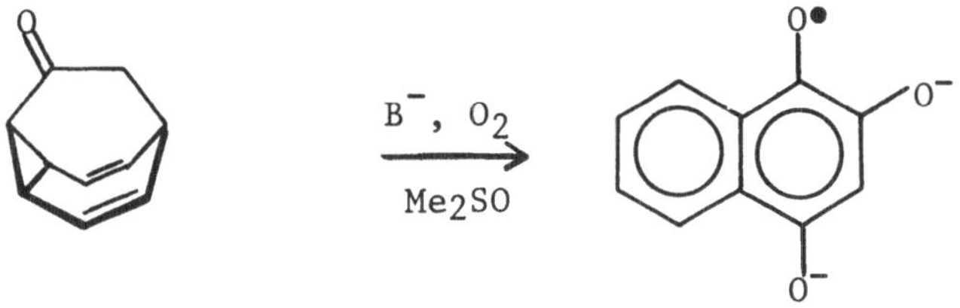

$$\xrightarrow[\text{Me}_2\text{SO}]{\text{B}^-,\ \text{O}_2}$$

(27)

45   26

$$\xrightarrow[\text{Me}_2\text{SO}]{\text{B}^-,\ \text{O}_2}$$ 26   (28)

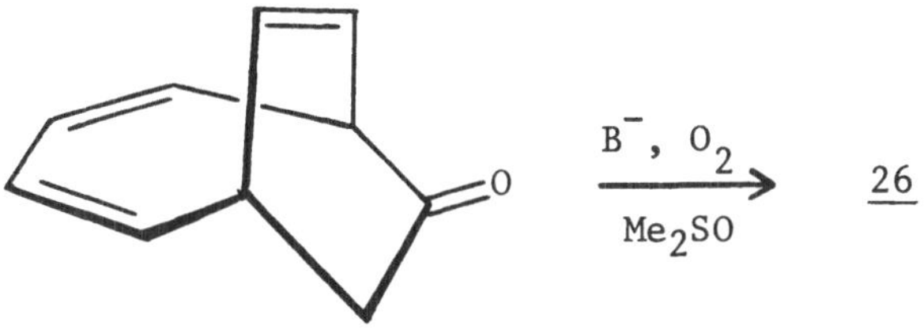

46

   The overall course of the reaction scheme seems to involve
isomerization at the dione stage as shown in Scheme XXIII. The
1,2-naphthosemiquinone[31] can be detected as a precursor to 26. The
interesting conversion of 45 to 47, and the exact structure of 47
is still under investigation.[42] Presumably the bullvalene semidione (49) is involved in the conversion of 45 to 47.

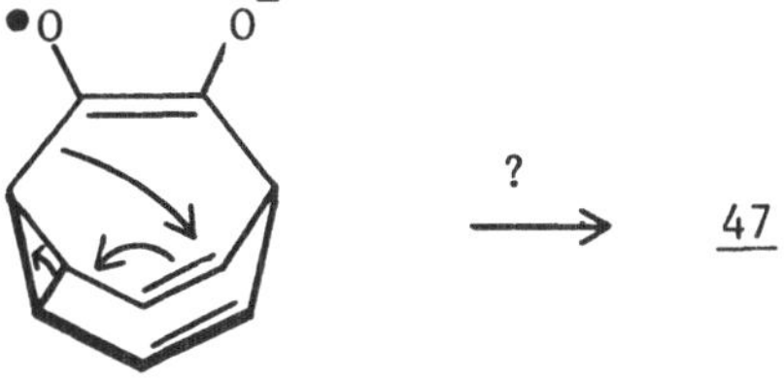

$$\xrightarrow{?}$$ 47   (29)

49

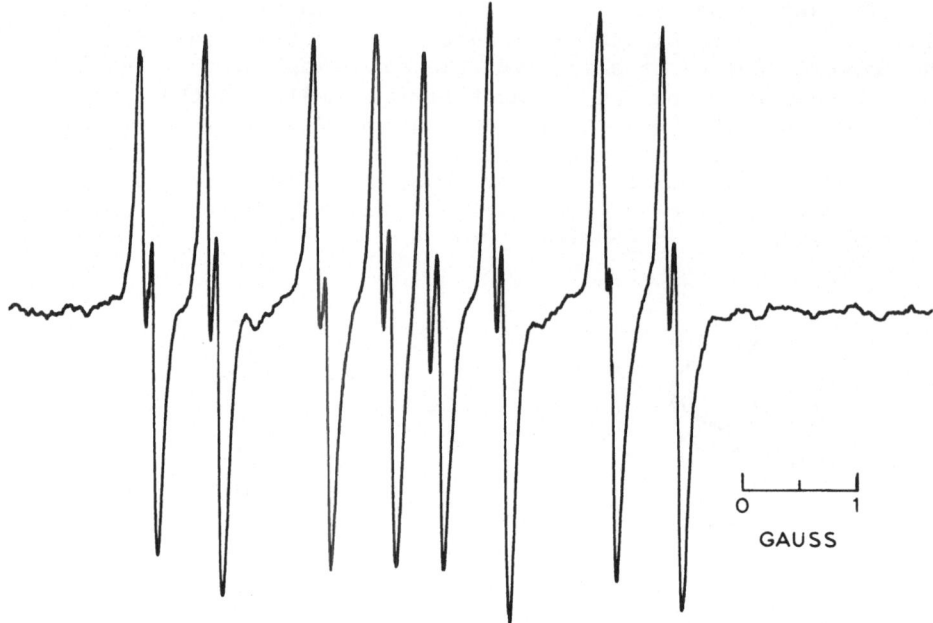

Figure 8. ESR spectrum of the semiquinone radical dianion of 2-
hydroxynaphtho-1,4-quinone in Me$_2$SO.

$\underline{a}^H$=5.46 (2H), 0.86 (2H), 0.27 (2H), 0.13 (2H)

$\underline{47}$

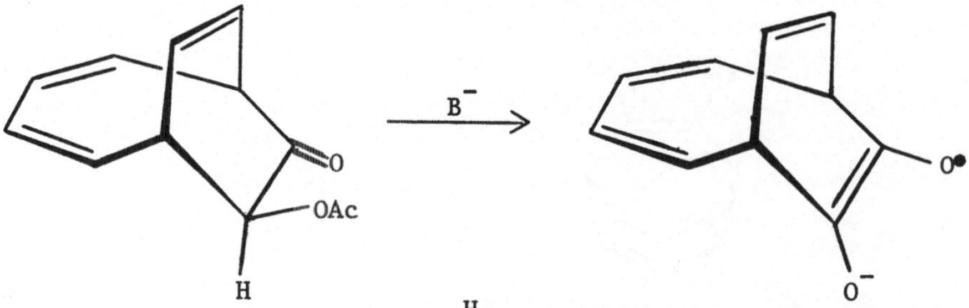

$\underline{a}^H$=2.95 (2H), 1.30 (2H), 0.33 (4H)

$\underline{48}$

Scheme XXII

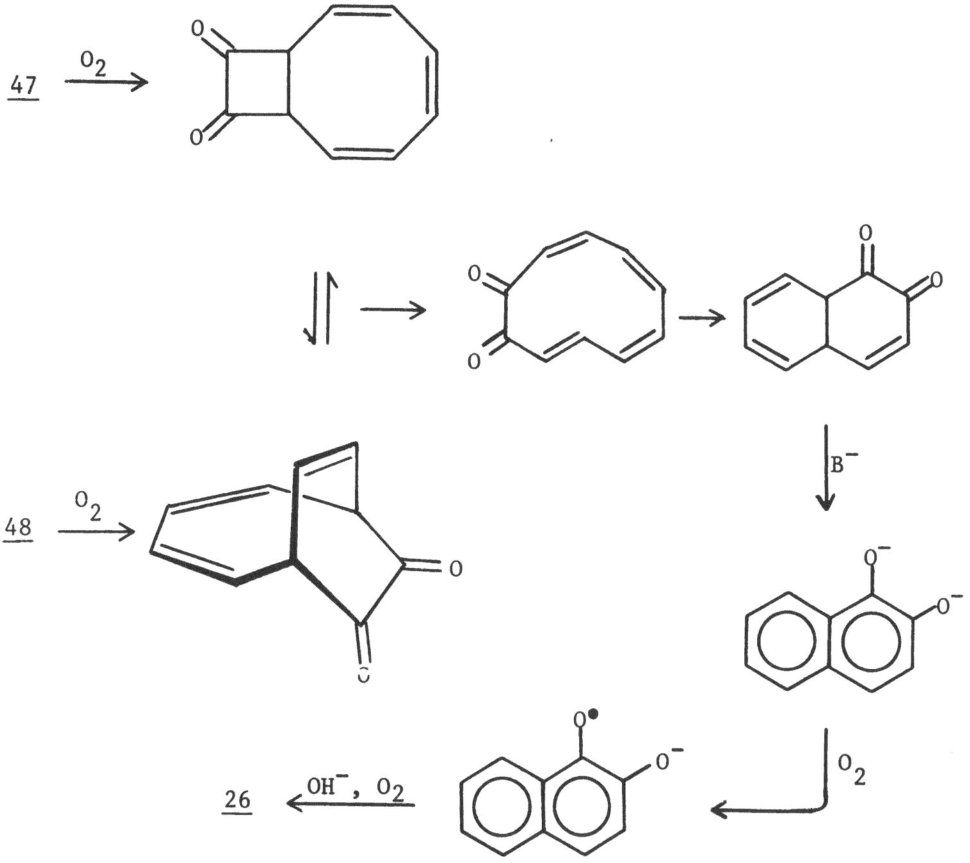

Scheme XXIII

SIGMATROPIC REARRANGEMENTS FOLLOWED
BY THE OBSERVATION OF 1,2-SEMIDIONES

Unsaturated bicyclic semidiones such as 50-52 show no evidence
of the 1,3-sigmatropic rearrangements indicated in equations 30-32,
even with UV irradiation. However, when *exo*- or *endo*-hydroxynor-
bornene-2-one is treated with base in DMSO the first ESR signal
that can be detected (~0.1 sec after mixing) is that of the rearranged semidione, bicyclo[3.2.0]hept-2-ene-6,7-semidione (53).[43]
On the other hand when 3-alkoxy-norbornene-2-one is oxidized in
basic solution, formation of the unrearranged semidione is observed.
The oxidation of the alkoxy ketone can form 50 without going through
the enediol dianion, whereas the hydroxy ketone in basic solution
must be in equilibrium with the dianion (Scheme XXIV). We conclude
that 54, but not 50, undergoes a facile thermal 1,3-sigmatropic

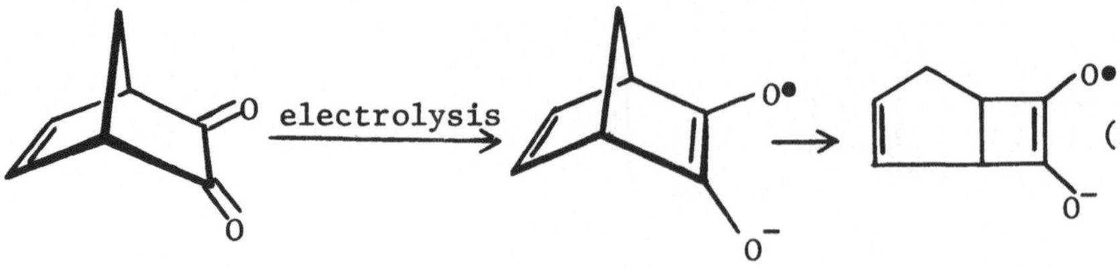

(30)

$\underline{50}$ $\underline{53}$

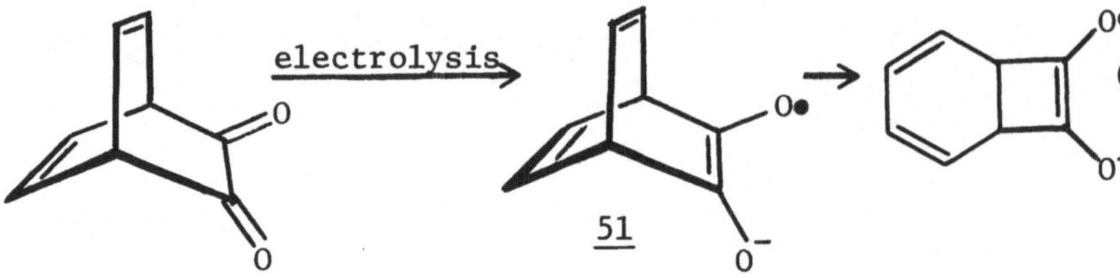

(31)

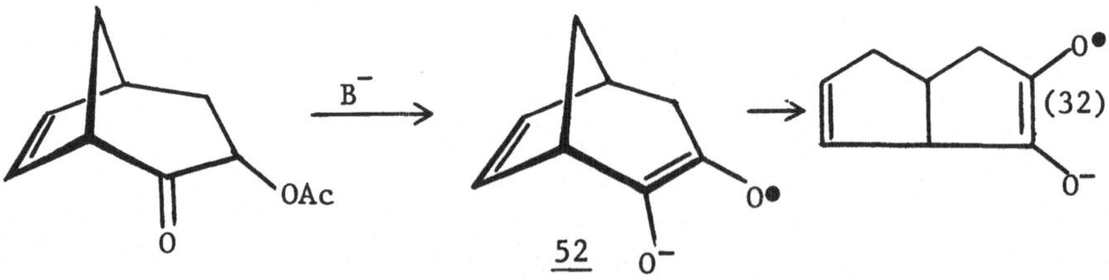

$\underline{51}$

(32)

$\underline{52}$

rearrangement (part b of Scheme XXIV). The rearrangement of $\underline{54}$ to $\underline{55}$ is reversible. Thus, all three monomethyl derivatives of $\underline{54}$ and $\underline{55}$ as shown in Scheme XXV produced roughly the same ratio of the isomeric semidiones $\underline{56}$ and $\underline{57}$.[43] The rearrangement, $\underline{54} \rightleftharpoons \underline{55}$, may or may not be concerted. A possible stepwise rearrangement is shown in Scheme XXVI, which nicely predicts that the dianion should rearrange more readily than the radical anion because formation of the substituted ketene intermediate relieves electrostatic repulsion.

In the bicyclo[3.2.1]octene system, oxidation in basic solution of ketones $\underline{58}$ and $\underline{59}$[44] gives rise to the unrearranged isomeric bicyclo[3.2.1]hept-2-ene-6,7-semidiones without any indication of isomerization (Scheme XXVII). In these cases the intermediate dianions do not rearrange.

On the other hand, ketone $\underline{60}$ gives rise to a semidione whose well-resolved hyperfine splitting pattern can only be rationalized as being that of the rearranged structure $\underline{61}$ (Scheme XXVIII). It is easy to see why the bridgehead carboalkoxy group in $\underline{60}$ would

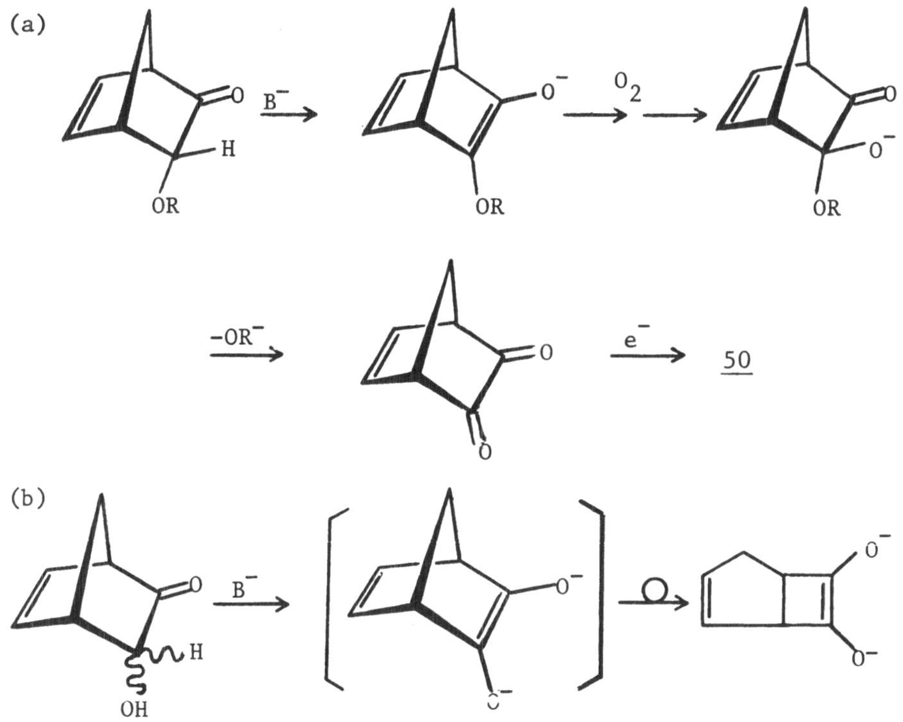

(a)

(b)

Scheme XXIV

facilitate the bicyclic ⇌ monocyclic rearrangement in the dianion. In 61 the ester function is conjugated with the double bond, which may explain why the rearrangement proceeds to yield exclusively 61.

In one case we have observed a photochemical 1,3-sigmatropic migration, presumably at the radical anion stage. Semidione 62 gives an equilibrium mixture of 62 and 63 when irradiated with UV light (Scheme XXIX).

A 1,4-sigmatropic rearrangement has been found to occur readily when semidiones of the general structure 64 are exposed to oxygen.[40] The semidione ESR signal disappears and a new ESR signal, due to 65, gradually grows (eq 33). This process appears to involve a 1,4-sigmatropic rearrangement to form a cyclopropanol anion or radical, which then undergoes ring opening (Scheme XXX).

VALENCE ISOMERIZATION OF 1,4-SEMIDIONES

The valence isomerizations shown in equations 34-36 have been examined. The isomerization seen in equation 34 is a paramagnetic

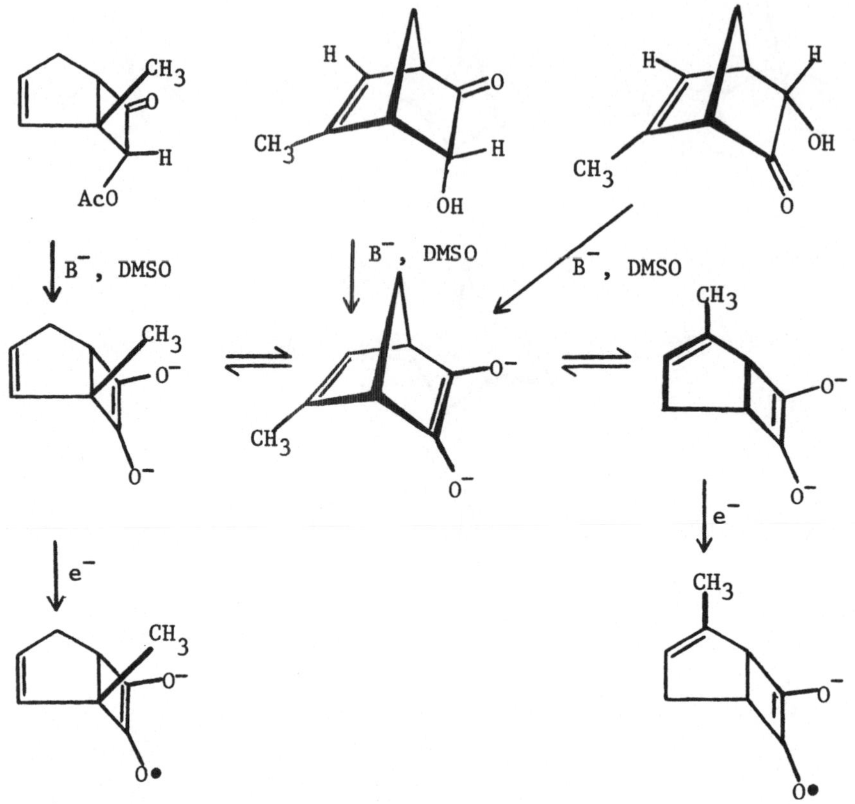

Scheme XXV

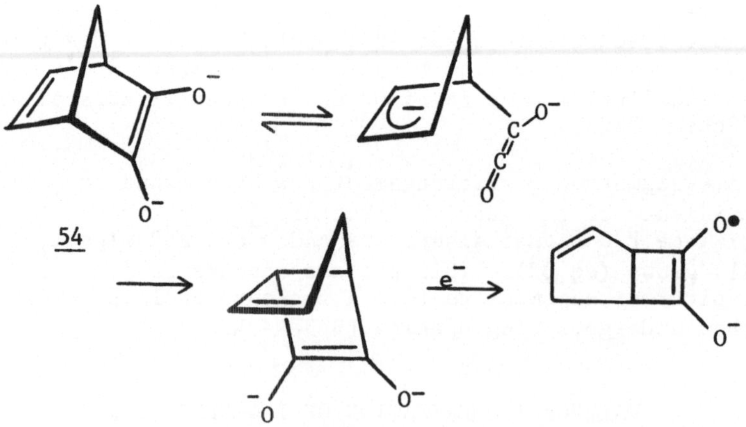

Scheme XXVI

R = CH$_3$

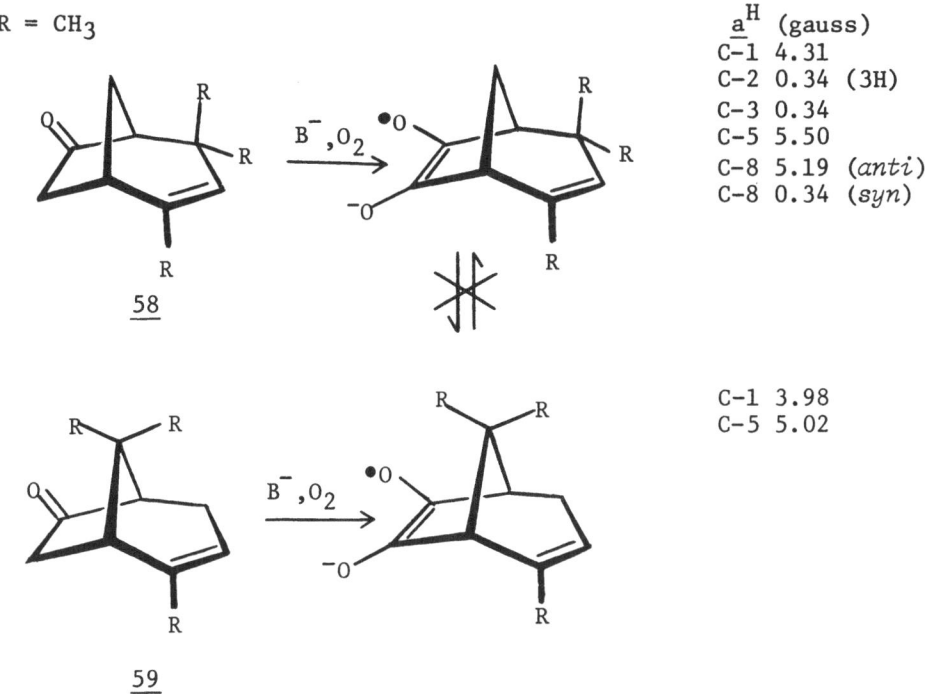

$\underline{a}^H$ (gauss)
C–1 4.31
C–2 0.34 (3H)
C–3 0.34
C–5 5.50
C–8 5.19 (*anti*)
C–8 0.34 (*syn*)

C–1 3.98
C–5 5.02

Scheme XXVII

analogue of the norcaradiene–cycloheptatriene valence isomerization, which, for the parent system, lies far on the side of cycloheptatriene. On the other hand, the 1,4-semidione 66, prepared as shown in Scheme XXXI, is preferred to the valence isomer 67. This is easily ascertained by an examination of the analogue of 66 having methyl groups at the bridgehead positions. For the monocyclic structure 67 it is expected that $\left|\underline{a}^H\right| \cong \left|\underline{a}_{CH_3}^H\right|$, since it is well known (e.g., for the ethyl radical) that $\left|Q_{\circ}^H\right| \cong \left|Q_{\circ}^H\right|$. Instead, the values of the hyperfine splitting $\underset{CH}{CH} \quad \underset{CCH_3}{CCH_3}$ constants listed in Chart 1 were observed, as expected for the bicyclic structure.

The reason for the preference of structure 66 over 67 is undoubtedly that the 1,4-semidione is stabilized by the same factors which stabilize the corresponding enedione. The enedione in the bicyclic system is a stable molecule, whereas in the monocyclic system a rational valence bond structure cannot be drawn.

We next asked whether 67 (R = CH$_3$) might be thermally accessible and thus allow the *syn–anti* isomerization of 66 ⇌ 68 to occur, as was observed in the bicyclo[3.1.0]hexane system (Scheme XXI). The enedione precursors of 66 (R = *exo*-CH$_3$) and 68 (R =

E = EtO$_2$C-, R = CH$_3$

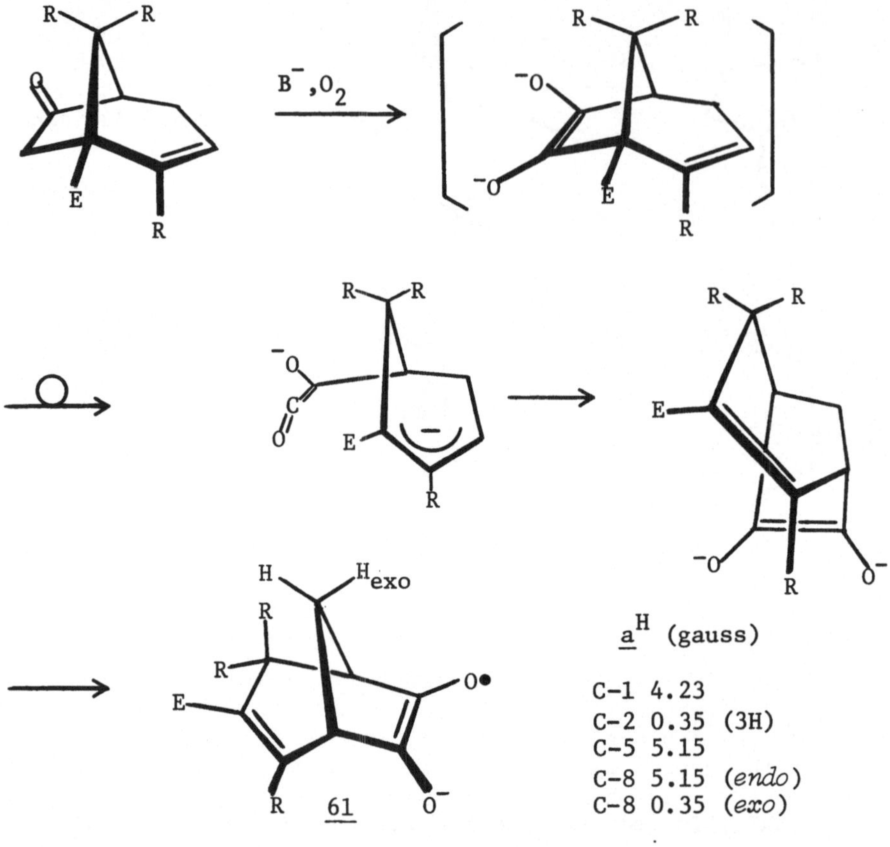

$\underline{a}^H$ (gauss)

C-1 4.23
C-2 0.35 (3H)
C-5 5.15
C-8 5.15 (*endo*)
C-8 0.35 (*exo*)

Scheme XXVIII

R = CH$_3$

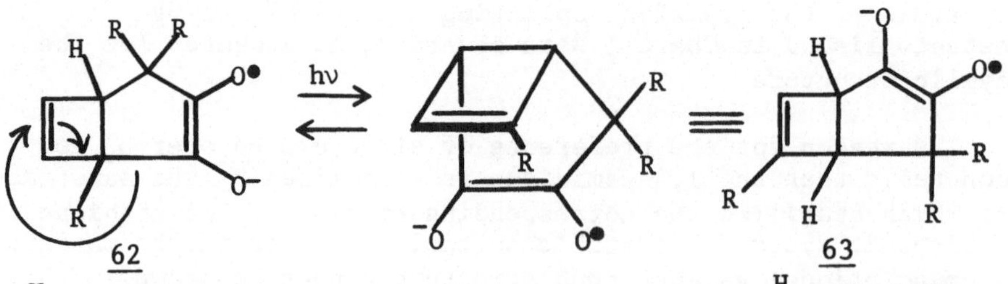

$\underline{62}$

$\underline{a}^H$=0.91 (1H),
   0.40 (3H),
   0.14 (2H) G

$\underline{63}$

$\underline{a}^H$=8.20 (1H),
   0.40 (3H),
   0.21 (1H),
   0.14 (1H) G

Scheme XXIX

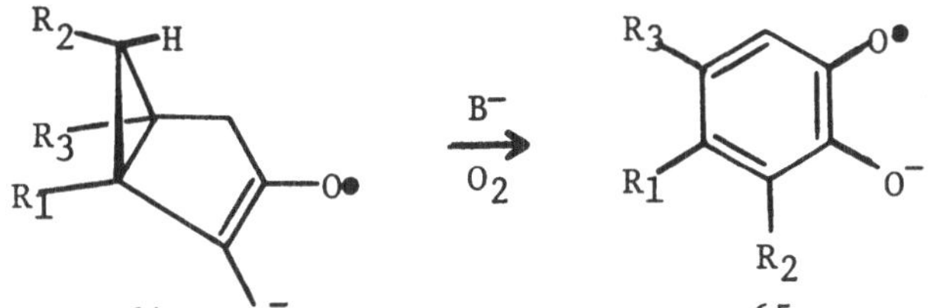

(33)

– – – – – – – –

* = radical or anion

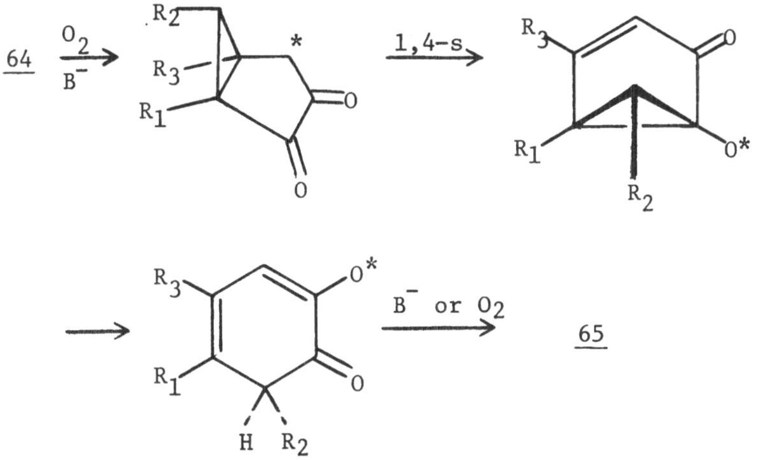

Scheme XXX

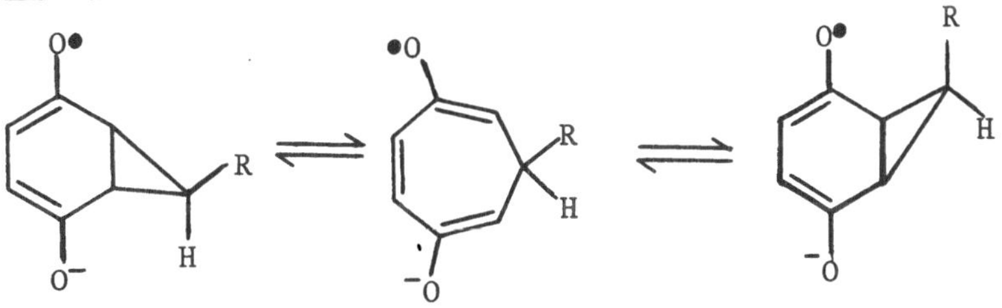

– – – – – –

(34)

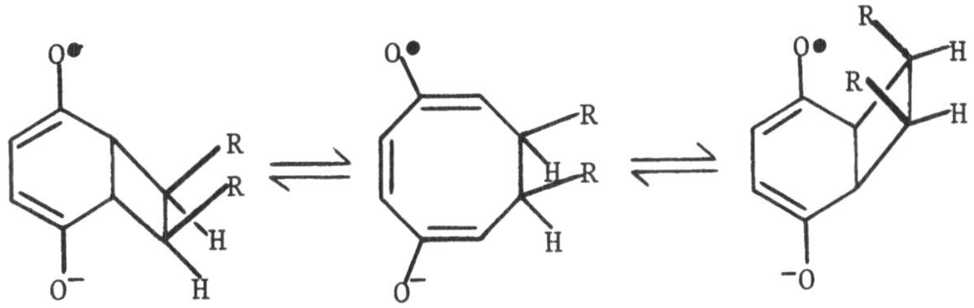

(35)

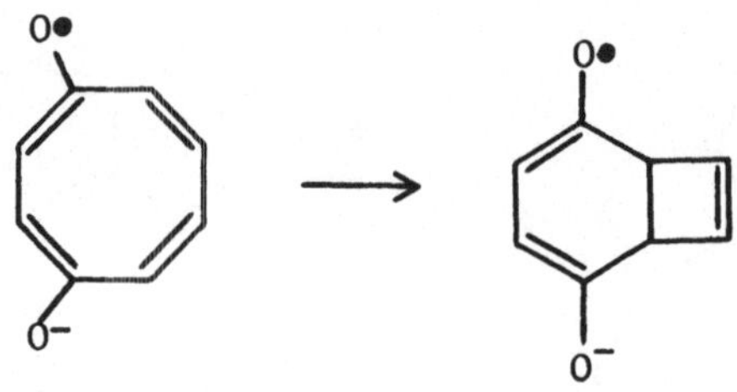

(36)

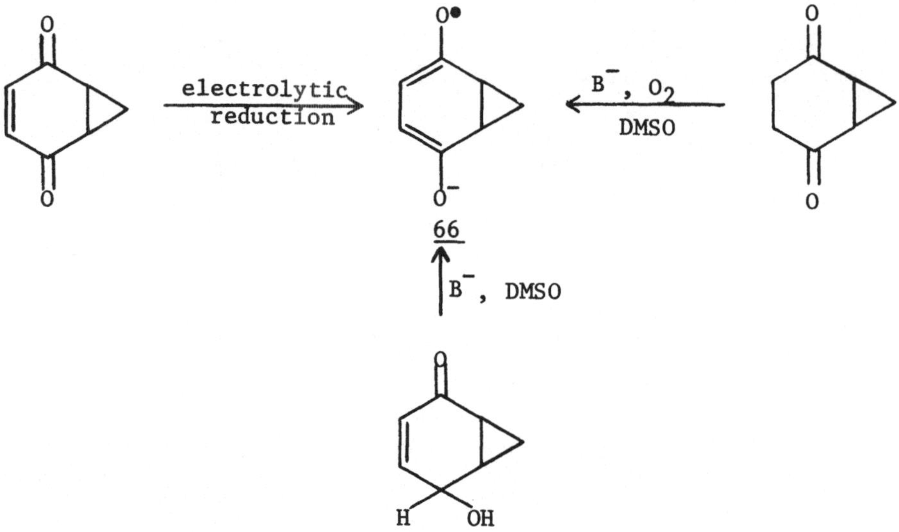

Scheme XXXI

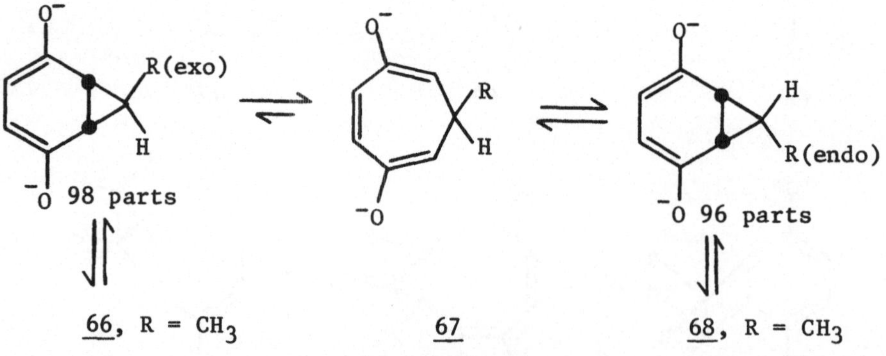

Scheme XXXII

Chart 1

Hyperfine Splitting Constants (gauss) for
Bicyclo[4.1.0]hept-3-ene-2,5-semidiones

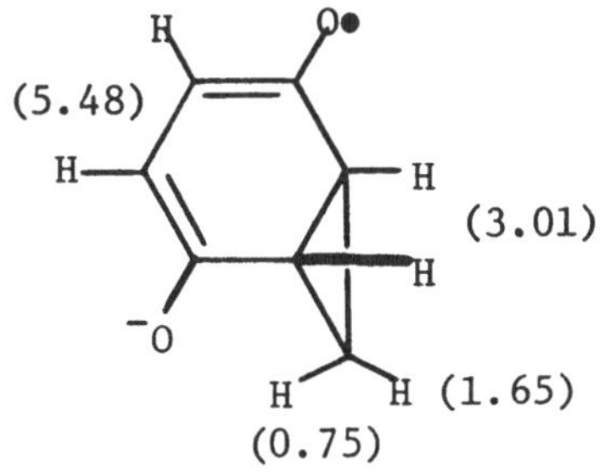

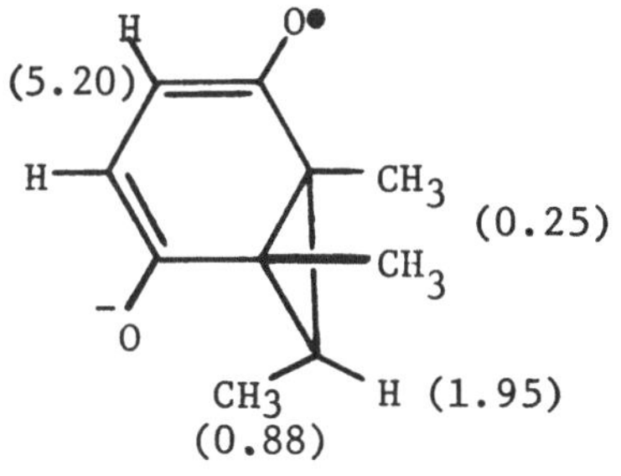

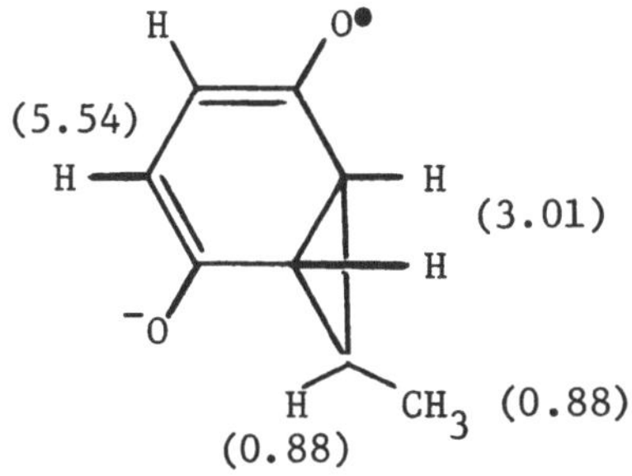

66, R = CH₃

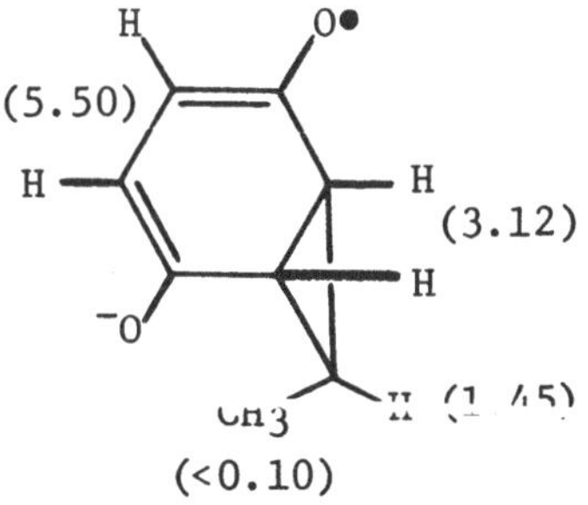

68, R = CH₃

endo-CH₃) were synthesized and were found to be stable. Upon electrolytic reduction each dione initially formed its own 1,4-semi-
dione; no evidence for the isomeric semidione could be observed.
The equilibrium shown in equation 34 is either excluded or else
must occur very slowly. However, upon extensive reduction, either
of the 1,4-enediones gave the same ratio of 66/68 = 96 parts/4 parts.
The simplest explanation is that upon extensive reduction the
dianions have an appreciable concentration. Although equilibrium
34 must occur slowly, the analogous equilibrium involving the dianions apparently occurs quite readily at 25° (Scheme XXXII).

The formation of the 2,5-semidiones in the bicyclo[4.2.0]-
octane system (eq 35) has also been investigated.[33] The two isomeric saturated diones shown in Scheme XXXIII gave the same semidione upon base-catalyzed oxidation. Perhaps again valence isomerization at the dianion stage has occurred, although in this case a
base-catalyzed stepwise epimerization of the bridgehead hydrogen
atoms in the semidione may be responsible for the isomerization
assumed to occur for the syn,syn-7,8-dimethyl isomer. Hydrogen-

deuterium exchange at the bridgehead positions occurs readily under
the reaction conditions. There is no epimerization at C-7,8 since
the known semidione isomeric to 70 with *trans*-methyl groups is not
formed under the reaction conditions.[33]

R = CH$_3$

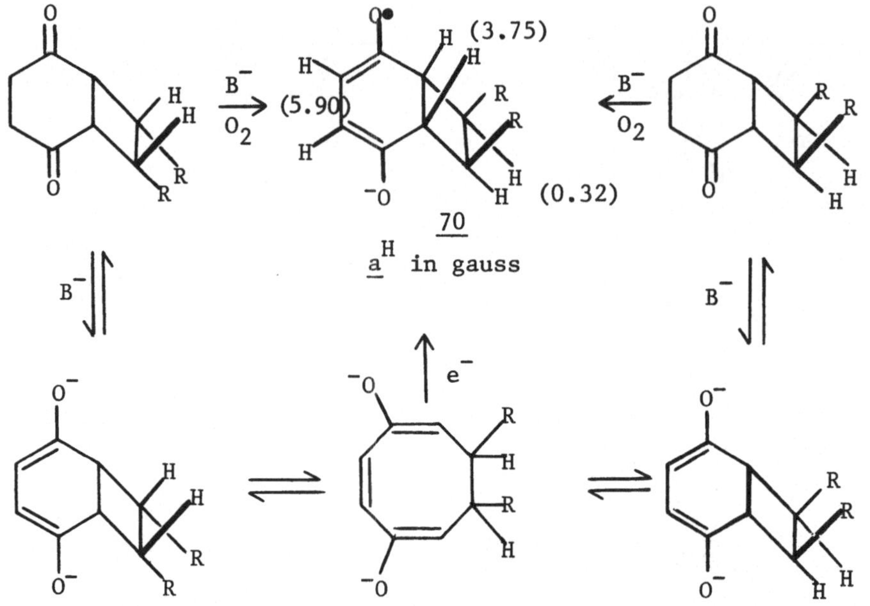

Scheme XXXIII

A consideration of equilibrium 36 takes us back to cycloocta-
traene valence isomerization. The structure of the only 1,4-semi-
dione observed in this system is obviously bicyclic, as evidenced
by the observed hyperfine splitting constants for 71 and the different values for a_H^H and $a_{CH_3}^H$ at C-7,8. The synthesis of 71 and
the observed hyperfine splitting constants are given in Scheme
XXXIV. As mentioned previously (Scheme XVI), 71 readily rearranged
in the presence of oxygen to form the *p*-semiquinone of the benzo-
cyclobutene.

Attempts to reduce the known cycloocta-1,3,6-triene-5,8-dione
45 electrochemically failed to produce any interpretable ESR signals.[32] The polarographic reduction involved a 2-electron wave
(-1.0 v relative to s.c.e.), apparently because the monocyclic
radical anion is more easily reduced than the parent dione. Base-
catalyzed oxidation of cycloocta-1,3-diene-5,8-dione produced semidione 71 and a new species, which is assigned structure 72 (eq 37).[32]

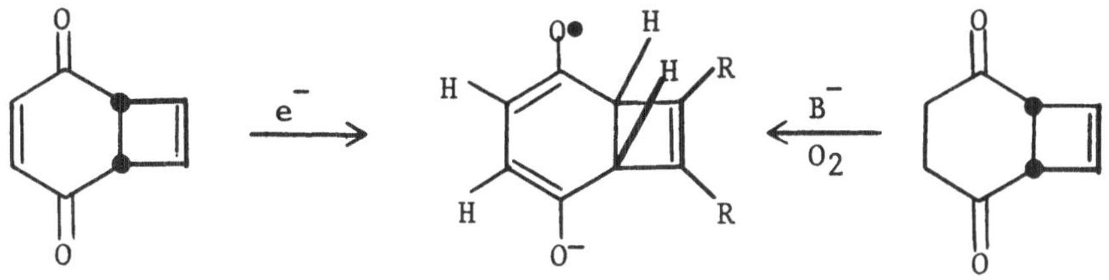

$$\underline{71}$$

R = H, $\underline{a}^H$=5.68 (2H),
         5.05 (2H),
         0.46 (2H)

R = CH$_3$, $\underline{a}^H$=5.75 (2H),
            5.10 (2H),
            0.14 (6H)

Scheme XXXIV

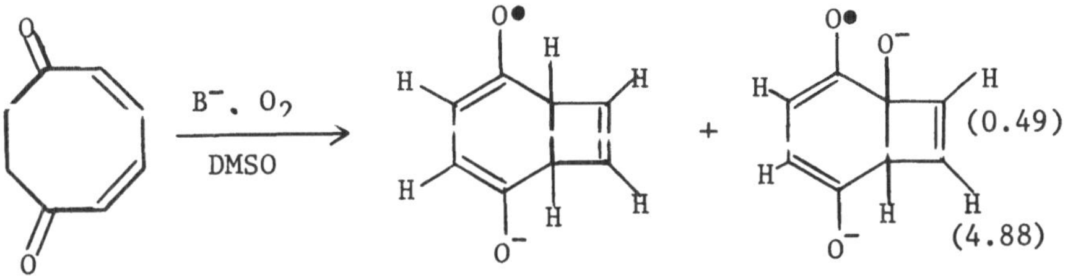

(0.49)   (37)

(4.88)

$$\underline{71} \qquad \underline{72}$$

$\underline{a}^H$=5.88 (1H), 5.38 (1H),
      4.88 (1H), 0.49 (2H)

The formation of these semidiones requires valence isomerization between the monocyclic and bicyclic structures. Semidione 72 is not formed by oxidation of 71 or its bicyclic precursors. Scheme XXXV presents a possible rationalization for the formation of 72.

Electrolytic reduction of cycloocta-1,3,6-triene-5,8-dione, followed by exposure to oxygen generated p-benzosemiquinone.[32] p-Benzosemiquinone is also the final oxidation product in basic solution of cycloocta-1,3-diene-5,8-dione. p-Benzoquinone can also be formed by autoxidation in basic solution of 5-hydroxy-α-tropolone. All of these reactions apparently involve a benzylic acid type of rearrangement which leads to decarbonylation (Scheme XXXVI), as previously observed for 1,2,3-triketones.[18]

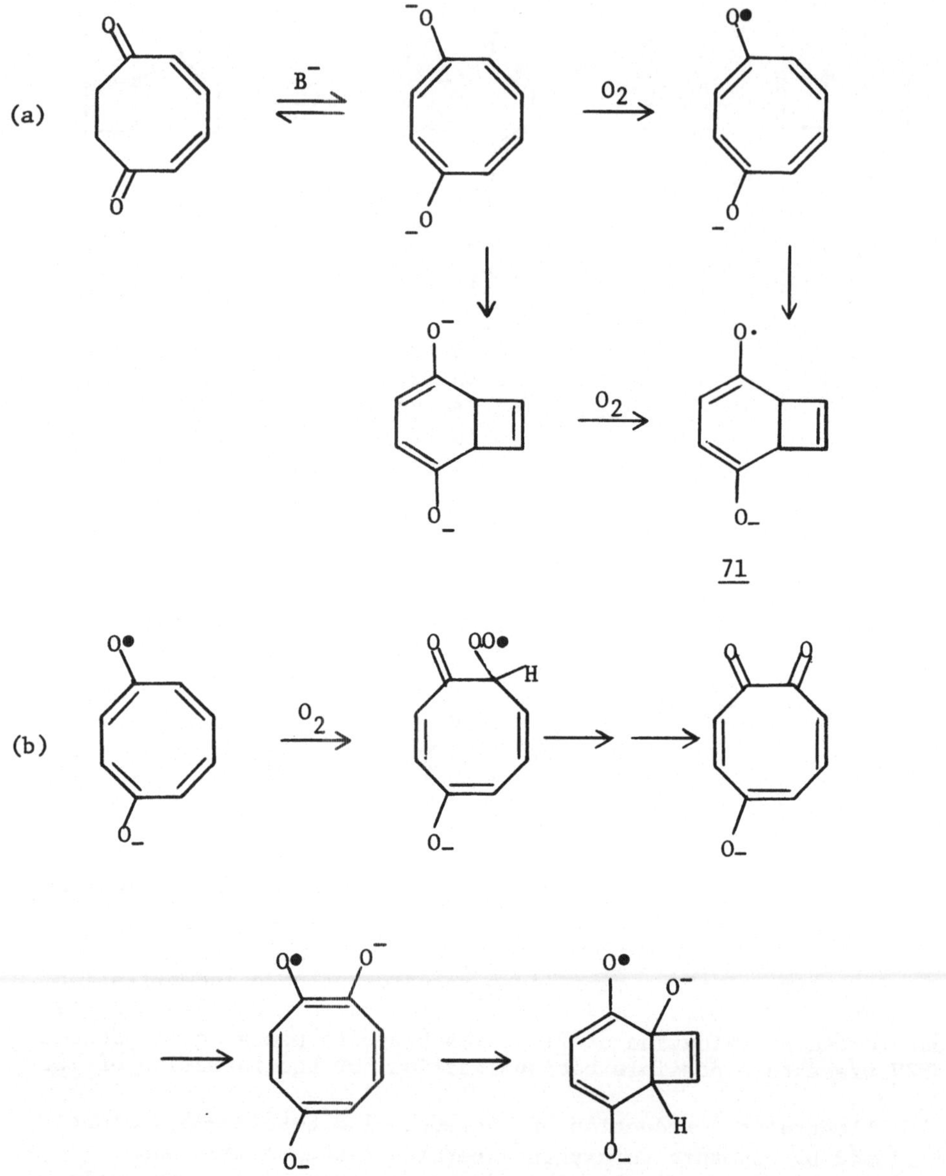

Scheme XXXV

CONCLUSIONS

Our studies of molecular rearrangements of unsaturated and
strained ring systems by observation of semidiones are complicated

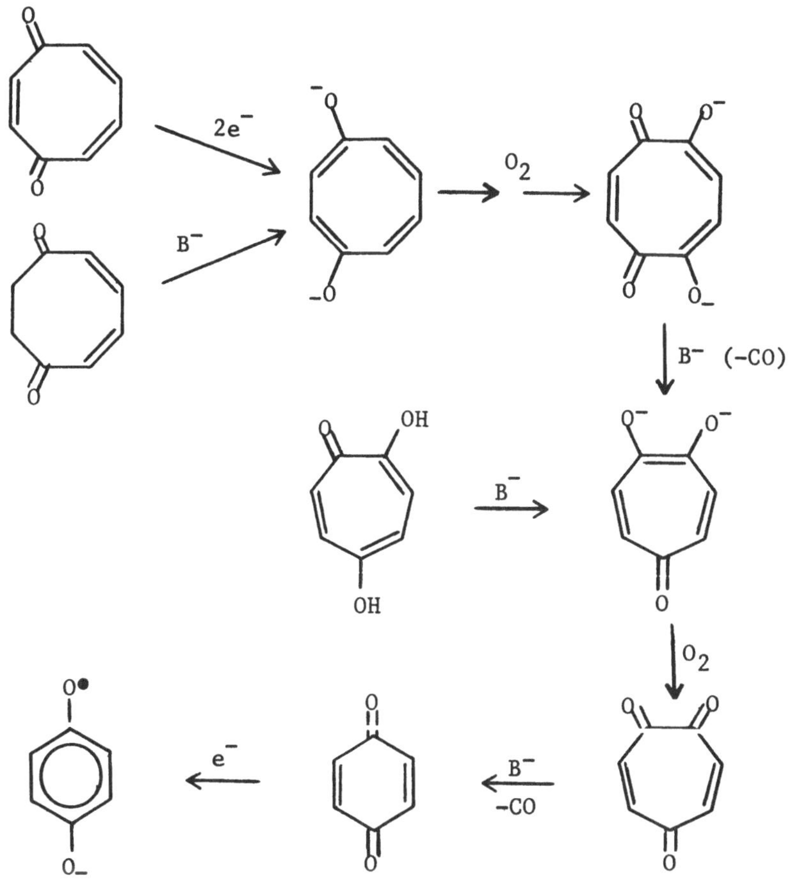

Scheme XXXVI

by the disproportionation reaction (eq 38). In general we have

$$2 \; \underset{\substack{| \;\; | \\ \cdot O \;\; O^-}}{RC{=}CR} \; \rightleftharpoons \; \underset{\substack{\| \;\; \| \\ O \;\; O}}{RC{-}CR} \; + \; \underset{\substack{| \;\; | \\ {}^-O \;\; O^-}}{RC{=}C{-}R} \tag{38}$$

observed either rearrangements of the dianions (Schemes XII, XXV, XXVIII, XXXII) or the diones (Scheme XXXIII), although in the case of bicyclo[3.1.0]hexane-2,3-semidione system (Scheme XXI) we have observed the rearrangement of a paramagnetic species. The dianion rearrangements observed may be concerted (Schemes XII, XXXII) or, in the case of 1,3-sigmatropic rearrangements, they may involve intermediates (Schemes XXVI, XXVIII). An overall view suggests that the presence of the paramagnetic center in the molecule does not lead to particularly facile molecular rearrangement, and, in fact, in most or all cases one of the diamagnetic analogues of the semidione (the enediol dianion or the diketone) rearranges more easily. This may be the result of orbital symmetry control, for if a given reaction of the dione involves 4n electrons in a cyclic transition state the enediol dianion will involve 4n + 2 electrons. Thus, one or the other of these substances will be able to satisfy the orbital requirements for a Möbius (4n) or Hückel (4n + 2) transition state for a concerted process, while the semidione spin probe will possess something in between (~4n + 1 electrons) and will not lead to a particularly favorable transition state, whether a sign inversion is or is not required in the basis orbital set.

ACKNOWLEDGEMENT

I am personally indebted to the graduate students and post-doctoral fellows who have worked in various areas of electron transfer processes and semidione chemistry and who are acknowledged in the references cited. Major support was supplied by the National Science Foundation and by the Petroleum Research Fund. An unrestricted grant from the latter agency in 1964 was instrumental in our early investigation of electron transfer processes by the use of ESR spectroscopy.

REFERENCES

1. G. A. Russell and E. T. Strom, *J. Am. Chem. Soc.*, <u>86</u>, 744 (1964).
2. G. A. Russell, E. G. Janzen and E. T. Strom, *J. Am. Chem. Soc.*, <u>84</u>, 4155 (1962).
3. G. A. Russell and R. D. Stephens, *J. Phys. Chem.*, <u>70</u>, 1320 (1966).
4. G. A. Russell, E. G. Janzen, A. G. Bemis, E. J. Geels, A. J. Moye, S. Mak and E. T. Strom, "Selective Oxidation Processes," Vol. 51 in the Advances in Chemistry Series published by the American Chemical Society, pp. 112-171.
5. E. T. Strom, G. A. Russell and R. D. Stephens, *J. Phys. Chem.*, <u>69</u>, 2131 (1965).
6. G. A. Russell, R. D. Stephens and E. R. Talaty, *Tetrahedron Lett.*, 1139 (1965).
7. G. A. Russell and P. R. Whittle, *J. Am. Chem. Soc.*, <u>89</u>, 6781 (1967).
8. R. O. C. Norman and R. J. Prichett, *J. Chem. Soc. B*, 378 (1967).
9. G. A. Russell, D. F. Lawson and L. A. Ochrymowycz, *Tetrahedron*, <u>26</u>, 4697 (1970).
10. G. A. Russell and R. L. Blankespoor, *Tetrahedron Lett.*, 4573 (1971).
11. G. A. Russell, V. Malatesta and R. L. Blankespoor, *J. Org. Chem.*, <u>43</u>, 1837 (1978).

12. G. A. Russell and K. Y. Chang, *J. Am. Chem. Soc.*, 87, 4381 (1965).
13. G. A. Russell, E. G. Janzen and E. T. Strom, *J. Am. Chem. Soc.*, 86, 1807 (1964).
14. G. A. Russell and E. G. Janzen, *J. Am. Chem. Soc.*, 84, 4153 (1962).
15. G. A. Russell and G. Hamprecht, *J. Org. Chem.*, 35, 3007 (1970).
16. G. A. Russell, M. Ballenegger and H. L. Malkus, *J. Am. Chem. Soc.*, 97, 1900 (1975).
17. G. A. Russell, R. L. Blankespoor, S. Mattox, P. R. Whittle, D. Symalla and J. R. Dodd, *J. Am. Chem. Soc.*, 96, 7249 (1974).
18. G. A. Russell and S. A. Weiner, *J. Am. Chem. Soc.*, 89, 6623 (1967).
19. G. A. Russell, D. F. Lawson, H. L. Malkus, R. D. Stephens, G. R. Underwood, T. Takano and V. Malatesta, *J. Am. Chem. Soc.*, 96, 5830 (1974).
20. G. A. Russell, G. Wallraff and J. L. Gerlock, *J. Phys. Chem.*, 82, 1161 (1978).
21. G. A. Russell and C. E. Osuch, *J. Am. Chem. Soc.*, in press.
22. G. A. Russell and J. L. Gerlock, *J. Am. Chem. Soc.*, 96, 5838 (1974).
23. Unpublished results with Mr. T. Takano.
24. G. A. Russell, G. R. Underwood and D. C. Lini, *J. Am. Chem. Soc.*, 89, 6636 (1967).
25. F. A. L. Anet and N. Z. Hag, *J. Am. Chem. Soc.*, 87, 3147 (1965).
26. G. A. Russell, R. G. Keske, G. Holland, J. Mattox, R. S. Givens and K. Stanley, *J. Am. Chem. Soc.*, 97, 1892 (1975).
27. Unpublished results with Mr. C. E. Osuch.
28. G. A. Russell and P. R. Whittle, *J. Am. Chem. Soc.*, 91, 2813 (1969).
29. Unpublished results with Dr. P. Riedl.
30. Unpublished results with Dr. E. Goettert and Mr. J. Siddens.
31. G. A. Russell, R. L. Blankespoor, K. D. Trahanovsky, C. S. C. Chung, P. R. Whittle, J. Mattox, C. L. Myers, R. Penny, T. Ku, Y. Kosugi and R. S. Givens, *J. Am. Chem. Soc.*, 97, 1906 (1975).
32. G. A. Russell, C. M. Tanger and Y. Kosugi, *J. Org. Chem.*, in press.
33. G. A. Russell, J. R. Dodd, T. Ku, C. Tanger and C. S. C. Chung, *J. Am. Chem. Soc.*, 96, 7255 (1974).
34. M. Oda, M. Oda, S. Miyaroshi and Y. Kitahara, *Chem. Letters*, 293 (1977).
35. G. A. Russell, P. R. Whittle, R. G. Keske, G. Holland and C. Aubuchon, *J. Am. Chem. Soc.*, 94, 1693 (1972).
36. R. F. Heldeweg, H. Hogeveen, *Tetrahedron Lett.*, 1517 (1975).
37. G. A. Russell, K. Schmidt, C. Tanger, E. Goettert, M. Yamashita, Y. Kosugi, J. Siddens and G. Senatore, "Forty Years of Free Radicals," Advances in Chemistry Series published by the American Chemical Society, pp. 376-399.

38. G. A. Russell, P. R. Whittle and R. G. Keske, *J. Am. Chem. Soc.*, 93, 1467 (1971).
39. Work in progress with Mr. G. Senatore.
40. G. A. Russell, J. J. McDonnell, P. R. Whittle, R. S. Givens and R. G. Keske, *J. Am. Chem. Soc.*, 93, 1452 (1971).
41. R. Sustmann and F. Lübbe, *J. Am. Chem. Soc.*, 98, 6037 (1976).
42. Work in progress with Dr. T. Morita.
43. G. A. Russell, K. D. Schmidt and J. Mattox, *J. Am. Chem. Soc.*, 97, 1882 (1975).
44. A. Stockis and E. Weissberger, *J. Am. Chem. Soc.*, 97, 4288 (1975).
45. M. Oda, Y. Kayama, H. Miyazaki and Y. Kitahari, *Angew. Chem. Int. Ed. Eng.*, 14, 418 (1975).

EFFECTS OF STERIC STRAIN ON THE PREPARATION, STABILITY AND

REACTIONS OF ARYL ANIONS AND AMINES

Shelton Bank

Department of Chemistry, State University of New York

at Albany, Albany, New York 12222

Steric effects as important contributors to the behavior of organic compounds were first identified by Kehrmann less than 100 years ago.[1] The subsequent major developments were outlined by Brown in 1956.[2] Since that time steric effects have become an important part of the fabric of organic chemistry and recent advances are highlighted in chapters in advanced textbooks.[3,4] Steric effects have not always occupied a position of high regard, and the classic work since 1940 has altered indelibly the then prevailing view. The present-day organic chemist values and utilizes steric factors along with polar and resonance factors to account for chemical effects.

Perhaps the most useful modern steric concept is that of strained homomorphs. The proposal "that molecules having the same or closely similar molecular dimensions be termed 'homomorphs'"[5] can now be viewed as a simple and profound insight that stands the demanding tests of both time and clarity. It is one of the very few simple ideas that has survived and remained valuable for more than a quarter of a century. Among its most useful features is that information obtained with, for example, molecular addition compounds provides a quantitative approach to the evaluation of steric strain in structurally similar hydrocarbons or other organic molecules. The steric strain of a variety of related homomorphs has been estimated in this way (Table 1).[2,5,6] Interestingly, the rigidity of fused rings provides many comparable steric situations and[7,8], therefore, further useful analogies, as summarized in Table 2. These principles have been used for estimating steric effects in a great variety of species.[9]

Steric effects on reactivity and stability have been considerably less well studied with carbanions than with other species. For example, the following statement is found in the major work summarizing steric effects in 1956: "very little is known

Table 1. Steric Strains in Homomorphs[a]

Structures			Strain in kcal/mole
			1.4
			2.7
			6.0

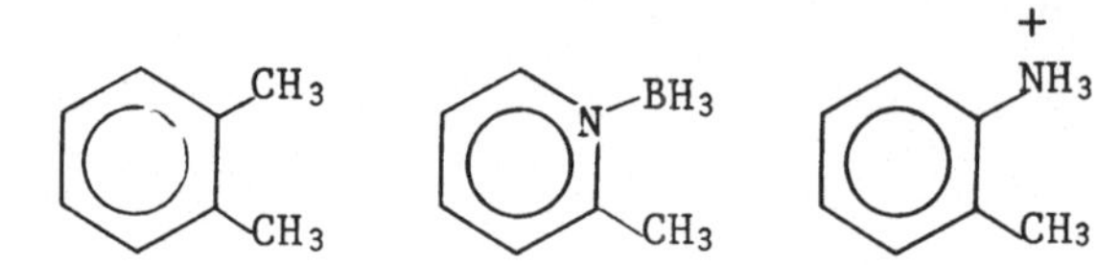

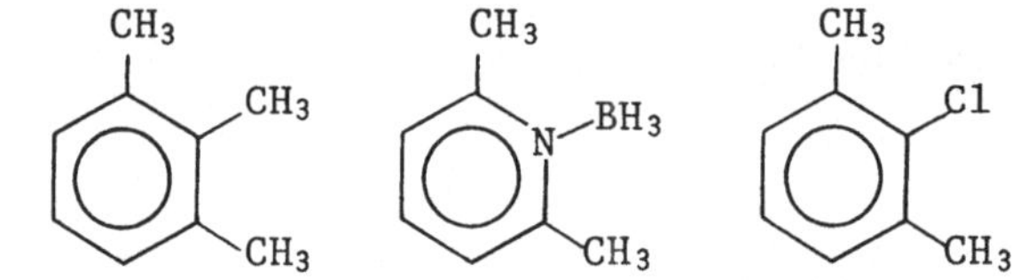

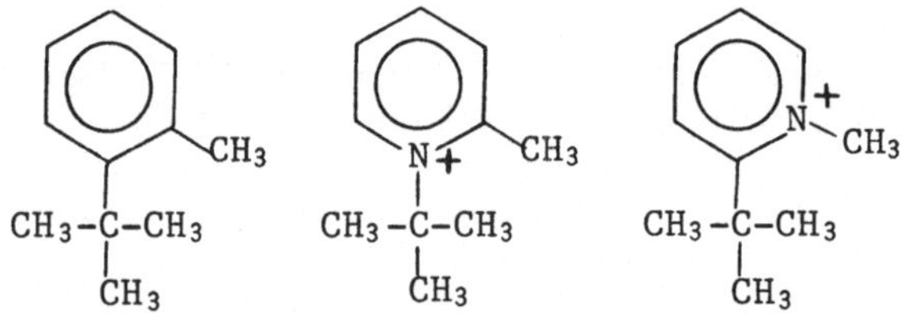

[a]References 2 and 6.

regarding steric effects on the reactivity of carbanions".[10] Although the intervening years have brought considerable understanding to the chemistry of carbanions, there has been no systematic study of steric effects and they are not considered in any major sense in monographs on carbanions.[11] Indeed Eliel's comment of 1956 could be accurately restated today. The work described in this chapter attempts to provide some quantitative estimates of steric effects on the formation, stability and reactivity of aryl carbanions and some isoelectronic amines. Three types of aryl system and varying measures of kinetic, structural and thermodynamic effects are considered. Attention is focused first on diarylmethyl carbanions and changes brought about by substitution of methyl for hydrogen at the central carbon atom.

Table 2. Steric Strains in Fused Ring Homomorphs

Structures (Strains in kcal/mole)

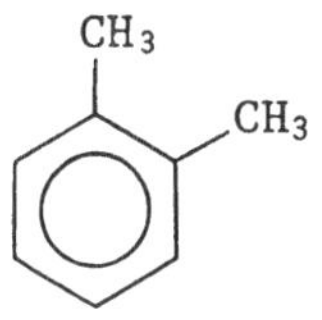

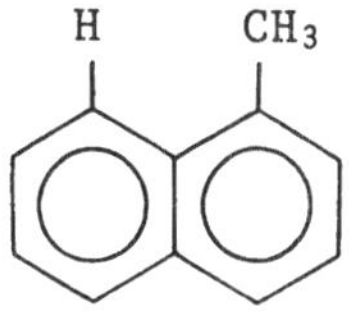

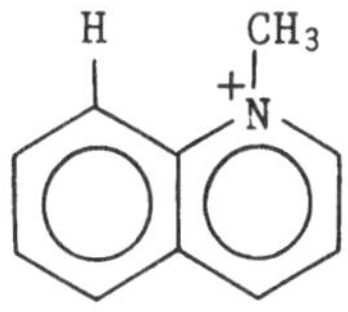

(1.4)[a]   (1.6)[b]

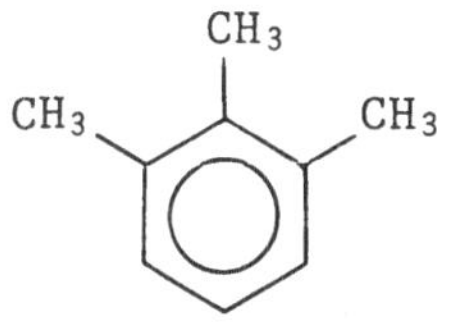

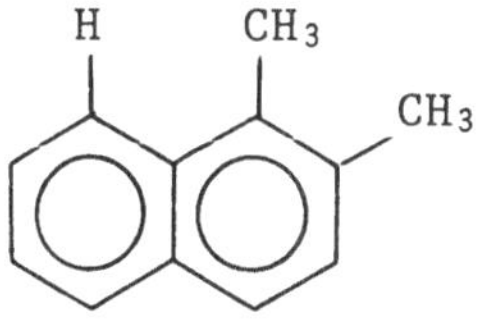

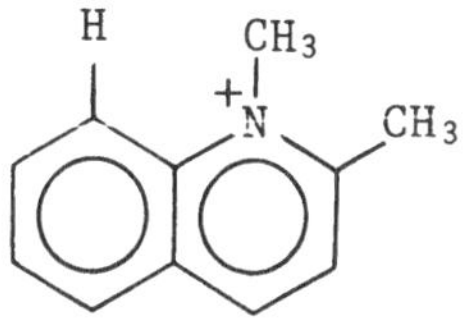

(2.7)[a]   (3.4)[b]

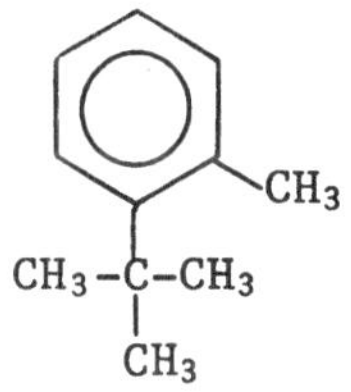

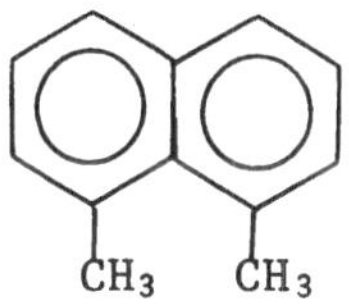

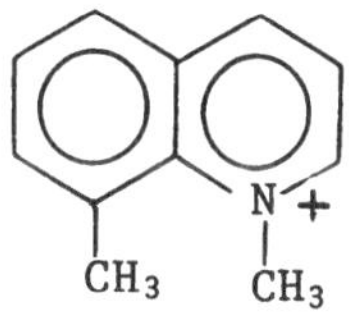

(6.0)[a]   (7.6)[b]

[a]Reference 6.   [b]Reference 7.

Next is a spectral and structural study of substituted naphthyl-amines which are isoelectronic analogs of anions. Finally, a kinetic study of the reactivity of 9-alkyl-10-lithio-dihydroan-thracenyl anions is described.

For diarylmethyl anions of the type:

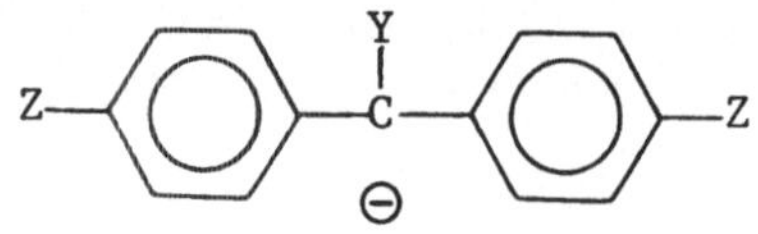

	Z	Z	Y
I	H	H	H
II	CH₃	CH₃	H
III	H	H	CH₃
IV	CH₃	CH₃	CH₃

there are dramatic differences in the thermodynamic and kinetic
formation of III and IV as compared to I and II. As will be seen,
substitution of a methyl for hydrogen at the central carbon atom
brings about a large effect which is not accounted for by electronic effects but is rationally accommodated by a steric effect.

In thermodynamic measurements of the acidity of hydrocarbons
using cesium cyclohexylamide in cyclohexylamine, Streitwieser
found the following values for the equilibrium constants: I-H,
33.4; II-H, 35.1; III-H> 35; toluene (V), 41 pKa units.[12]
Accordingly each methyl in the *para*-position has an acid-weakening
effect of 0.85 pKa units. This result is readily accounted for
on the basis of the electronic effect of the methyl group. In
contrast, substitution of a single methyl at the *alpha*-position
has a large effect that is not readily accounted for in terms of
an electronic effect.

In the present study we utilize nuclear magnetic resonance
(NMR) spectroscopy to further define the pKa values of III and
IV and also related compounds. Figure 2 (p 118) shows the aromatic
region of the proton NMR for the independently prepared anions of
III, IV, p-Xylyl (VI) and 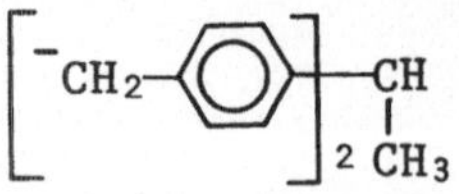   (VII) (lithium coun-

terion). As can be seen for most of these, the spectra are
sufficiently different that it is possible to see which one (or
both) are formed when both precursors are present. The effect
of aryl and methyl substitution on pKa, from these NMR equilibrium
measurements and from the spectral equilibrium data of Streitwieser,
is shown in table 3. It is seen that the quantitative effect of

an α-methyl group on pKa is ~4 pKa units (5.5 kcal/mole). This effect is largely steric in origin since in the delocalized anion the change from hydrogen to methyl brings about considerable steric interactions with the *ortho*-hydrogens:

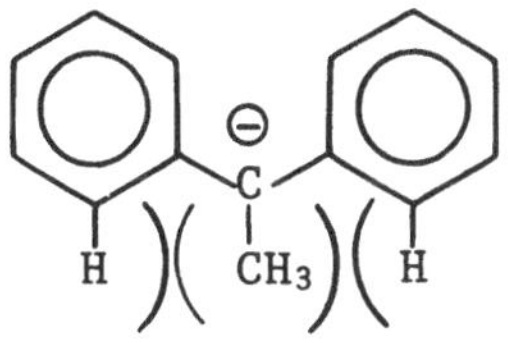

Table 3. Aryl and Methyl Substituent Effect on pKa[a]

41[b]

42-3[c]

33[b]

35[b]

37[d]

42-3f

38-9[e]

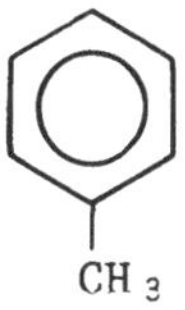

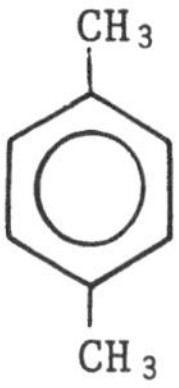

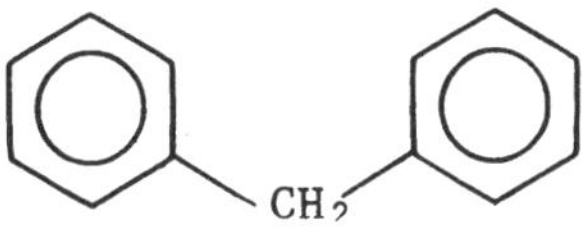

[a]Per hydrogen

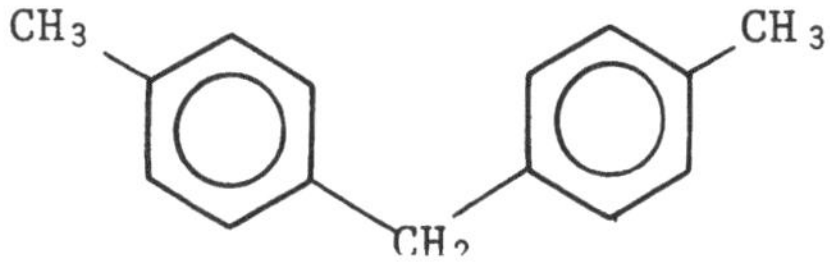

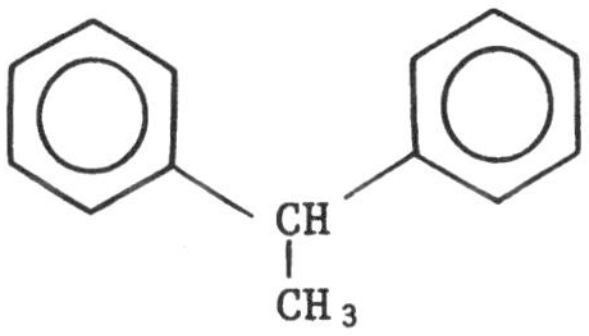

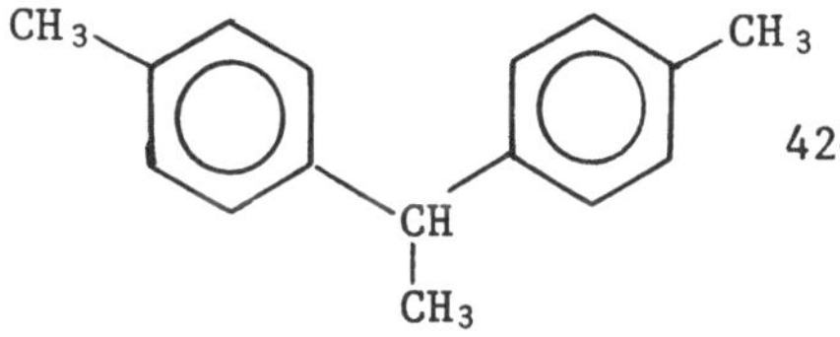

[b]Equilibrium acidities determined relative to 9-phenylfluorene toward cesium cyclohexylamide in cyclohexyl amine. (Reference 12).

[c]Estimated by proton NMR from equilibrium experiments with benzyllithium in THF.

[d]Determined by proton NMR equilibrium studies using p-xylyllithium and III-H in THF.

[e]Estimated from proton NMR equilibrium studies with p-xylyllithium and VII-H in THF.

[f]Estimated from proton NMR equilibrium studies with p-xylyllithium in THF.

An estimated upper limit of ~3.0 kcal/mole may be calculated for this effect from the homomorphs, hemimellitene or 1,2 dimethylnaphthalene.

These conclusions are supported and amplified by related changes in pKa in the isoelectronic and homomorphic diarylamines as well as corresponding changes in the *ortho* hydrogen proton chemical shifts. The pKa values of the relevant aryl ammonium ions[13,14] are recorded in table 4. Thus, introduction of a *para*-methyl group in aniline has the same base-strengthening effect

Table 4. Effect of Methyl Substituents on the pKa Values of Anilines and Diphenylamines

Compound	pKa	ΔpKa
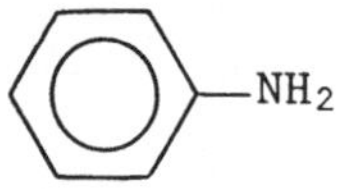 ⟨⟩—NH$_2$	4.62[a]	
 CH$_3$—⟨⟩—NH$_2$	5.07[a]	.45
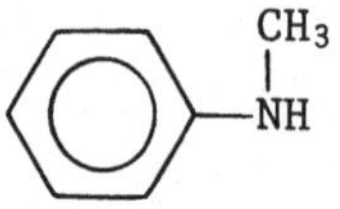 ⟨⟩—NH—CH$_3$	4.85[a]	.23
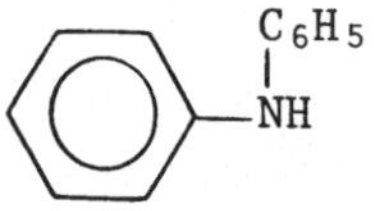 ⟨⟩—NH—C$_6$H$_5$	0.76[b]	
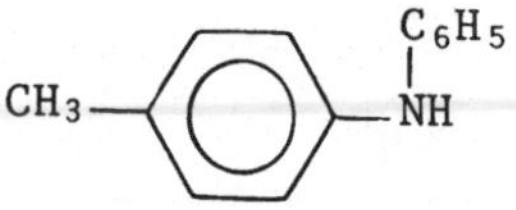 CH$_3$—⟨⟩—NH—C$_6$H$_5$	1.21[b]	.45
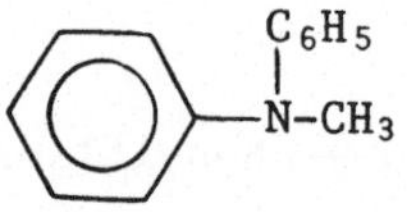 ⟨⟩—N(CH$_3$)—C$_6$H$_5$	1.76[b]	1.00

[a] Reference 14.   [b] Reference 13.

of 0.45 units as introduction of a *para*-methyl group in diphenylamine. In contrast, replacement of an α-hydrogen by methyl in aniline brings about a change that is considerably smaller (0.23) than the corresponding increase in base strength of 1.0 pKa in diphenylamine. The effect is neither a polar effect nor that

expected for B-strain, but is, in fact, the effect expected if the
α-methyl group decreases nitrogen lone-pair delocalization in the
same way as the methyl-*ortho*-hydrogen interaction does with anions
(discussed above). Although directionally similar, the magnitude
of the effect in the amine is smaller. This is likely due to
opposing steric effects on solvation and delocalization for the
amines and reinforcing steric effects for the anions. The steric
effect on solvation will be greatest for the charged species, which
is tetrahedral for nitrogen and trigonal for the carbanion. On
the other hand, the steric effect on delocalization contributes
only in the trigonal species for both. Additionally, we note that
the base-strengthening effect of a *para*-methyl group is the same
for both aryl amines (0.45 units) and approximately half that for
the isoelectronic anions. It is likely that this is the result
of greater delocalization in the charged anions and we will
return to this point in a later section.

   In further support, steric interaction with neighboring hydrogens brings about magnetic deshielding[15] and this effect is seen
in the *ortho*-hydrogens of the *alpha*-methyl substituted carbanions.
Table 5 records the proton chemical shifts. For compounds with
steric interactions the *ortho*-protons shift by 0.33 and 0.22 ppm
respectively whereas the *meta*-protons are essentially unchanged.
Further, the magnitude of the deshielding shift is in agreement
with other values and with that expected in these systems.[15]
Finally, the low temperature spectra of II[18] and IV[19] (Figure 1),
where rotation is frozen on the NMR time scale, magnifies and
thus more clearly reveals the effect. The finding that one *ortho-*

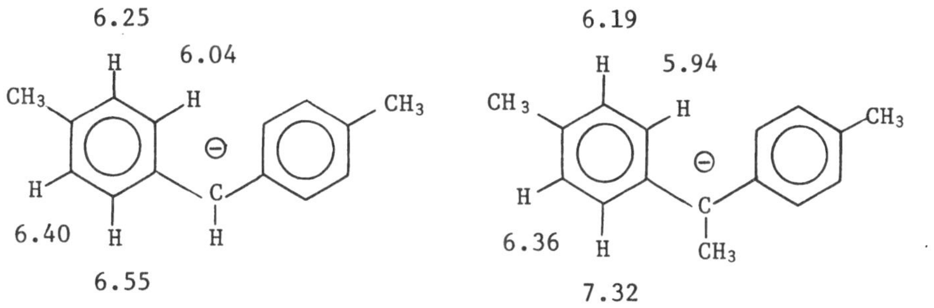

Figure 1. NMR chemical shifts of aromatic protons at low temperatures.

proton has a downfield shift of 0.77 ppm, while the other *ortho-*
proton and the *meta*-protons have similar chemical shifts, con-
firms the conclusion that steric strain between the *alpha*-methyl
group and the *ortho*-hydrogens is important. Thus, the structure

Table 5.   NMR Data of Aromatic Protons of Lithioarylmethyl Anions
           in THF.[a]

Compound	H_o	H_m	H_p
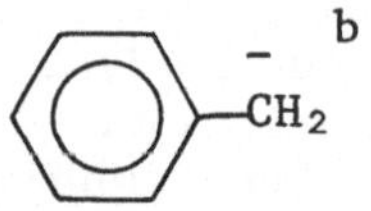 phenyl–CH_2^- [b]	6.09	6.30	5.50
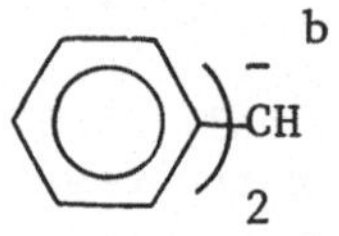 (phenyl)$_2$–CH^- [b]	6.51	6.54	5.65
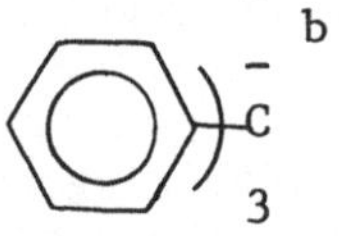 (phenyl)$_3$–C^- [b]	7.31	6.52	5.96
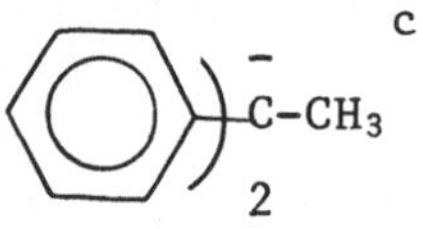 (phenyl)$_2$–C^-–CH_3 [c]	6.84	6.51	5.60

[a]Chemical shifts in ppm downfield from tetramethylsilane.

[b]Reference 16.

[c]Reference 17.

and thermodynamic acidity are strongly affected by the *alpha*-methyl group. As discussed below, the kinetic acidity of the methine proton is even more strongly affected.

Reaction of diphenylmethane with *n*-butyllithium in tetrahydrofuran (THF) at 0–25° to give the anion I is fast and quantitative. In sharp contrast, reaction of IVH with *n*-butyllithium is very slow and the spectrum is complicated by formation of significant amounts of acetaldehyde enolate and ethylene. These products of the reaction of THF with *n*-butyllithium[20] are observed in the absence of hydrocarbon or, when metalation is slow, in its presence.

Reaction of IVH with the stronger base, *t*-butyllithium in THF, at 0–25° for 15 minutes produced a red anion (VII). The ¹H NMR (Figure 2) resembles that of VI more than those of III or IV. The anion solution was stable for several hours at room temperature and at least 4 days at –10°. Quenching with ethanol gave the parent hydrocarbon IV-H as the only product.

Two remarkable aspects of the reaction are that t-butyl-lithium reacted with the methyl protons rather than with the central methine proton and that when reaction was complete there were no discernible aromatic protons ascribable to a neutral ring. This could be explained by extensive formation of the dianion (VII). Methylation of the anion solution with methyl iodide

VII

provides 1,1-di(4-ethylphenyl)ethane as the sole methylated product. Thus, with the powerful metalating agent, t-butyl-lithium, IVH gives dianion VII as the product.

The nature of the anion produced from IVH depends upon the metalating agent; the diarylmethyl anion (IV) was formed when benzyllithium was used. Reaction of toluene with t-butyllithium in THF for 15 minutes at -60° to -70° produced the characteristic benzyl anion. The reaction of this anion with IVH in THF was followed by use of proton NMR. After one and three-quarter hours the spectrum of the benzyl anion was replaced by that of a diphenylmethyl type anion (IV). These shifts agree quite well with those of anion III and are distinct from those of anions VI or VII. Confirmation of the anion structure was obtained by reaction with methyl iodide and isolation of 2,2-di(p-tolyl) propane.

It is seen that the metalation and subsequent methylation of IVH takes two distinct paths, depending on the nature of the base used. A reasonable explanation of these results is that the benzyl-type anion is the product of kinetic control and that the diarylmethyl anion is the product of thermodynamic control. In reaction with an excess of lithium reagent, rapid metalation at a benzyl site should strongly affect the position at which further metalation would occur. The methine position is now far less acidic due to the presence of the negative charge in one of the aromatic rings. On the other hand, the kinetic and thermo-dynamic acidity of the other $para$-methyl group is largely un-affected. Confirmation of kinetic vs. thermodynamic control was provided by anion interconversion.

The benzyl-type dianion VII was generated in ~75% yield; then less than an equivalent of toluene was added, and the reaction mixture was observed as a function of time by proton NMR. The spectra in Figure 3 depict the changes. Initially, the peaks of toluene, unreacted IVH and VII are present. After several hours, the toluene peak at 7.2 ppm is gone, the *para*-proton of

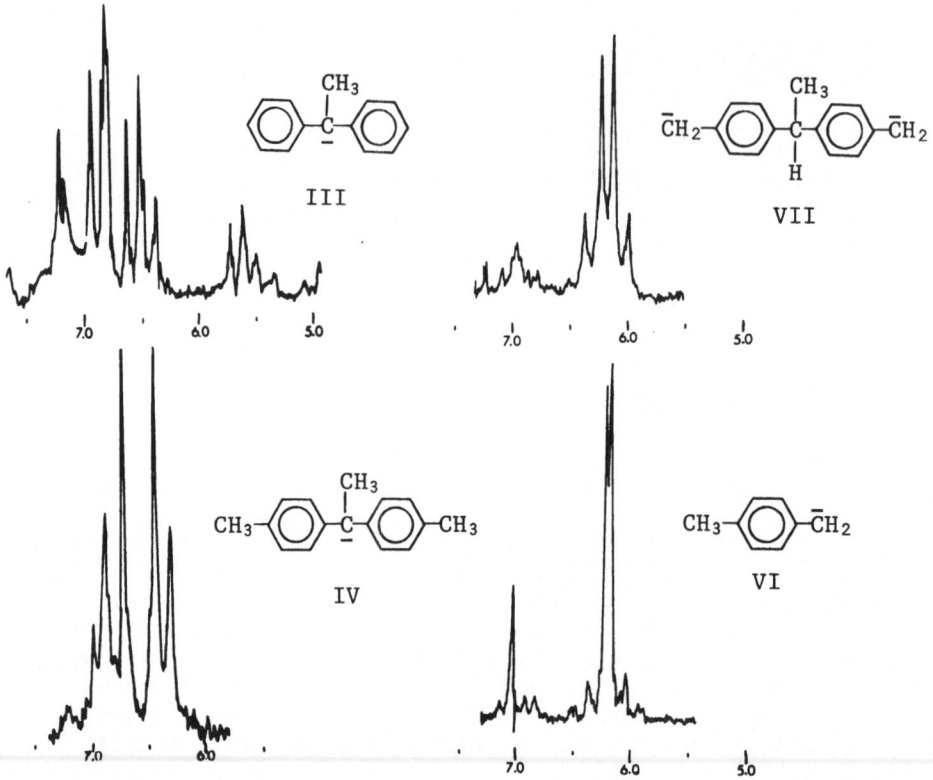

Figure 2: - Aromatic proton NMR spectra of lithium arylmethyl anions in THF.

the benzyl anion at 5.50 ppm is present and there is evidence of anion IV. Still later, anion VII is gone, toluene has reappeared and the amount of anion IV has increased. At the conclusion of the experiment only toluene and anion IV are present. This shows that the anion VII and IV are equilibrated *via* the benzyl anion, which is formed *in situ* from VII and toluene. This leads to the order of anion stability: IV > benzyl > VII in full accord with expected thermodynamic stabilities.

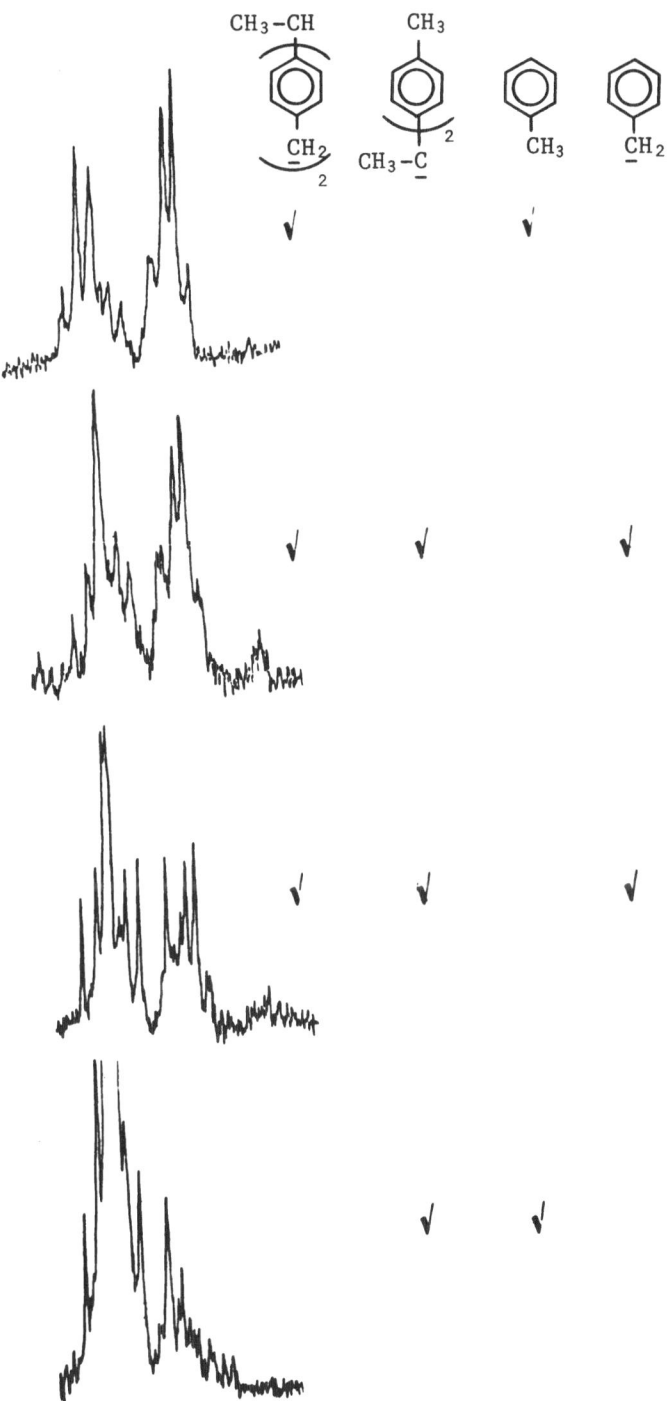

Figure 3. - Aromatic proton NMR spectra of anion equilibration *via* the benzyl anion.

It is probable that several factors contribute to the reversal of kinetic and thermodynamic acidity of the two reaction sites. First, the reference base is considerably more basic than that formed by reaction at either site[21]. Additionally, for the reacting hydrocarbon acid there is considerably greater F-strain[2] for attack at the single methine position than at the six methyl positions. The steric bulk of the reacting base is large and aggregation, which is likely, would increase the size even more[22]. The increased steric demands of the base will serve to increase the relative steric requirements at the two sites. As demonstrated so clearly by H. C. Brown, the magnitude of steric effects in acid-base reactions are a function of the steric requirements of the reference acid or base[2,6]. Accordingly, an exceedingly potent and large base selects the sterically less hindered site. The less powerful and smaller base, benzyllithium, selects the thermodynamically more favorable site.

In summary, the seemingly small structural change from hydrogen to methyl at the *alpha*-position of a diarylmethyl anion has a large demonstrable effect on both the kinetic and thermodynamic acidities of the precursor hydrocarbons. The magnitude and direction of the effects are readily accounted for on the basis of the quantitative steric effect studies of H. C. Brown and, in particular, by use of the concept of homomorphs.

The second area of investigation of the effect of steric consideration on stability and structure concerns species which are isoelectronic with anions, namely, naphthylamines. In particular, we consider the implications that inspection of homomorphs suggest relative to the conjugation of lone pair electrons with a naphthyl ring, and then we consider structural and spectral data that demonstrate the utility of the model.

In a classic study, Brown and Cahn[23] determined the rates of reaction of substituted pyridines with methyl iodide. Inspection of the data, as displayed in Figure 4, reveals a steric effect for the 2-substituted compounds. Most importantly, the pattern that the effect follows is characteristic, namely: Me- < Et- < *i*-Pr << *t*-Bu. The large break between *t*-butyl and *i*-propyl is attributed to the ability of the ethyl and isopropyl groups to rotate to minimize the steric interaction. In this way, a hydrogen atom is positioned in the area of greatest interaction and the effect of these groups is similar to that of a methyl substituent. In contrast, a *t*-butyl group cannot minimize its steric requirements through rotation since it is spherically symmetrical.

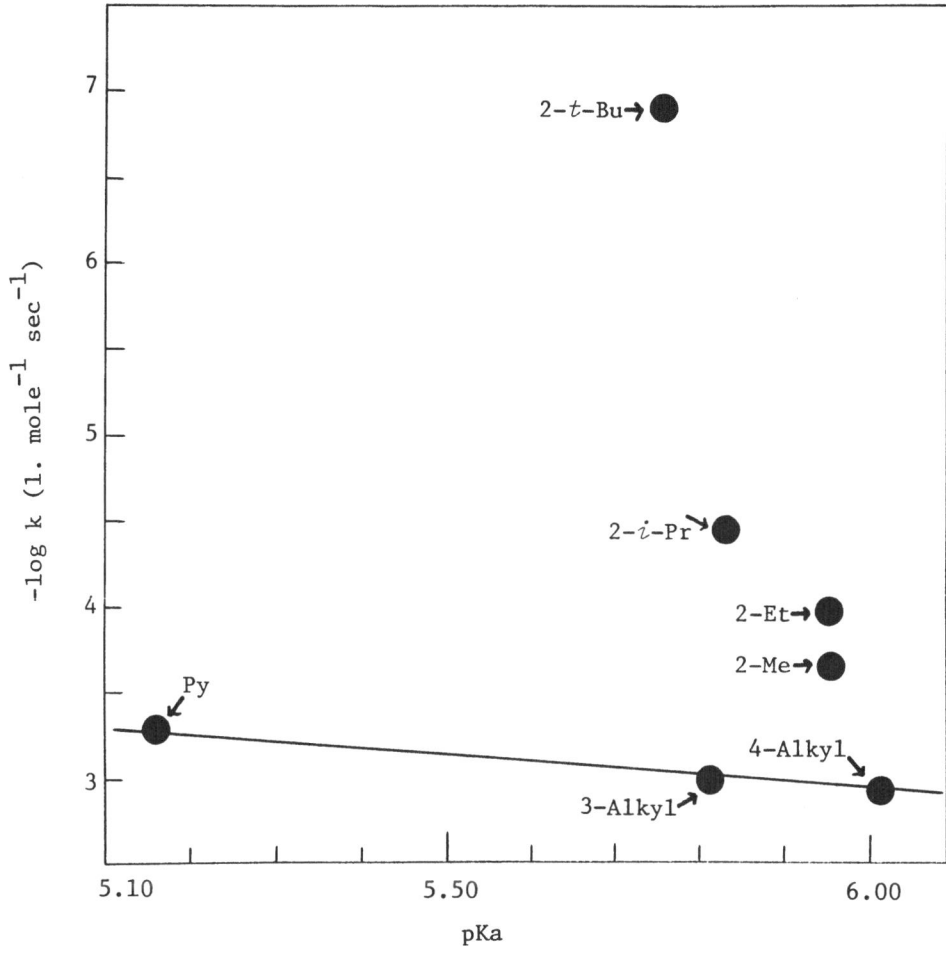

Figure 4. Relationship between the strengths of alkylpyridine
bases and their rates of reaction with methyl iodide.

     Figure 5 depicts the most likely minimum strain conformation
of 2-isopropyl methyl pyridinium iodide and related forms in homomorphic species. Of central interest to this work is the point
that the hydrogen atom or lone pair of electrons will occupy
the position of minimal steric interaction. In dimethylaminonaphthalene the implication is that in this conformation the
lone pair is essentially unavailable for conjugation with the aryl
system. From the difference in strain between the 2-isopropyl
and the t-butyl compounds some 4.5 kcal/mole of strain would be
introduced by the rotation that would place the lone pair in
conjugation, with the methyl group and the *peri*-hydrogen abutting.
Interestingly, this is the same order of magnitude (~4.0 kcal/mole)

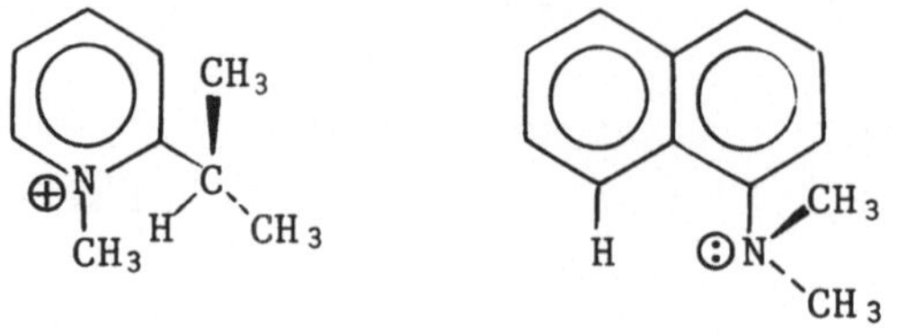

Figure 5. Most likely minimum energy conformations of 2-isopropyl-
N-methyl pyridinium ion and 1-N-N-dimethylaminonaphthalene.

as the value for the resonance energy of the nitrogen lone pair
conjugation.[24]

Conjugation of the lone pair with the aryl ring has important
consequences with regard to the energy of the molecule. In particular, the orbitals most affected are the highest occupied molecular orbital (HOMO) and the lowest vacant molecular orbital
(LVMO). Experimentally, these changes are reflected in oxidation
and reduction potentials, respectively. Figure 6 displays the
relevant potentials[25] for the naphthalene series.

The effect of nitrogen lone pair conjugation is seen in the
changes brought about in going from naphthalene to α-naphthyl-
amine. Thus, introduction of the α-nitrogen brings about an increase in reduction potential of ~0.2 ev and a decrease in oxidation
potential of ~0.5 ev. The former is measured in solution and the
latter in the gas phase which accounts in part for its greater
magnitude. The direction of both effects is consonant with the
increased electron density brought about by conjugation with the
lone pair.

For dimethylaminonaphthalene, where conjugation is expected
to be greatly diminished, both the reduction and oxidation potentials are closer in magnitude to naphthalene than to α-naphthyl-
amine. Estimates of the magnitude can be obtained from the differences between naphthalene and naphthylamine. On this basis,
30 ± 5% of the resonance interaction of the lone pair would
appear to be present in the dimethylamino compound. Finally,
we note the expected mirror-image nature of reduction and
oxidation curves.

The steric effect on resonance has been elegantly explored
via use of electronic spectra.[24] For the amines, a most general
consequence of deviations of planarity of the conjugating groups
is a decrease in the absorption intensity of the K-bands, i.e.,
those bands that are ascribable to the fully conjugated species.
For these systems, the equation $\varepsilon/\varepsilon_0 = \cos^2 \psi_a$, where ψ_a is the
angle of twist, is valuable when changes in λ_{max} are small.

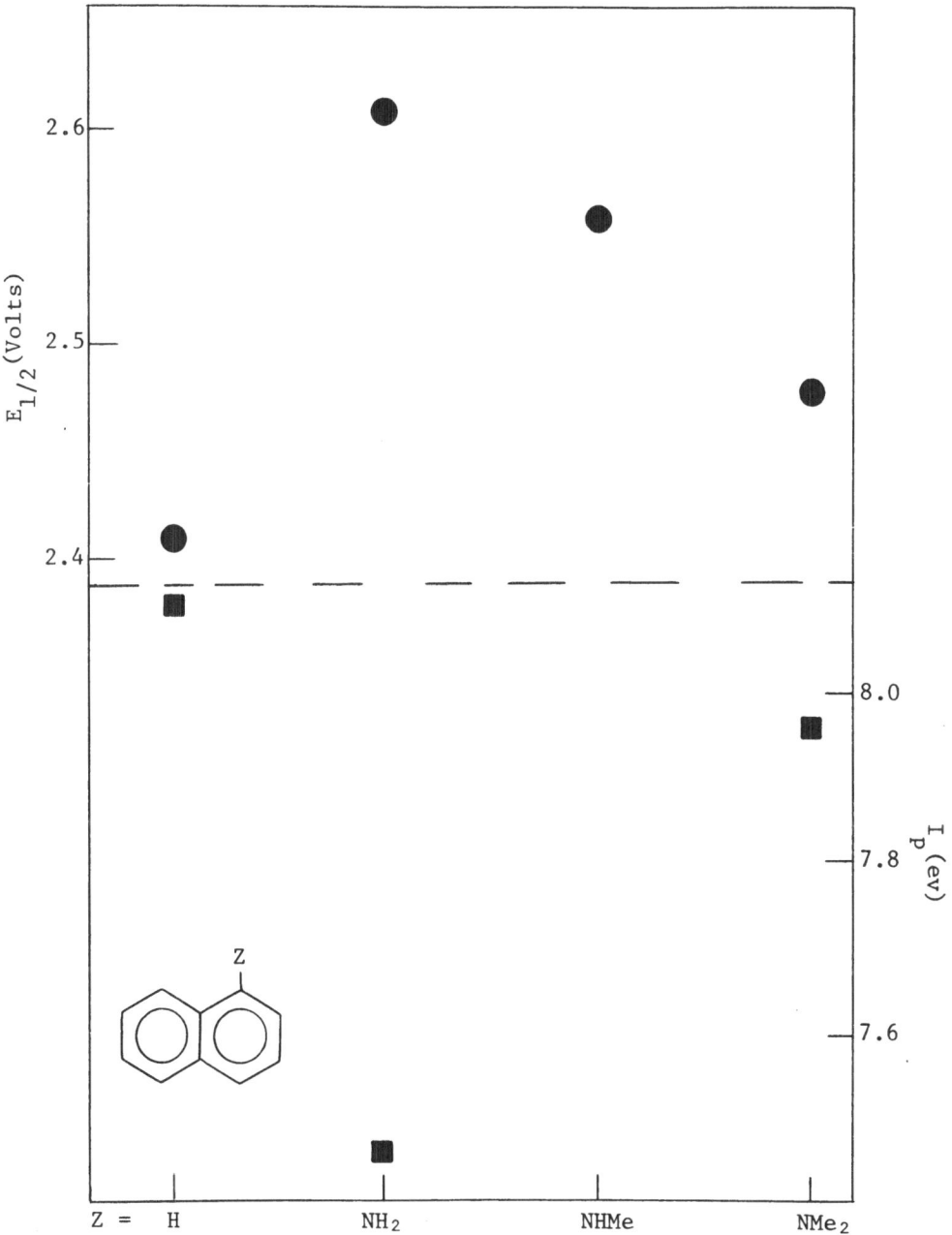

Figure 6. Oxidation and reduction potentials for 1-substituted naphthalenes.

Table 6 and Figure 7 depict the absorption of the substituted naphthyl- and phenyl-amines. For the aniline compounds which are free from a steric effect there is an increase in the extinction

Table 6. Ultraviolet Absorption of Naphthylamines and Anilines[a]

Aromatic	Substituent		
	NH$_2$	NHMe	NMe$_2$
1-Naphthyl	25,000[b]	18,500[b]	10,800[b]
2-Naphthyl	59,957[c]		50,110[d]
Phenyl	9,130[e]	13,200[e]	15,500[e]
o-Tolyl	8,800[e]	9.980[f]	6,300[e]

[a]Extinction coefficient of bands at λ_{max} 240 ± 10 nm. [b]Reference 26. [c]Sadtler Index, UV spectrum #22. [d]Reference 25. [e]Reference 24. [f]Sadtler Index, UV spectrum #23052.

coefficient with increasing methyl substitution on nitrogen. In contrast, for the amines in which increasing methyl substitution on the nitrogen interferes with lone pair conjugation there is a decrease in the extinction coefficient. Interestingly, for the homomorphic α-naphthyldimethylamine and *ortho*-methyldimethyl-aniline the calculated nitrogen lone pair resonance effect is the same and equal to 42% of that for the unhindered compounds. That value agrees very well with the resonance contribution estimated from the reduction and oxidation potentials. Finally, the angle of twist is estimated as 56°.

The proton and ^{13}C NMR are particularly informative for the conformation of the lone pair electrons in these molecules. Whenever a lone pair is directed toward a *peri*-proton, the proton is strongly deshielded. The magnitude of this effect, which has been established with anthracene derivatives, is of the order of 0.4 – 1.0 ppm[28]. It is important to note that this effect is larger than a steric deshielding.

Figure 8 depicts the chemical shifts of the proton NMR of the hydrogen on C-8 as a function of the substituent. Central to the understanding is the observation that for the amino-, N-methylamino- and dimethylammonium-naphthyl compounds the chemical shift is similar. Only for the dimethylamino compound is there a substantial downfield shift. Since the steric

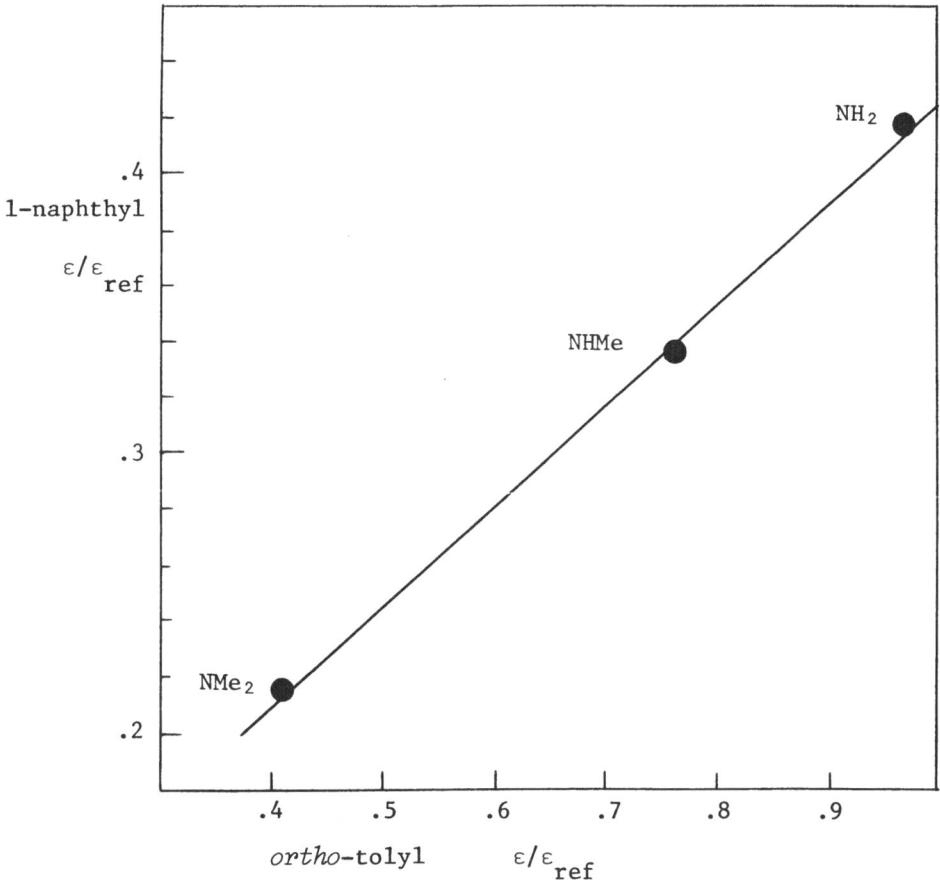

Figure 7. Correlation of the ultra-violet absorption spectra of
1-naphthyl- (ε_{ref} 2-naphthyl) and *ortho*-tolyl amines (ε_{ref} phenyl).

requirement of the dimethylammonium compound is equal to or greater
than that of the dimethylamino compound, this is strong evidence
for the lone pair shift as discussed above. We conclude that
the electron pair on nitrogen in dimethylaminonaphthalene is
pointed towards the *peri*-hydrogen and would therefore be unavailable for conjugation. This result supports the previous
findings.

     Figure 8 also depicts the relationship between the chemical
shifts of the methyl protons of similarly substituted *ortho*-
tolyl derivatives. As discussed previously, these form homomorphic
sets and the same pattern is noted. The chemical shifts for the
amino and N-methylamino compounds are essentially the same whereas

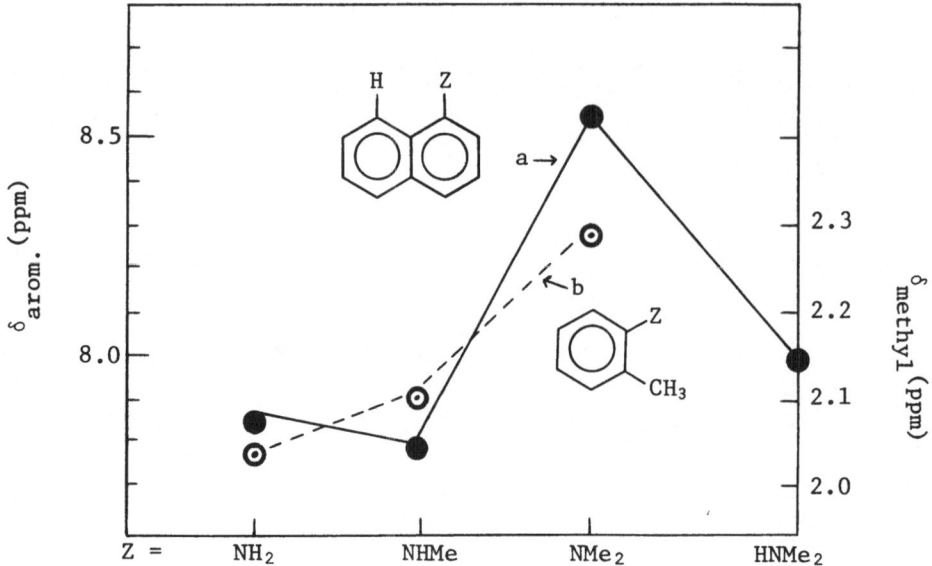

Figure 8. (—●—) Proton chemical shifts of *peri* hydrogens in
1-naphthylamines.   (--⊙--) Methyl proton chemical shifts of
ortho tolylamines.

the dimethylamino compound shows a downfield shift. As expected,
the magnitude of the shift is less since there are three methyl
protons and there are hybridization differences between the *peri*-
hydrogen and the hydrogens on methyl; nevertheless, the agreement is substantial.

Figure 9 records the ^{13}C chemical shifts of the 4 and 8
carbon atoms for the series of naphthylamines. For carbon atom
4, the chemical shifts are sensitive to the substituent at C-1
and are correlated as are benzene compounds with σ^+ constants.[26,29]
In actual fact, the modest decrease in going from the amino substituent to the N-methylamino substituent is in accord with the
correlation. The deviation of the dimethylamino group signifies
a marked decrease in the resonance contribution of the nitrogen
atom in this compound.

The same pattern, albeit for different reasons, is observed
for the chemical shift at C-8. The chemical shifts at position
8 for naphthalene rings are considerably less sensitive to the
electronic effects of substituents, but do show steric effects.[27,29]
The magnitude and the direction of the shift (~4.0 ppm) for the
dimethylamino compound is consistent with such an effect. Thus
the proton and ^{13}C NMR, in three distinct kinds of effects, support

and amplify the conclusions gathered from the spectral and electro-chemical studies.

Attention is now focused on the basicity of the phenyl- and naphthylamines and the implications of conjugation loss. Table 7 records the pKa values of similarly substituted compounds. Thus, substitution of methyl for hydrogen on the amino group has

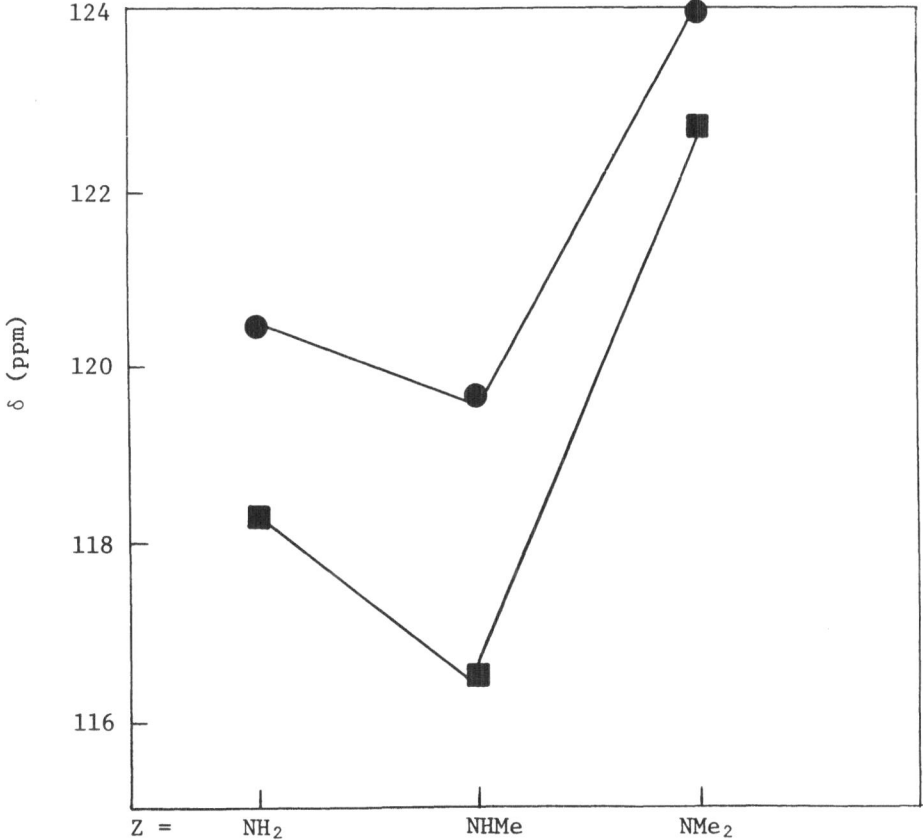

Figure 9. Carbon chemical shifts of C-8 (circles) and C-4 (squares) of 1-naphthylamines.

an additive base-strengthening effect of 0.22 units in both the unhindered aniline and the 2-naphthyl compounds. For the 2-methylphenyl and 1-naphthyl compounds the situation is not so clear. Substitution of a single hydrogen by methyl actually

Table 7.   Effect of Methyl Substitution on pKa Values[a] of Anilines
           and Naphthylamines.

Aromatic	Substituent		
	NH_2	NHMe	NMe_2
Phenyl	4.62	4.85	5.06
2-Methylphenyl	4.39	4.62	5.94
4-Methylphenyl	5.08	5.36	5.56
1-Naphthyl	3.96	3.67	4.83
2-Naphthyl	4.14		4.57

[a]Reference 14.

brings about a decrease in basicity in the 1-naphthyl compounds.
For the disubstituted compounds, the basicity is increased and
the effect is larger than that seen with the unhindered compounds.

The increase in basicity of dimethylaminoaphthalene and 2-
methylphenyldimethylamine is expected on the basis of a decrease
in lone pair conjugation. However, the magnitude of the effect,
as measured by ΔpKa of ~1.2 units, is, perhaps, smaller than
expected. From the earlier estimates, the decrease in the resonance effect of nitrogen of ~65% would bring about a ΔpKa of
~2 units. Quite possibly steric hindrance to solvation, which
effect opposes that of conjugation, moderates the observed value.
This effect has direct analogy with that found in the diaryl
compounds discussed in the first section.

In that regard, a little is known about the strain in the
isoelectronic and homomorphous anions. Figure 10 displays the
proton NMR of naphthylamine and the lithium salt of 1-methyl-
naphthalene. The spectral similarities and differences are those
expected. Thus, the gross patterns are very similar, but the
electronic effects on hydrogens 2 and 4, as well as the steric
effect on hydrogen 8, are greater for the anion than for the
amine. Delocalization is, as expected, greater in the charged
species.

Interestingly, attempts to prepare the anion, which is isoelectronic and homomorphic with dimethylaminonaphthalene, have
been unsuccessful to date. Thus, reaction of 1-isopropyl-
naphthalene with n-butyllithium, t-butyllithium and sodium metal

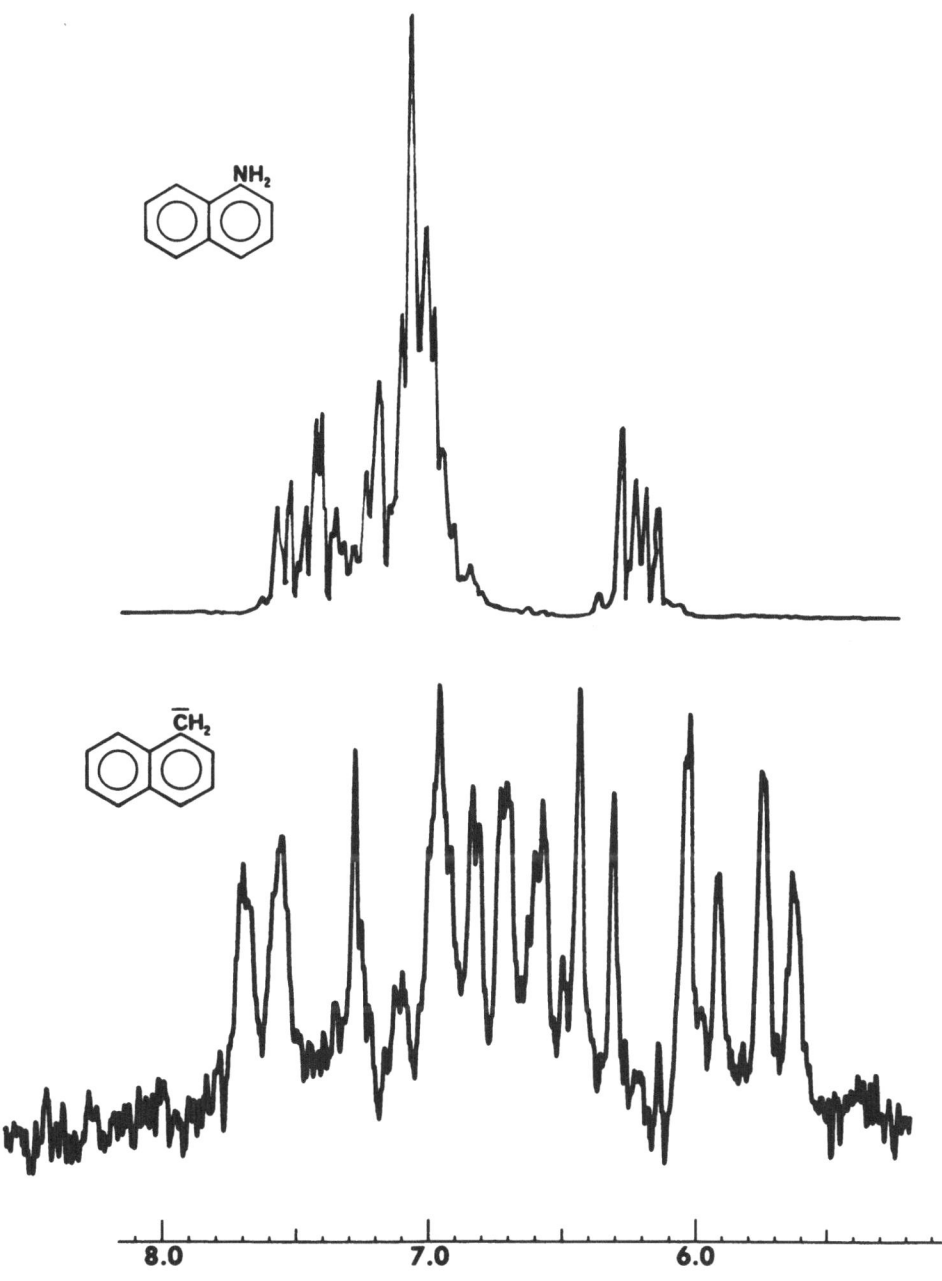

Figure 10. Aromatic proton NMR spectra of 1-naphthylamine and 1-naphthylmethyllithium.

in THF do not lead to the α-anion. While this evidence is not
conclusive, it is noteworthy that the 1-naphthylmethyl anion
is easily prepared under those conditions. Our previous work
suggests that effects in anions are greater than in amines, but
additional work is clearly required to demonstrate that effect
here.

Finally, there has been some work on the corresponding
cations. Surprisingly, the results are quite contrary to those
for the anions. Zaharadnik[30] is in fact able to prepare the cation
of the dimethylcarbinyl compound under conditions where the monomethyl and unsubstituted precursor are inert. Contributing
factors to this reversal may include the stabilizing, rather than
destabilizing, effect of methyl on cations, the greater effect
of resonance in cations relative to anions, and hybridization
differences. Nevertheless, the ready formation of what appears
to be a highly strained cation is puzzling if delocalization is
extensive. It may be that the driving force for cation formation
derives most from the fact that the system is tertiary, whence
resonance does not contribute importantly.

The third area of investigation concerns steric effects
on the rate and products of anion reactions. This study focuses
on providing quantitative data for the rate-determining step in
reactions of 9-alkyl-10-lithio-9,10-dihydroanthracenyl anions with
organic halides. The relevant steric factors are homomorphic
with known systems and are shown in Figure 11.

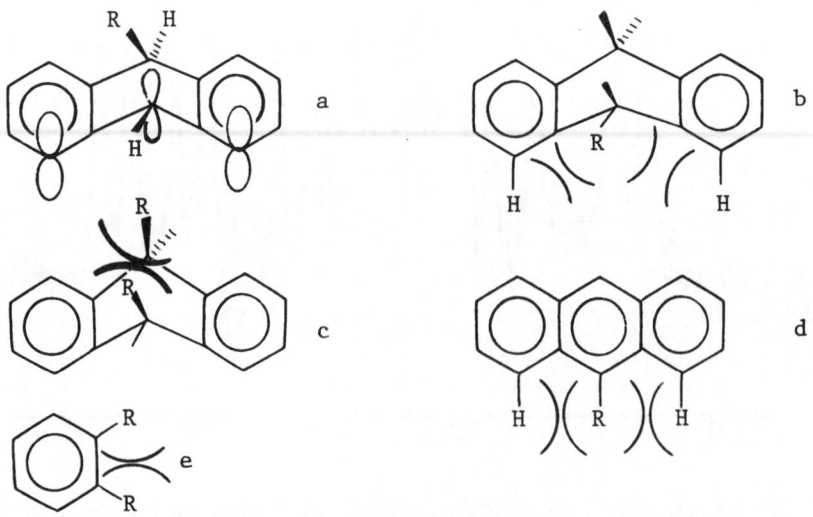

Figure 11. Steric Effects in 9-Alkyl-10-lithio-9,10-dihydro-
anthracenyl anions and related homomorphs.

For the anion, it is likely that the 9-alkyl group occupies a pseudo-axial position (Figure 11a) since in the pseudo-equatorial position (Figure 11b) there will be significant *peri*-like interactions with the aromatic hydrogens. Additionally, the anion probably perfers a pseudo-axial orientation in order to maximize delocalization with the aromatic rings. For the incoming alkyl group, attack at the pseudo-axial position minimizes *peri*-interactions but maximizes 1,4-dialkyl interactions (Figure 11c).

The reactions of the alkylanthracenyl anion with halides have, in fact, fascinating stereochemical results. In the main, reaction with primary halides leads to a predominance of the *cis* product, while reaction of the secondary halides leads to the *trans*.[31] We have used kinetic measurements to sort out the factors influencing the transition states for *cis* and *trans* products.[32]

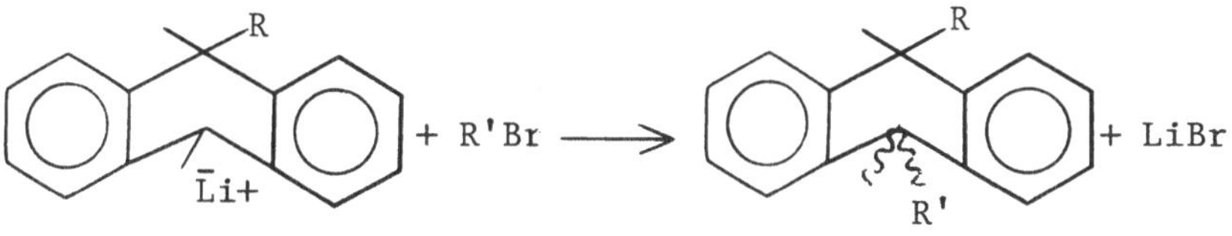

Table 8. Second-Order Rate Constants for Reaction of 9-Alkyl-10-Lithio-9,10-Dihydroanthracene with Alkyl Bromides[a]

Bromide	9-Alkyl	k(corr.)	*cis*/*trans*	k(*cis*)	k(*trans*)
n-hexyl	H	2673			
	Et	1473	76/24	1124	346
	i-Pr	574	59/41	337	237
	t-Bu	312	10/90	31	280
i-Propyl	H	62			
	Et	40	25/75	10	30
	i-Pr	37	13/87	5	32
	t-Bu	36	2/98	0.7	35

[a]Measured with stopped-flow apparatus[32] in THF at 20° at 400 nm.

[b] Absolute rate constants were corrected for the elimination reaction (<15%).

The rate and product data in Table 8 indicate that, for both primary and secondary halides, the transition state leading to *cis* product is markedly influenced by the size of the substituent in the 9-position. In direct contrast, the transition state leading to the *trans* product is insensitive to the substituent.

These data are depicted graphically and correlated with a reaction known to be sensitive to strain in Figures 12 and 13. Here we correlate the free energies of activation ($\Delta G^{\pm}$) of the

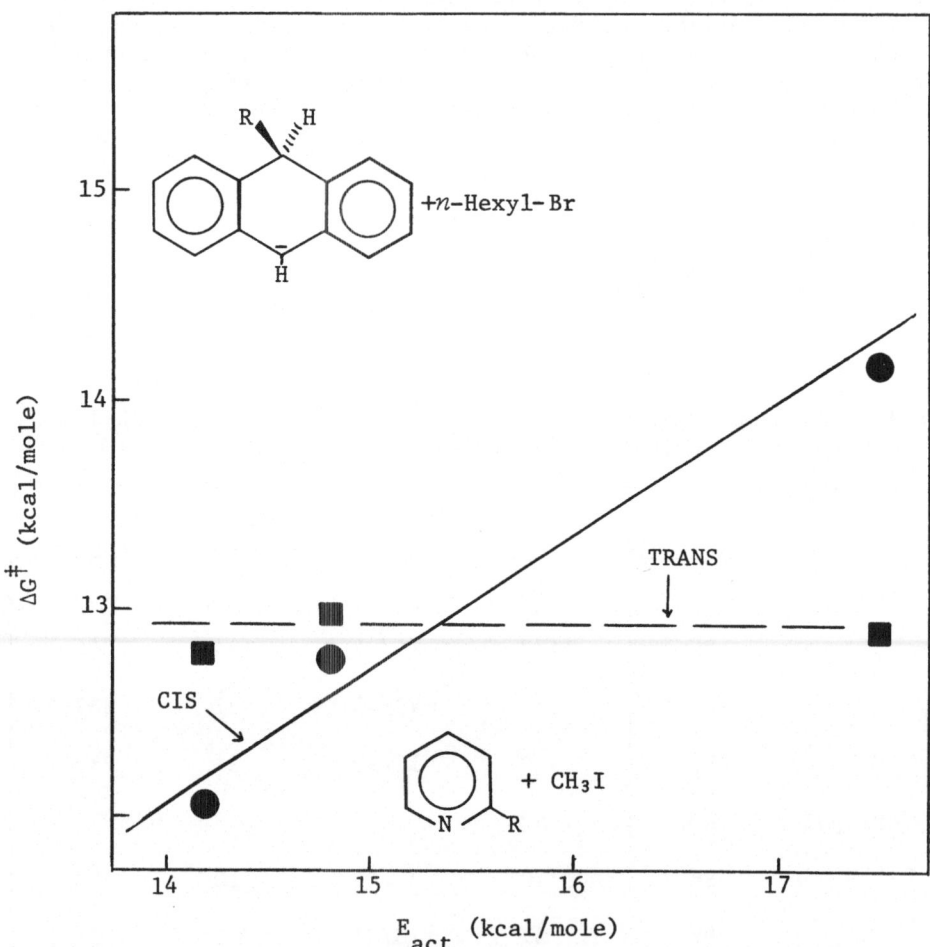

Figure 12. Correlation of free energies of activation of reactions of 9-alkyl-10-lithio-9,10-dihydroanthracenes with *n*-hexyl bromide and activation energies of reactions of 2-alkyl pyridines with methyl iodide.

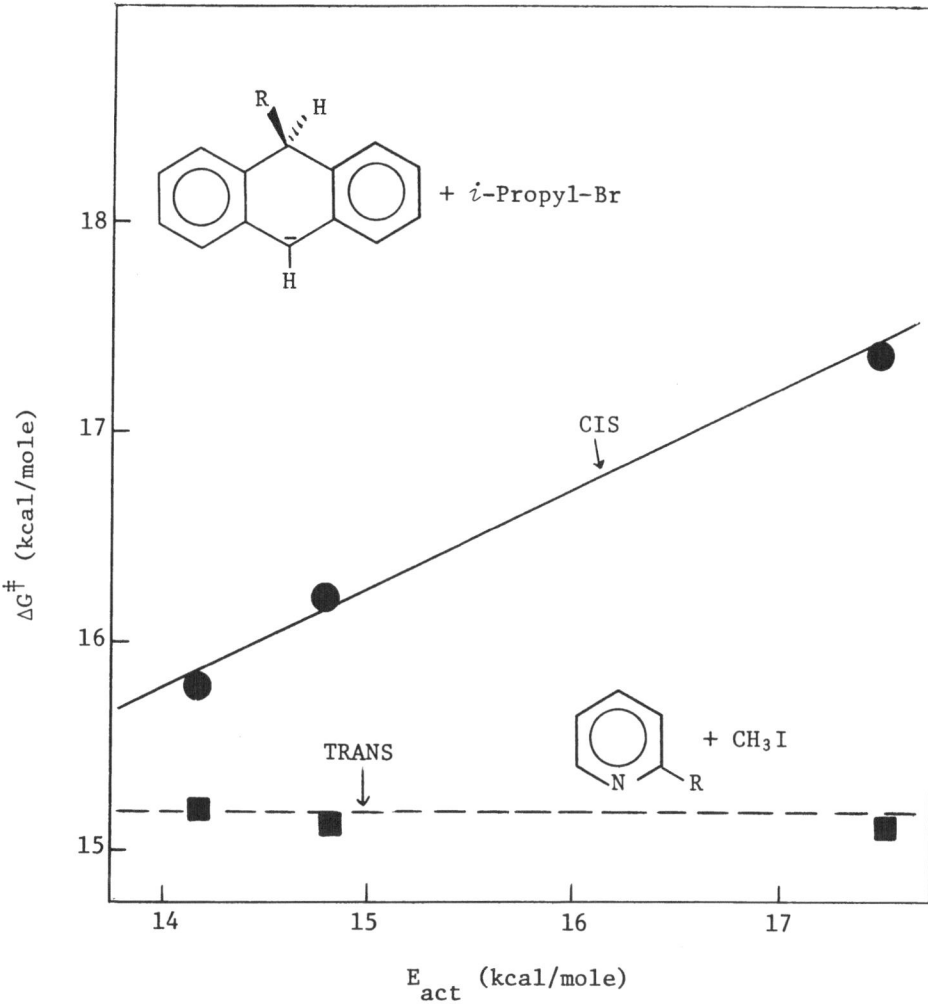

Figure 13. Correlation of free energies of activation of reactions of 9-alkyl-10-lithio-9,10-dihydroanthracenes with *i*-propyl bromide and activation energies of reactions of 2-alkyl pyridines with methyl iodide.

anthracenyl anions with primary and secondary halides, respectively, and the activation energies (E_A) for reactions of 2-substituted pyridines with methyl iodide, studied by Brown and Cahn.[23] Inspection of these figures reveals several interesting features. First, for primary and secondary halides, the formation of the *cis* compounds quantitatively correlates with the reaction known to be sensitive to steric effects. Additionally the slopes are similar (~0.5-0.6), indicating comparable steric effects in the

two anthracenyl anion reactions. Secondly, the rate of formation
of the *trans* compounds is, in both series, totally independent
of the steric bulk of the alkyl group.

There is some surprise in the fact that the reactions of
the 2-alkylpyridines with methyl iodide serves as a quantitative
model for the reactions of the 9-alkyl-10-lithio-9,10- dihydro-
anthracenyl anions. However, scale drawings (Figure 14) reveal
the similarities of the two systems. Essentially what they share
is a similar distance of the two interacting alkyl groups. Thus,
in spite of the significant differences in the hybridization of
the lone pairs, the basicities of the lone pairs, and the bond
angles connecting the interacting species, the homomorph concept
is exceedingly useful in providing a quantitative model for steric
effects.

The importance of this is evidenced further in the comparison of rate ratios for primary and secondary halides with a
sterically hindered nucleophile. While there is ample precedent
for a rate ratio ~15 with typical nucleophiles,[31] we observed
much larger values with the 9-alkyl-10-lithio-9,10- dihydro-
anthracene compounds. As expected, a sterically hindered base
gave a larger rate ratio, but we sought a quantitative model.
In actual fact, Brown and Cahn had studied a comparable halide
variation with 2-alkypyridines[23] which leads to the desired
quantitative estimate.

The relevant data, as recorded in Table 9, reveal that replacement of a hydrogen atom by methyl has a larger effect on
the reaction with isopropyl iodide than with ethyl iodide. This
fact is accommodated by the concept that the steric strain is
a function of both the reference acid and the reference base.[2,6]
Thus, increasing the size of the reference acid (halide) brings
about a greater effect for the larger base. This is precisely
what is observed, in magnitude and direction, for the same structural variation with the alkylanthracenyl anions and halides.
The more bulky 9-ethyl compound has a larger effect in reaction
with isopropyl bromide than in reaction with *n*-hexyl bromide.
Again, the concept of homomorphs and steric effects as a function of reference size of reactants has proven quite valuable.

Finally attention is focused on the possible geometries of
the two transition states. The geometry of the anion is not
known with certainty and, in fact, it might be a mixture of
conformers. Nevertheless, the geometry of the transition state
leading to the *cis* product most likely contains an axial-axial
orientation as shown in Figure 15a. This orientation accommodates
the steric effects and the stereochemical preferences and allows
anion stabilization by orbital overlap. In the case where the
trans product is formed, whereas there are several possible

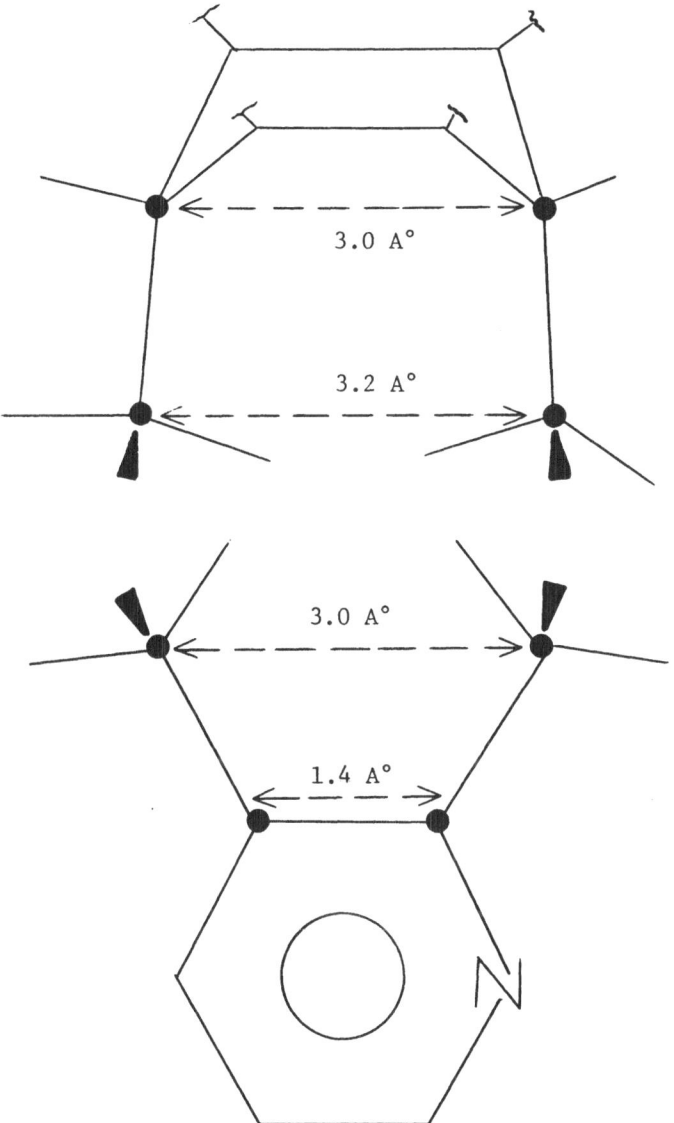

Figure 14.   Scale drawings of homomorphic bases.

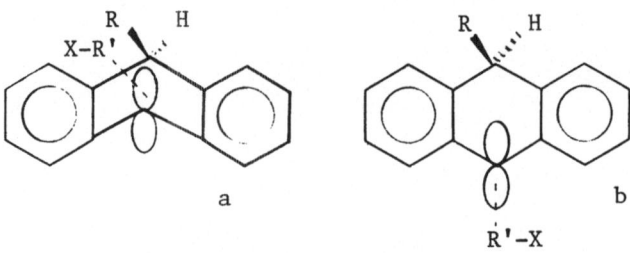

Figure 15. Possible transition states for formation of *cis* and
trans dialkyldihydroanthracenes.

Table 9. Log Rate Constant for Reactions of Primary and Secondary
          Halides with Hindered Bases.

Base	1°	Δ	2°	Δ
pyridine (N–H)	-3.51[a]		-4.00[b]	
		.60		1.09
2-methylpyridine (N–CH₃)	-4.11[a]		-5.09[b]	
	3.43[c]		1.79[d]	
		.38		.79
	3.05		1.00	

[a]Reaction with ethyl iodide at 60° (Reference 23).

[b]Reaction with isopropyl iodide at 80° (Reference 23).

[c]Reaction with *n*-hexyl bromide at 20° (Reference 32).

[d]Reaction with isopropyl bromide at 20° (Reference 32).

conformers which could lead to product, it is clear that the 9-substituent has little effect on the reaction rate. A reasonable model for the transition state is depicted in Figure 15b. In this scheme a flattening of the ring and attack from the underside leads to the *trans* product.

In summary, those concepts of a quantitative model for steric interactions developed more than 25 years ago with weakly basic amines have been found to be highly useful in accounting for the behavior of strongly basic organic carbanions. Thus, while the absolute rates of reaction often span differences of six orders of magnitude, the quantitative steric effects for homomorphs are the same. The utility is obvious. It permits the vast accumulated data on the one type of system to be used in other systems. In this way we and others[9] continue to reap benefits from the careful and creative work on homomorphs and molecular addition compounds.

In conclusion, some comment on the similarity of steric effects in systems of greatly differing stability and reactivity is in order. Two things are suggested by these observations. First, it is useful to separate the energy feature contributing to reactivity. Then, polar and electronic factors can dramatically shift the energy scale but the differences between homomorphs will remain the same. Second, the mechanism of the several effects must be essentially different and therefore the size of a substituent is not a marked function of the charged energy surface. This makes steric effects all the more powerful and it was an important moment in time when H. C. Brown saw this and chose to demonstrate it.

ACKNOWLEDGEMENTS

It is a pleasure to acknowledge the many contributions of my co-workers in this work. Mr. John Sturges, Dr. Matthew Platz and Mrs. Janet Frost Bank have capably and enthusiastically performed the many experiments discussed herein. Finally NATO provided a research grant and I am grateful for the support.

REFERENCES

1. F. Kehrmann, *J. prakt. Chem.*, **40**, 257 (1889).
2. H. C. Brown, *J. Chem. Soc.*, **1956**, 1248.
3. W. J. le Noble, "Highlights of Organic Chemistry," Marcel Dekker, Inc., New York, 1974, Chap. 7.
4. F. A. Carey and R. J. Sundberg, "Advanced Organic Chemistry," Plenum Press, New York, 1977, Chap. 3.

5. H. C. Brown, G. K. Barbaras, H. L. Berneis, W. H. Bonner,
    R. B. Johnanneson, M. Grayson and K. LeRoi Nelson, *J. Am.
    Chem. Soc.*, 75, 1, (1953).
6. H. C. Brown, *J. Chem. Educ.*, 36, 427 (1959).
7. J. Packer, J. Vaughan and E. Wong, *J. Am. Chem. Soc.*, 80,
    905 (1958).
8. M. S. Newman and W. H. Powell, *J. Org. Chem.*, 26, 812 (1961).
9. Y. Okamoto and K. I. Lee, *J. Am. Chem. Soc.*, 97, 4015 (1975).
10. E. L. Eliel in M. S. Newman, Editor "Steric Effects in Organic
    Chemistry," John Wiley and Son, Inc., New York, 1956, p. 161.
11. D. J. Cram, "Fundamentals of Carbanion Chemistry," Academic
    Press, New York, 1965.
12. A. Streitwieser, Jr., J. R. Murdoch, G. Häfelinger and C. J.
    Chang, *J. Am. Chem. Soc.*, 95, 4248 (1973).
13. N. S. Frumina and E. G. Tregub, *Zh. Analiticheskol Khim.*,
    26, 669 (1971).
14. N. F. Hall, M. R. Sprinkle, *J. Am. Chem. Soc.*, 54, 3469
    (1932); D. D. Perrinn, "Dissociation Constants of Organic
    Bases in Aqueous Solution" Butterworths, London, 1965.
15. B. V. Cheney, *J. Am. Chem. Soc.*, 90, 5386 (1968).
16. V. R. Sandel and H. H. Freedman, *J. Am. Chem. Soc.*, 85,
    2328 (1963).
17. S. Bank and J. S. Sturges, *J. Organometal. Chem.*, in press
    (1978).
18. C. H. Bushweller, J. S. Sturges, M. Cipullo, S. Hoogasian,
    M. W. Gabriel and S. Bank, *Tetrahedron Lett.*, 1359 (1978).
19. S. Bank and J. S. Sturges, research in progress.
20. R. B. Bates, L. M. Kiroposki and D. E. Potter, *J. Org. Chem.*,
    37, 560 (1972).
21. R. Breslow and R. Goodin, *J. Am. Chem. Soc.*, 98, 6076 (1976).
22. P. West and R. Waack, *J. Am. Chem. Soc.*, 89, 4395 (1967),
    R. A. H. Casling, A. G. Evans and N. H. Rees, *J. Chem. Soc.*,
    B, 1966, 519.
23. H. C. Brown and A. Cahn, *J. Am. Chem. Soc.*, 77, 1715 (1955).
24. B. M. Wepster in W. Klyne and de la Mare, Editors, "Progress
    in Stereochemistry," Butterworths, London, 1958, p. 150.
25. C. Zener, *Phys. Rev.*, 36, 51 (1930). The reduction poten-
    tials were measured in DMSO with TBAP electrolyte using SCE
    as reference at 21°. The oxidation potentials are computed
    from the frequencies of the charge transfer complexes of DDQ
    and 1-substituted naphthalenes in methylene chloride
    (Reference 26) and the relationship between ionization
    potentials and charge transfer frequencies (Reference 27).
26. S. Bank and M. Platz, manuscript in preparation.
27. R. Foster, "Organic Charge-Transfer Complexes." Academic
    Press, New York, 1969, pp. 60-62.
28. F. Gobert, S. Combrisson and M. Platzer, *Tetrahedron*, 30,
    2919 (1974).
29. P. R. Wells, D. P. Arnold and D. Doddrell, *J. Chem. Soc.*,
    Perkin II, 1974, 1745.

30. R. Zahradnik, A. Kröhn, J. Pancis and J. Shobl., *Coll. Czech. Chem. Comm.*, 34, 2553 (1969).

31. (a) M. Daney, R. Lapouyade, M. Mary, and H. Bouas-Laurent, *J. Organometal. Chem.*, 92, 267 (1975); (b) C. Fabre, M. H. A. Salem, J. P. Mazaleyrat, A. Tchapla, and Z. Welvart, *ibid.*, 87, 9 (1975); (c) P. P. Fu, R. G. Harvey, J. W. Paschal, and P. W. Rabideau, *J. Am. Chem. Soc.*, 97, 1145 (1975); (d) H. E. Zieger and L. T. Gelbaum, *J. Org. Chem.*, 37, 1012 (1972).

32. S. Bank, J. Bank, M. Daney, B. Labrande, and H. Bouas-Laurent, *J. Org. Chem.*, 42, 4058 (1977).

33. A. Streitwieser, "Solvolytic Displacement Reactions," McGraw-Hill, New York, 1962, pp. 30-31.

COPPER(II)-INDUCED OXYGENOLYSIS OF o-BENZOQUINONES, CATECHOLS, AND PHENOLS: THE ACTIVE COPPER(II)-SPECIES, ROLE OF CUPRIC CHLORIDE, AND THE GENERAL QUESTION OF ACTIVATION OF MOLECULAR OXYGEN BY DIOXYGENASES

Milorad M. Rogić and Timothy R. Demmin

Corporate Research Center, Allied Chemical Corporation

Morristown, New Jersey 07960

The chemical energy required by biological systems to do work is derived primarily by oxidation of complex organic molecules to carbon dioxide and water. Since, in the overall process, oxygen acts as a final acceptor of electrons from the substrates and is converted to water, a special mechanism for the activation of molecular oxygen is not required. Nevertheless, it is still not known with certainty whether the required four electrons are transferred between the last few members of the electron-transporting chain in pairs or singly, nor is it known precisely how molecular oxygen accepts the electrons from the last member of this chain (the cytochrome c oxidase).[1]

A relatively small fraction of the oxygen used in biological oxidations actually does appear in the reaction products, generally in the form of hydroxyl or carbonyl groups. The remarkable ability of certain monooxygenases and dioxygenases to incorporate one or both atoms of the O_2 in the appropriate oxidation products, particularly during the oxidative cleavage of various catechols and phenols, is well known.[2] Unfortunately, the mechanisms of these oxidations are still not well understood, but since, under ordinary conditions, these reactions would be spin forbidden processes,[3a] it is generally assumed that they involve "activated" molecular oxygen.[3a-g] Presumably, "activation" of molecular oxygen could be achieved by formation of a "ternary complex" by an allowed reaction between the O_2, the enzyme molecule, EM, and the substrate species, SH_2, as indicated in I and II, respectively. The role of the "ternary

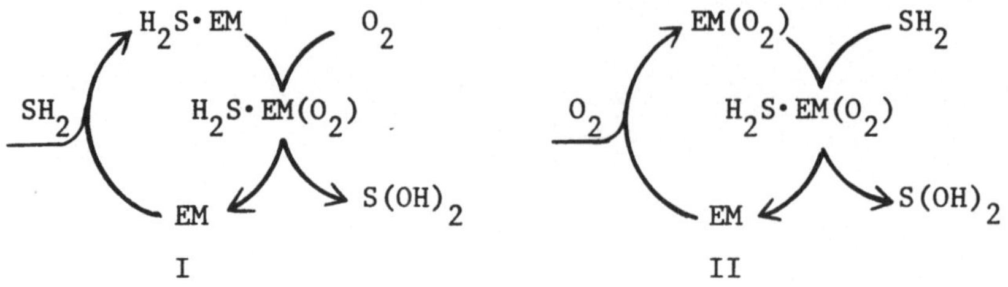

complex", $H_2S \cdot EM(O_2)$ in activation of molecular oxygen is to provide
a particular mechanism for a direct transfer of electrons from the
substrate to the complexed molecular oxygen in such a way that one
or both atoms of the O_2 would be incorporated in the resulting oxidation products.[4]

Our interest in new synthetic approaches to caprolactam led us
to explore various methods for cleaving carbon-carbon bonds in cyclic
C_6-systems[5-8] (eq 1), and some time ago we began investigating possibilities for a direct oxidative cleavage of catechols and phenols

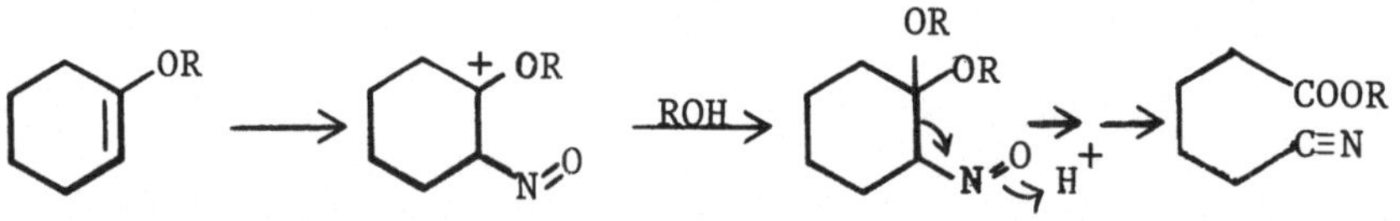

with molecular oxygen in the presence of certain transition
metals[9,10] (eq 2).

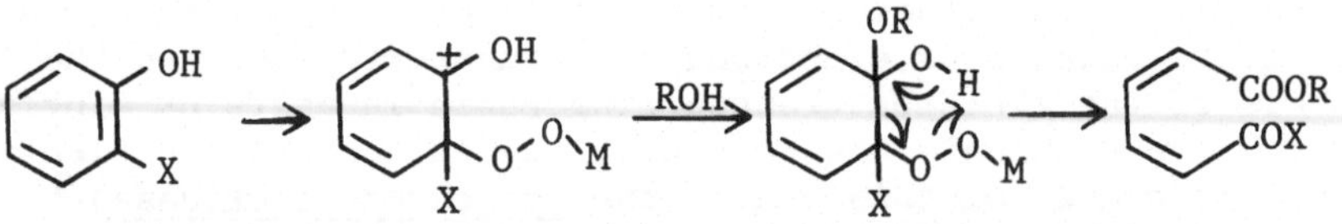

These efforts are described in the present paper. We discuss
the chemistry associated with copper(II)-induced carbon bond cleavage reactions in *o*-benzoquinones, catechols, and phenols; further,
we elucidate the nature of the "activated" oxygen-copper species
and consider the overall role of molecular oxygen in these transformations.

COPPER(I) AS A CATALYST FOR "ACTIVATION" OF MOLECULAR OXYGEN

In an attempt to learn more about the mechanism of oxidation of
phenols by the enzyme tyrosinase, Brackman and Havinga carried out

an extensive study of phenol oxidation using copper/amine catalysts.[11] At about the same time, Terentiev and co-workers investigated the oxidation of aromatic amines in pyridine solution containing cuprous chloride,[12] and Kinoshita reported that benzil undergoes carbon-carbon bond cleavage to benzoic acid under similar conditions.[13] In this country, Hay, Blanchard, Endres, and Eustance developed an oxidative polymerization of 2,6-disubstituted phenols catalyzed by cuprous chloride in pyridine/nitrobenzene solution,[14,15] and demonstrated that the pyridine cupric methoxy chloride complex, A, was an active catalyst for these oxidative polymerizations.

$$
\begin{array}{c}
\text{Me} \\
| \\
\text{Py} \diagdown \quad \diagup \text{O} \diagdown \quad \text{Cl} \\
\text{Cu} \quad \text{Cu} \\
\text{Cl} \diagup \quad \text{O} \diagdown \quad \text{Py} \\
| \\
\text{Me}
\end{array}
$$

A

We have demonstrated that addition of catechol under oxygen to the "Cu-Reagent" in pyridine, prepared as described below, resulted in consumption of approximately 1 molar equivalent of oxygen and led, after evaporation of solvent, hydrolysis and extraction, to an 80-85% yield of the *cis,cis*-muconic acid monomethyl ester (1), mp 80-81° (eq 3).

$$
\text{catechol} + \text{"Cu-Reagent"} + O_2 \xrightarrow{\text{Py}} \text{product} \underline{1} \tag{3}
$$

The "Cu-Reagent" was prepared either by reaction of oxygen with four molar equivalents of cuprous chloride in pyridine solution containing five equivalents of methanol (method A), or by reaction of oxygen with four molar equivalents of cuprous chloride in pyridine, followed by addition of the required amount of methanol (method B). Alternatively, the reaction of complex A with an equivalent amount of water in pyridine also provided the same reagent (method C).

$$
\begin{array}{l}
4CuCl + O_2 + 4MeOH \xrightarrow[\text{Py}]{A} \\
4CuCl + O_2 \xrightarrow{\text{Py}} [\ ] \xrightarrow[\text{4MeOH}]{B} \\
2(PyCuClOMe)_2 + 2H_2O \xrightarrow[\text{Py}]{C}
\end{array}
\Biggr\} \longrightarrow \text{"Cu-Reagent"} \tag{4}
$$

When ethanol, *n*-butyl alcohol, or isopropyl alcohol were used
instead of methanol to prepare the "Cu-Reagent" by methods A and B,
followed by reaction with catechol under similar conditions, the
corresponding *cis,cis*-muconic acid monoalkyl esters were obtained:
R = Et, mp 101-102°, 63%; *n*-Bu, mp 54.5-56°, 76%; i-Pr, mp 51-71°
(mixture of double bond isomers), 26% yield.[10]

Oxidation of catechol to *cis,cis*-muconic acid monomethyl ester
(1) is a 4-electron oxidation, and in principle the reaction could
involve stepwise oxidations of catechol to *o*-benzoquinone, followed
by the oxidative cleavage of *o*-benzoquinone to the observed monomethyl ester 1. Consequently, it was of interest to establish whether *o*-benzoquinone was a reaction intermediate, and more importantly,
whether phenol itself could be used as a catechol precursor (Scheme I).

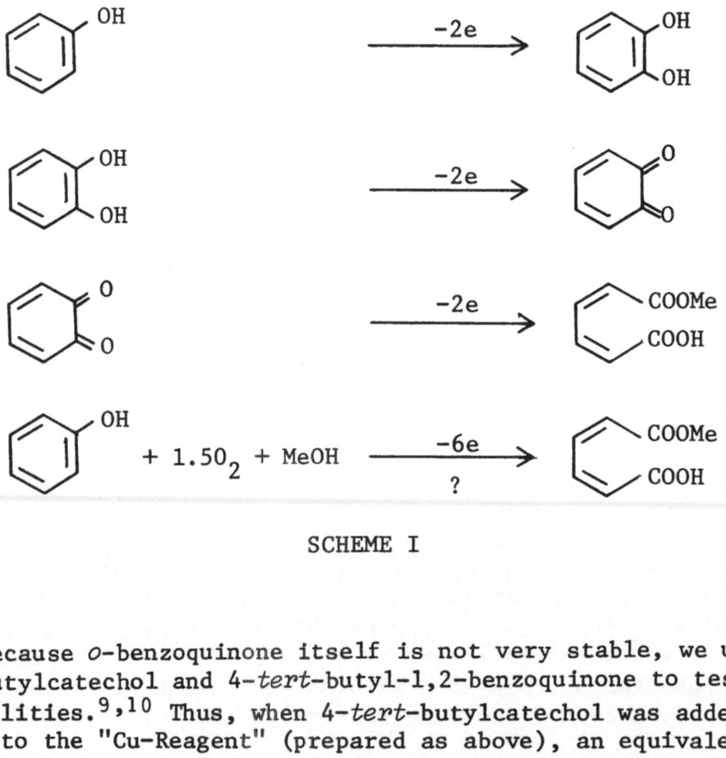

SCHEME I

Because *o*-benzoquinone itself is not very stable, we used 4-
tert-butylcatechol and 4-*tert*-butyl-1,2-benzoquinone to test these
possibilities.[9,10] Thus, when 4-*tert*-butylcatechol was added under
oxygen to the "Cu-Reagent" (prepared as above), an equivalent amount
of oxygen was consumed. Standard workup gave a mixture of the isomeric 3- and 4-*tert*-butylmuconic acid monomethyl esters (2) (55%)
and (3) (40%) (eq 5, X = OH, n = 1). The reaction with 4-*tert*-
butyl-1,2-benzoquinone (eq 5, X = 0, n = 1/2), required one-half
molar equivalent of oxygen and afforded essentially identical yields
of the isomeric monomethyl esters 2 and 3.

$$\text{(structure)} + n\text{"Cu-Reagent"} + n\text{O}_2 \xrightarrow{\text{Py}} \text{(structure 2)} + \text{(structure 3)} \quad (5)$$

$$\underline{2} \qquad \underline{3}$$

The oxidation of phenol under similar reaction conditions was significantly slower. For example, addition of phenol to the "Cu-Reagent" under oxygen led to a slow oxygen uptake to give, after one day, about 50-60% of the monomethyl ester $\underline{1}$ (eq 6).

$$\text{(structure, phenol)} + \text{"Cu-Reagent"} + \text{O}_2 \xrightarrow{\text{Py}} \text{(structure 1)} \quad (6)$$

$$\underline{1}$$

Since both 4-*tert*-butyl derivatives gave essentially the same mixture of the isomeric monomethyl esters $\underline{2}$ and $\underline{3}$, it seems reasonable to conclude that corresponding *o*-benzoquinone is indeed the intermediate in the overall oxidation of the catechol. Moreover, since the oxidation of phenol also provided the monomethyl ester $\underline{1}$, it appears that the overall oxidation of phenol might also proceed in a stepwise manner as suggested in Scheme I (but see Note 16).

DOES OXIDATION OF CATECHOL TO *CIS,CIS*-MUCONIC ACID MONOMETHYL ESTER REQUIRE MOLECULAR OXYGEN?

The salient feature of the pyrocatechase catalyzed carbon-carbon bond cleavage of catechol in the reaction with $^{18}\text{O}_2$ in the presence of H_2O, or with O_2 in the presence of $\text{H}_2{}^{18}\text{O}$, is that one of the oxygen atoms in each carboxylic group of the *cis,cis*-muconic acid produced came from the molecular oxygen and not from the added water.[2,3] Tsuji carried out the oxidation of catechol[17] with the $^{18}\text{O}_2$ enriched oxygen[18] in the presence of "Cu-Reagent" prepared by method A (*vide supra*), and found that one atom of the $^{18}\text{O}_2$ becomes incorporated in the carboxylic group of the *cis,cis*-muconic acid monomethyl ester ($\underline{1}$), while the other becomes converted to water.[18] This experimental observation led him to propose that cuprous chloride in pyridine/methanol is a good system for activating molecular oxygen,[17,18] and hence that it can be used as a nonenzymatic model for pyrocatechase. To account for the incorporation of molecular oxygen in the reaction product, Tsuji proposed the following mechanism[18] (eq 7).

$$\text{(7)}$$

Since Tsuji's copper reagent[17] was evidently the same as our "Cu-Reagent", one may be tempted to conclude that the "activation" of molecular oxygen in our systems was achieved similarly as in Tsuji's system, namely by reaction of oxygen with the "Cu-Reagent". However, from the observed stoichiometries between cuprous chloride and oxygen (eq 4), it is evident that during the preparation of the "Cu-Reagent" only the amount of oxygen required to oxidize the available copper(I) to copper(II) is consumed. Furthermore, the generated "Cu-Reagent" does not show any tendency toward further reaction with oxygen. Consequently, it is hard to see how the required "activation" of molecular oxygen could take place when the substrate is introduced to the reaction system. It is possible, however, that the "activation" of oxygen and subsequent transformations involve a "ternary complex" $H_2S \cdot EM(O_2)$ which could be formed analogously to enzymatic systems,[4] e.g., I or II, by interaction of molecular oxygen with the intermediate $H_2S \cdot EM$ resulting from reaction of the substrate SH_2 with the copper(II)-reagent EM.

Alternatively, it is possible that the various copper(II)-species present in the "Cu-Reagent" are the actual oxidizing agents, capable of bringing about the oxidative carbon-carbon bond cleavage of catechol or o-benzoquinone even in the absence of oxygen (see note 19). To distinguish between these possibilities we carried out the following experiments under strictly anaerobic conditions. For example, addition of a freeze-pump-thaw degassed solution of catechol in pyridine/methanol to a similarly degassed "Cu-Reagent" prepared as above, followed by the standard workup, but under an inert atmosphere, gave the cis,cis-muconic acid monomethyl ester (1) in essentially the same yield as when the reaction was carried out in the presence of oxygen

$$\text{(8)}$$

(eq 8). Both 4-tert-butylcatechol and 4-tert-butyl-1,2-benzoquinone reacted with the "Cu-Reagent" under nitrogen to give the same mixture of monomethyl esters 2 and 3 (eq 9; X = OH, n = 2; X = O, n = 1) that was produced in the presence of oxygen.   Reaction of catechol with

$$\text{[structure]} + n\text{"Cu-Reagent"} \xrightarrow[\text{Py}]{\text{N}_2} \text{[structure]}\begin{array}{l}\text{COOMe}\\\text{COOH}\end{array} \qquad (9)$$

$$\underline{2} + \underline{3}$$

pyridine cupric methoxy chloride complex A in pyridine/methanol
under anaerobic conditions, <u>but in the absence of water</u>, led predomi-
nantly to polymeric material.   However, a similar reaction using the
4-*tert*-butyl derivatives produced, after evaporation of solvent, a
brown oily material.   Pentane extraction, followed by evaporation of
the solvent and distillation, gave 2,2-dimethoxy-6-carbomethoxy-4-
tert-butyloxacyclohexa-3,5-diene $\underline{4}$ in 65% yield (eq 10; X = OH, n =
6; X = O, n = 3).   Acid hydrolysis of this orthoester-ester in chloro-
form gave, quantitatively, 6-carbomethoxy-4-*tert*-butyloxacyclohexa-
3,5-diene-2-one ($\underline{5}$), mp 84-85° (eq 10).

$$\text{[structure]} + n\text{PyCuClOMe} \xrightarrow[\text{Py/MeOH}]{\text{N}_2} \text{[structure]} \longrightarrow \text{[structure]} \qquad (10)$$

$$\underline{4} \qquad\qquad \underline{5}$$

     Clearly, the oxidative carbon-carbon bond cleavage in catechols
and *o*-benzoquinone brought about by reaction with "Cu-Reagent" under
anaerobic conditions conclusively establishes that molecular oxygen
is not involved in the cleavage reaction.   It follows that the various copper(II)-species present in the "Cu-Reagent" are the actual
oxidizing agents capable of bringing about the observed conversion
of catechol and *o*-benzoquinone into muconic acid monomethyl ester
($\underline{1}$) even in the absence of oxygen.   Since, during this transformation, the copper(II)-species must be reduced to copper(I), it follows
that the role of oxygen must be to reoxidize the reduced copper
species back to copper(II)-reagent.   It is known that the oxidation
of cuprous chloride in methanol gives the insoluble cupric methoxy
chloride[15] and water (eq 11) and, in methanol containing pyridine,[15]
provides pyridine cupric methoxy chloride complex A and water (eq 12).

$$4\text{CuCl} + \text{O}_2 \xrightarrow{\text{MeOH}} 4\text{CuClOMe} + 2\text{H}_2\text{O} \qquad (11)$$

$$4\text{CuCl} + \text{O}_2 + 4\text{Py} \xrightarrow{\text{MeOH}} 2(\text{PyCuClOMe})_2 + 2\text{H}_2\text{O} \qquad (12)$$

A

Evidently, the oxygen atom incorporated in the carboxylic group of the half-ester 1, resulting from anaerobic reaction of catechol with "Cu-Reagent" (eq 8) [generated (eq 4) by the reaction of complex A with water], must come from the water. If, on the other hand, the "Cu-Reagent" was prepared by reaction of cuprous chloride with $^{18}O_2$, the active copper(II)-species would be now equivalent to pyridine cupric methoxy chloride and $2H_2{}^{18}O$, and, consequently, it is understandable that one of the $^{18}O_2$ atoms would be incorporated in the carboxylic group and the other in the $H_2{}^{18}O$ by-product (see note 20). These relationships between the oxygen, copper(I), copper (II), methanol and the substrate on the one hand, and the half-ester product and the water by-product on the other, are schematically illustrated in Figure 1.

We will now discuss the nature of the active copper(II)-species in the "Cu-Reagent" and the mechanism of the carbon-carbon cleavage reaction.

NATURE OF THE ACTIVE COPPER(II)-SPECIES IN THE "Cu-REAGENT"

It has been demonstrated that reaction between cuprous chloride and oxygen in pyridine provides bispyridine cupric chloride and some copper-oxygen species assumed[21] to be corresponding "copper(I)-peroxide" (eq 13).

$$4CuCl + O_2 \xrightarrow{Py} 2Py_2CuCl_2 + (Cu-O-O-Cu)\cdot nPy \qquad (13)$$

We, however, could not find any conclusive evidence that this copper-oxygen product was indeed the suggested copper(I)-peroxide.

Any mechanism for the oxidation of cuprous chloride in pyridine must take into consideration the following experimental observations:

(i) one equivalent of oxygen oxidizes four equivalents of cuprous chlorides;[15,21,22,23,24,25,26]

(ii) the reaction between oxygen and cuprous chloride is first order in each reactant;[24,26]

(iii) one of the reaction products is bispyridine cupric chloride, which contains all of the chlorine atoms and half of the total amount of the copper;[21]

(iv) the second product contains the other half of the original copper and both oxygen atoms and neither pyridine solutions of this product nor the amorphous solid remaining after evaporation of the solvent show any evidence for the peroxide structure;

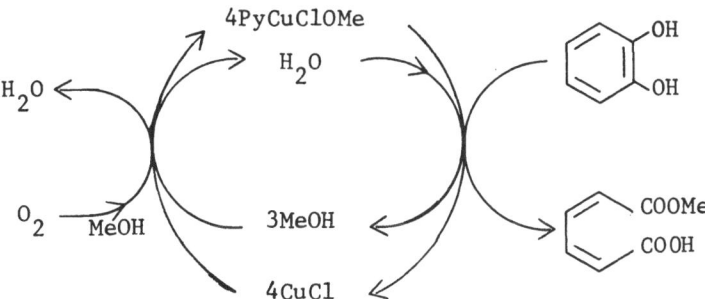

Figure 1 – Schematic representation of the oxidation of catechol
           with "Cu-Reagent" generated either by reaction of com-
           plex A with water, or by reaction of cuprous chloride
           with oxygen in the presence of methanol.

     (v) pyridine solution of the copper-oxygen species is stable
for an extended period of time and does not react with oxygen (see
Note 27);

     (vi) acid-catalyzed hydrolysis of the pyridine solution contain-
ing copper-oxygen species provides cupric oxide;[21]

     (vii) pyridine solution containing the copper oxygen species is
not ESR active.[21]

     The absence of an EPR signal in the pyridine solution of this
species[21] is certainly not evidence for the presence of copper(I),
for significant antiferromagnetic coupling of the unpaired electrons
on the two adjacent copper(II) centers would also eliminate the necessary condition for the observation of the EPR signal.[28,29] The absence of peroxide absorption in the Raman spectrum of the copper-
oxygen species, the stability of these species toward oxygen, and
the fact that on hydrolysis these species provide cupric oxide,
strongly argues against the suggested[21] copper(I)-peroxide structure.
Therefore, we must conclude that the oxygen-oxygen bond in this product has already been broken. Clearly, because of its particular
properties, the above copper-oxygen product cannot be a *bona fide*
cupric oxide.

     We propose that oxidation of cuprous chloride in pyridine pro-
vides a mixture of bispyridine cupric chloride and a copper(II)-
oxide that exists as a tight di-μ-oxo-bridged copper(II)-pyridine
dimer, oligomer or polymer effectively solvated with pyridine (eq 14).

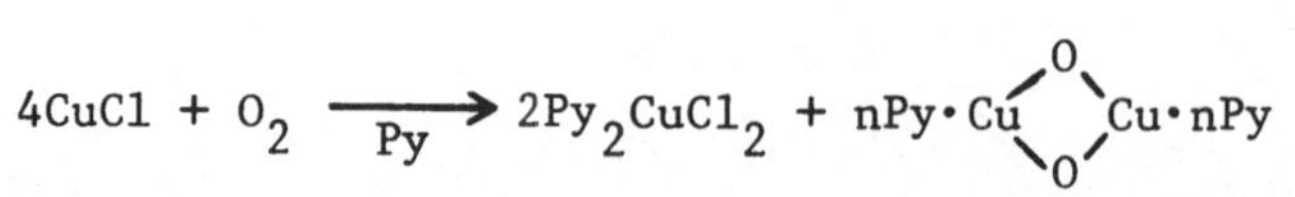

$$4CuCl + O_2 \xrightarrow{\text{Py}} 2Py_2CuCl_2 + nPy \cdot Cu \underset{O}{\overset{O}{\diamond}} Cu \cdot nPy \qquad (14)$$

The apparent chemical equivalency (eq 3) of the "Cu-Reagents" (eq 4) may formally be explained by assuming that the reaction of oxygen with cuprous chloride in pyridine produces the same products (eq 14), whether methanol is present or not. Further reaction of the copper-oxygen-pyridine species with methanol gives cupric methoxy hydroxide, the same species that could also be generated, perhaps reversibly, by reaction of cupric methoxide with water in the same solvent (Scheme II). In agreement with this picture, the "Cu-Reagent" can also be prepared by reaction of the complex A with water (eq 4), or by reaction of bispyridine cupric chloride with cupric methoxide and water in pyridine. Indeed, all properties of these "Cu-Reagents" were identical, regardless of the method of preparation (see Note 30).

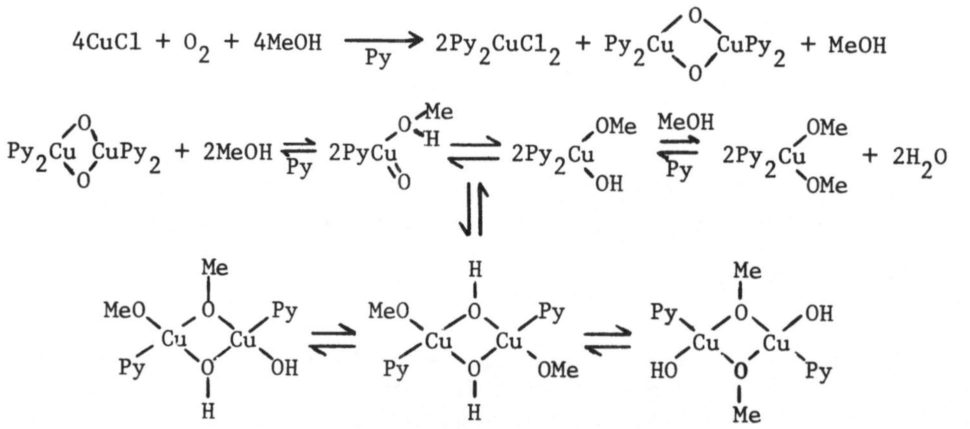

$$4CuCl + O_2 + 4MeOH \xrightarrow{\text{Py}} 2Py_2CuCl_2 + Py_2Cu \underset{O}{\overset{O}{\diamond}} CuPy_2 + MeOH$$

SCHEME II

Consequently, it is reasonable to conclude that the "Cu-Reagent" is a mixture of bispyridine cupric chloride and di-μ-methoxy or hydroxy-bridged copper(II)-pyridine dimers as indicated in Scheme II. However, it is quite likely that the di-μ-oxo-bridged copper(II)-dimers in pyridine could exist in equilibrium not only with the monomer but also with oligomeric or polymeric pyridine cupric methoxy hydroxide[31,32] (see Note 33).

Interestingly, filtration of the "Cu-Reagent" does provide approximately 50% of the total copper as bispyridine cupric chloride. While oxidation of catechol in pyridine/methanol solution containing this isolated bispyridine cupric chloride gave 4,5-dimethoxy-1,2-benzoquinone (eq 15), the same oxidation with the pyridine solution

containing the other 50% of the original copper ("CuO-Solution"),
provided the monomethyl muconic acid ester 1 (eq 16).

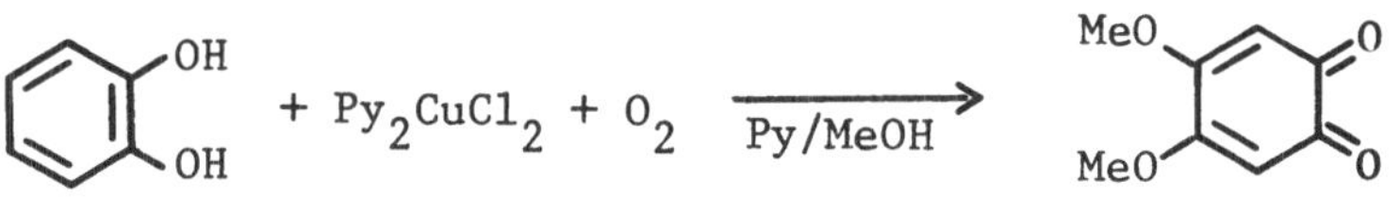

$$\text{(catechol)} + Py_2CuCl_2 + O_2 \xrightarrow{Py/MeOH} \text{(dimethoxy-o-benzoquinone)} \qquad (15)$$

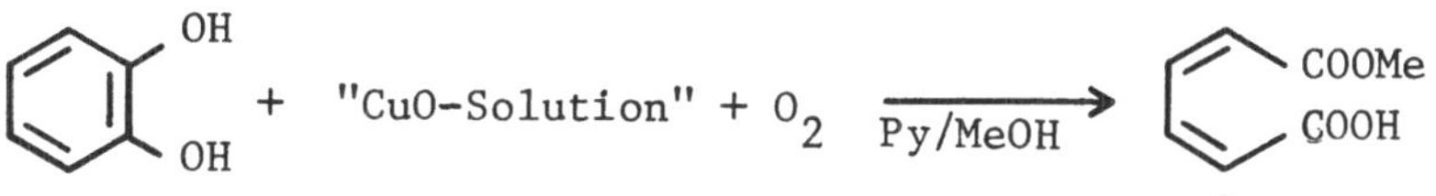

$$\text{(catechol)} + \text{"CuO-Solution"} + O_2 \xrightarrow{Py/MeOH} \text{(muconic ester 1)} \qquad (16)$$

These experiments clearly established that the di-μ-methoxy-hydroxy-
bridged copper(II)-pyridine species, rather than bispyridine cupric
chloride, are the active component of the "Cu-Reagent" responsible
for the carbon–carbon bond cleavage reaction. Furthermore, it also
follows that it should be possible to generate the active species by
reaction of cupric methoxide with water in pyridine, even in the
absence of bispyridine cupric chloride.

REACTION OF CUPRIC METHOXIDE/WATER IN PYRIDINE WITH CATECHOLS
AND *o*-BENZOQUINONES

The Cleavage of the Carbon–carbon Bond

Addition of a pyridine solution of catechol to a mixture of
cupric methoxide[34] and water in pyridine under anaerobic conditions,
followed by acid hydrolysis and the usual workup, gave a 15–20% yield
of the *cis,cis*-muconic acid monomethyl ester (1) and a larger quantity of the unreacted catechol (eq 17). A similar reaction with 4-
tert-butylcatechol gave the expected isomeric 4-*tert*-butylmuconic

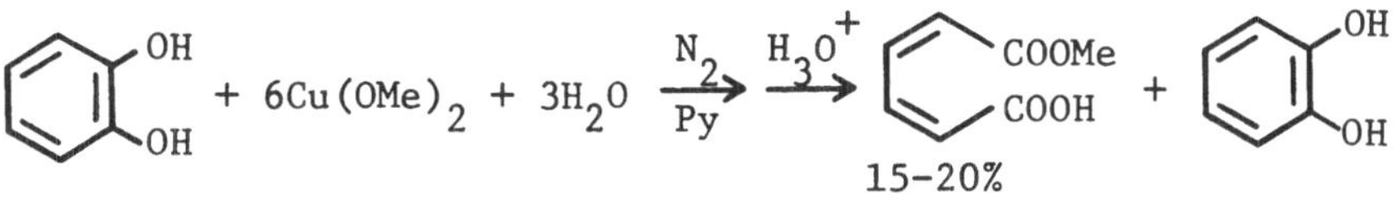

$$\text{(catechol)} + 6Cu(OMe)_2 + 3H_2O \xrightarrow[Py]{N_2} \xrightarrow{H_3O^+} \text{(muconic ester)} \underset{15-20\%}{} + \text{(catechol)} \qquad (17)$$

acid monomethyl ester 2 and 3 in a 30–35% yield (eq 18). Unexpectedly, the reaction of 4-*tert*-butyl-1,2-benzoquinone with an excess
of cupric methoxide/water in pyridine gave a mixture of the half
esters 2 and 3 (*ca* 60% in the usual ratio of 5:4.5), and 4-*tert*-
butylcatechol (*ca* 40%; eq 19).

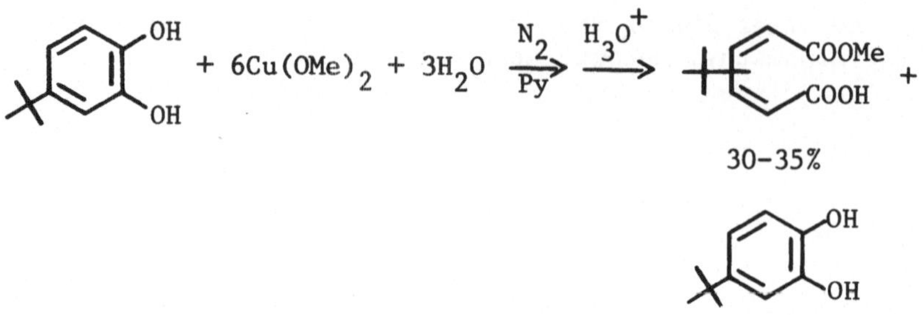

$$30\text{--}35\%$$ (18)

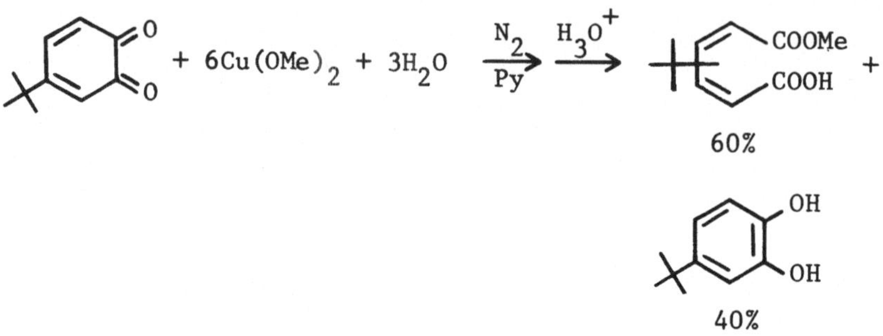

$$60\%$$ (19)

$$40\%$$

Clearly, the copper species generated by reaction of cupric methoxide with water in pyridine are indeed capable of bringing about the carbon-carbon bond cleavage in both catechols and o-benzoquinones.

The Reduction of o-Benzoquinone

The formation of the catechol in the reaction of 4-*tert*-butyl-1,2-benzoquinone with cupric methoxide and water in pyridine evidently results from a competitive reduction pathway. Since copper(I)-species produced in the cleavage reaction are the only reducing reagents which can be present in the system, it appears that this reduction is a very facile process that effectively competes with the cleavage reaction. Among the copper(I)-species that might be formed in the course of the cleavage reaction are cuprous hydroxide (or the corresponding oxide) and the copper(I)-salt of the muconic acids 2 and 3. Addition of 4-*tert*-butyl-1,2-benzoquinone to a pyridine solution containing cuprous hydroxide (generated *in situ* by reaction of cuprous-*tert*-butoxide[35] with water)[36] gave either the copper(II)-catecholate and cupric hydroxide (see Note 37), or the corresponding mixed complexes of copper(II)-catecholate and cupric hydroxide, which after acid hydrolysis afforded 4-*tert*-butyl-catechol quantitatively (Scheme III).

$$CuO\text{-}t\text{-}Bu \; + \; H_2O \xrightarrow{N_2,Py} CuOH + t\text{-}BuOH$$

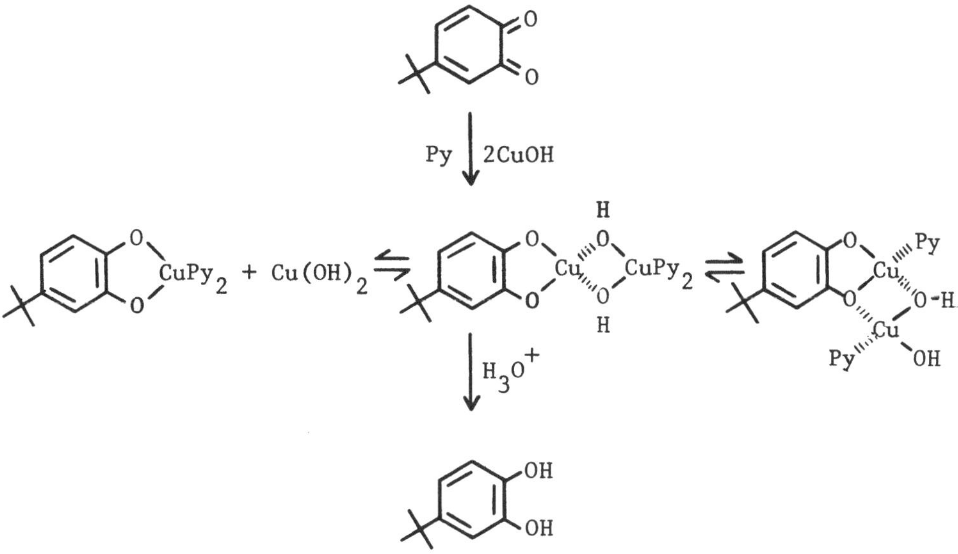

SCHEME III

Reaction of 4-*tert*-butyl-1,2-benzoquinone, first with cuprous hydroxide, and then with an excess of cupric methoxide/water, gave only trace amounts of the cleavage products. Similarly, if the cupric chloride was used instead of cupric methoxide in the last reaction, only small amounts of cleavage products were obtained. On the other hand, addition of the *o*-benzoquinone to a mixture of cuprous hydroxide, methanol, and cupric chloride in pyridine (see Note 38) led to clean carbon-carbon bond cleavage. Addition of the *o*-benzoquinone to a solution of cuprous chloride in pyridine containing methanol and water resulted in a rapid reaction, as indicated by the disappearance of the quinone absorptions in the IR spectrum. If this reaction mixture was quickly hydrolyzed in the presence of acid, 4-*tert*-butylcatechol was obtained quantitatively. However, if the reaction mixture was allowed to stand overnight before hydrolysis, or if the attempted reduction was carried out in the presence of cupric chloride and then hydrolyzed, only polymeric material was obtained.

The above experiments established that cuprous hydroxide, cuprous chloride, and probably other copper(I)-species (*e.g.*, cuprous monomethyl muconate), indeed react very readily with *o*-benzoquinone. Consequently, it is not surprising that reaction of the *o*-benzoquinone with an equivalent amount of cupric methoxide/water in pyridine gave almost equal amounts of the cleavage products, 2 + 3, and 4-*tert*-butylcatechol. The outcome of this particular reaction is remarkable, because it requires all of the resulting copper species to exist in a copper(II)-state. In other words, <u>the overall reaction occurred without a net change in the oxidation state of the copper reagent</u>. To account for the isolated products, it is then necessary to conclude that the *o*-benzoquinone was converted into copper(II)-catecholate and basic copper(II)-monomethyl muconate, which, before hydrolysis, were present in solution either as such, or as a mixed copper(II)-catecholate/basic copper(II)-monomethyl muconate complex[37,39] (Scheme IV).

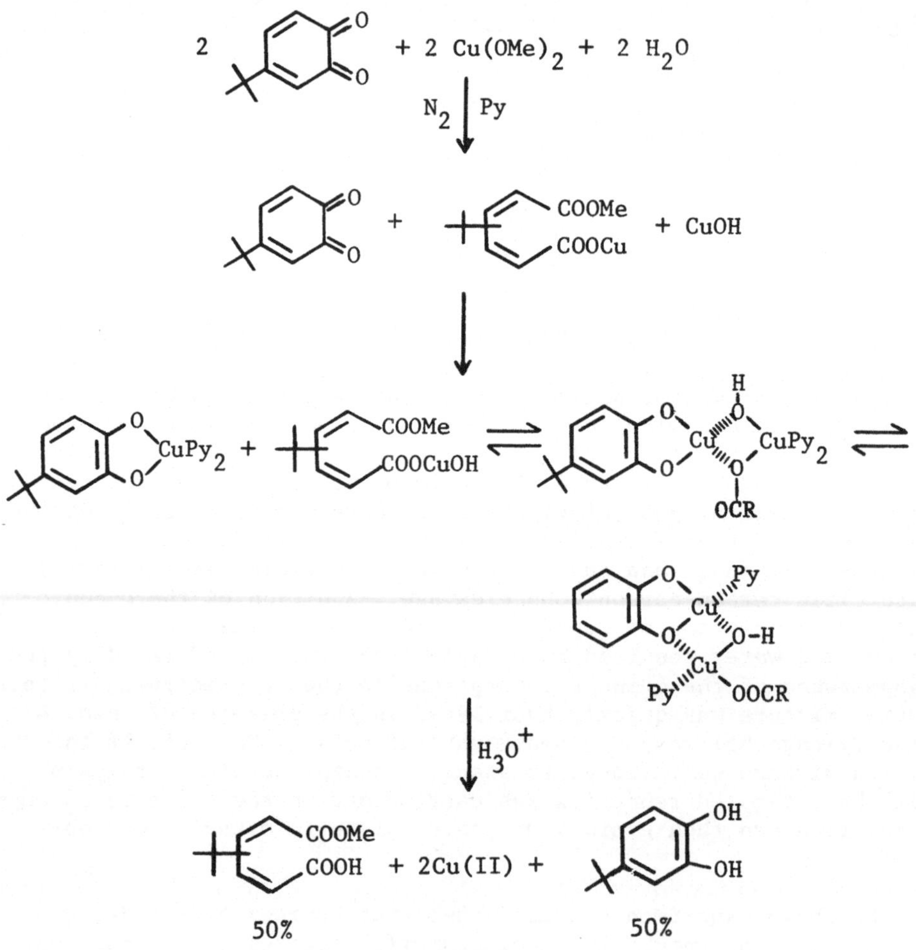

SCHEME IV

Copper(II)-catecholates as Intermediates in the Carbon-carbon Bond Cleavage Reactions of o-Benzoquinones and Catechols

Both the 4-*tert*-butylcatecholato-2,2'-dipyridyl copper(II) and 4-*tert*-butylcatecholatopyridine copper(II)-complexes[39] (see Note 40), as well as the catecholato-2,2'-dipyridyl complex[39] itself, were inert (see Note 41) to cupric methoxide water in pyridine (eq 20), and underwent polymerization in the presence of cupric chloride (eq 21), to provide the same polymeric material obtained by reaction of o-benzoquinone with cuprous chloride. Conversely, these catecholato copper(II)-complexes underwent clean carbon-carbon bond cleavage when treated with excess pyridine cupric methoxy chloride complex (complex A) in pyridine in the presence of water[43] (eq 22).

$$\text{(structure)} \ \text{CuX}_2 + 6\text{Cu(OMe)}_2 + 3\text{H}_2\text{O} \xrightarrow[\text{Py}]{\text{N}_2} \xrightarrow{\text{H}_3\text{O}^+} \text{(catechol)} \qquad (20)$$

$$\text{(structure)} \ \text{CuX}_2 + 6\text{CuCl}_2 + 3\text{H}_2\text{O} + 3\text{MeOH} \xrightarrow[\text{Py}]{\text{N}_2} \xrightarrow{\text{H}_3\text{O}^+} \text{POLYMER} \qquad (21)$$

$$\text{(structure)} \ \text{CuX}_2 + 6\text{PyCuClOMe} + 3\text{H}_2\text{O} \xrightarrow[\text{Py}]{\text{N}_2} \xrightarrow{\text{H}_3\text{O}^+} \text{(muconic ester)} \qquad (22)$$

R = H, *t*-Bu; X_2 = 2Py, Dipy

These experiments corroborate earlier observations with copper (II)-catecholate complex generated *in situ* and demonstrate (1) that the cupric methoxide/water reagent cannot effectively convert copper (II)-catecholate back to o-benzoquinone and cuprous hydroxide (eq 20), (2) that cupric chloride can reversibly oxidize copper(II)-catecholate to the same intermediate, presumably the semiquinone radical anion, that is formed by reduction of o-benzoquinone with cuprous chloride, and (3) that the "Cu-Reagent" generated by reaction of complex A with water can convert the copper(II)-catecholates cleanly to the muconic acid monomethyl esters (eq 22). The observed formation of polymer under thermodynamic conditions, either in reaction of the copper(II)- catecholates with cupric chloride (eq 21) or in reaction of o-benzo- quinone with cuprous chloride, is evidently a consequence of irre- versible polymerization of the semiquinone radical anion. The carbon- carbon bond cleavage of the o-benzoquinone that would result from the further 1-electron oxidation of the semiquinone radical anion by cup- ric chloride does not occur, simply because the system lacks the active copper(II)-species (see Scheme II) that could effect the oxi-

dative cleavage. However, in the reaction of the copper(II)-catecho-
late complexes with the "Cu-Reagent" (eq 22), all the requirements
for the carbon-carbon bond cleavage are met and this transformation
again becomes the exclusive reaction.

From the preceding discussion, it follows that the copper(II)-
complexes generated by reaction of cupric methoxide with water in
pyridine are indeed the active species capable of inducing the carbon-
carbon bond cleavage reaction. Furthermore, it is now evident that
copper(II)-catecholate can be formed as an intermediate from both
catechol and o-benzoquinone.

In the following section, we will first discuss the mechanism
of the two-electron oxidation/reduction of catechols and o-benzoqui-
nones, then we will briefly outline a possible mechanism of the car-
carbon-carbon bond cleavage of o-benzoquinone, and, finally we shall
discuss the details of the overall catechol transformation.

MECHANISM OF THE CLEAVAGE OF CATECHOL WITH "Cu-REAGENT"

Two Electron Oxidation Reduction of Catechol and o-Benzoquinone

The transformation of catechol with cupric methoxide/water in
pyridine, and of o-benzoquinone with cuprous hydroxide (or the oxide)
and cuprous monomethyl muconate strongly suggests that the semiquinone radical anion is not an important intermediate in these
reactions. If these reactions do, however, proceed by sequential
one electron processes,[44] then both electron transfers would have to
take place in very rapid succession. One of the ubiquitous characteristics of many copper compounds[45] is that they exist, in the
solid state as well as in solution, as polymers or highly clustered
aggregates.[28,29,31,32,46] We have already alluded to the possibility
that the cupric methoxy hydroxide, generated from cupric methoxide
and water in pyridine, may exist in solution as methoxy-hydroxy-
bridged copper(II)-dimers or oligomers in mobile equilibrium (see
Scheme II). An acid-base reaction between catechol and such a dimer
could produce the di-copper(II)-catecholate intermediate B. A direct
transfer of a pair of electrons from the substrate to each of the
two copper(II)-centers would generate o-benzoquinone and cuprous

(23)

C   B   D hydroxide, cuprous methoxide (or cuprous oxide) as in D (eq 23). Similarly, a direct transfer of two electrons, one from each of the copper(I)-centers of a dimeric copper(I)-oxygen species, as in D, would lead to the same intermediate B, which could either return to D or irreversibly collapse to copper(II)-catecholate complexed with cupric methoxy hydroxide, as in C.

Cleavage of o-Benzoquinone with Cupric Methoxide in Pyridine in the Presence of Water

Transfer of the methoxy and hydroxy groups from the cupric methoxy hydroxide dimer to the carbonyl groups of o-benzoquinone (Scheme V), would provide the intermediate organic-copper(II)-dimer. Oxidative cleavage of the carbon-carbon bond accompanied by other bond reorganization, as outlined in Scheme V, can generate directly the copper(I)-salt of muconic acid monomethyl ester, cuprous hydroxide and methanol. The two electrons provided by the substrate in a single step are distributed between the two copper(II)-centers, converting

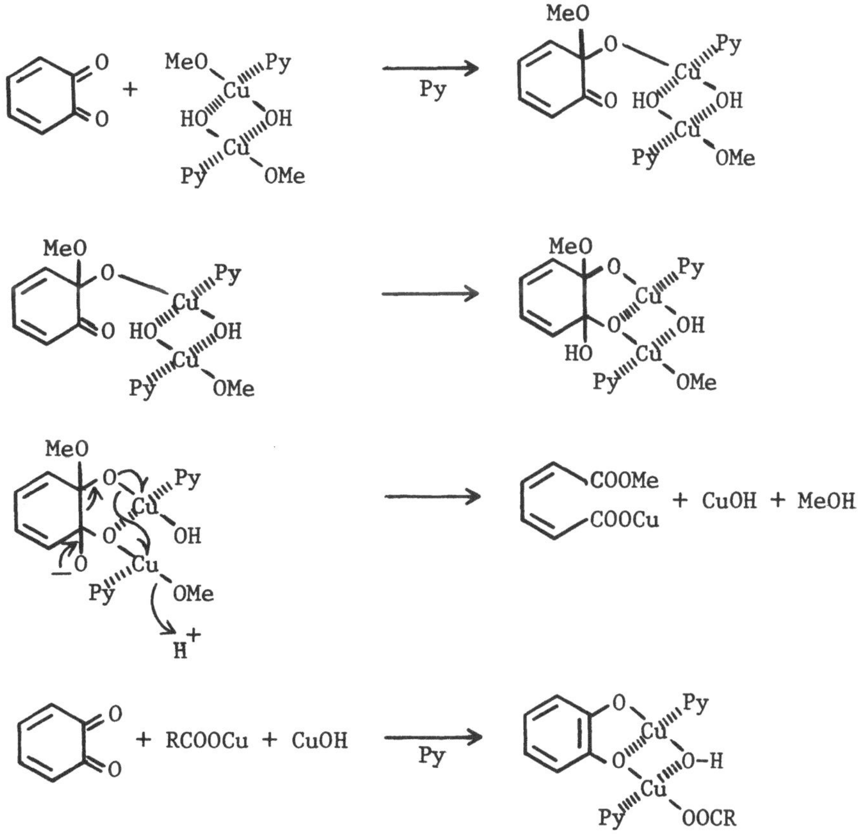

SCHEME V

them both to the copper(I)-state (see Notes 47 and 48). According
to experiments discussed earlier, the copper(I)-salt of muconic acid
monomethyl ester, together with cuprous hydroxide, can very effectively compete with cupric methoxy hydroxide dimer for o-benzoqui-
none and reduce it back to the above mentioned di-copper(II)-catecho-
late intermediate B and further to copper(II)-catecholate complexed
with basic copper(II)-monomethyl muconate (see Notes 37 and 50).
This mechanism satisfactorily explains both the exclusive formation
of the monoester and concomitant conversion of o-benzoquinone into
copper(II)-catecholate in the absence of cupric chloride. More importantly, the removal of two electrons from the organic substrate
in a single step[47],[48] does not violate the well-established propen-
sity for single electron oxidations by copper(II).[44] For example,
it is well known that four copper(II)-centers in the copper containing oxidase, laccase,[52],[53] remove electrons from the substrate in
pairs and transfer them further to oxygen, which is reduced to water.
Two of the four copper(II)-centers in the fully oxidized laccase
molecule exist as an "EPR-non-detectable copper(II)-pair"[52],[54],[56],
analogous to our proposed dimeric cupric methoxy hydroxide system in
pyridine (Scheme II). The absence of an EPR signal in this case was
also attributed to a total antiferromagnetic coupling of the unpaired
spins on the two adjacent copper(II)-centers.[52],[54],[56]

The Role of Cupric Chloride in the Overall Transformation of Catechol to Muconic Acid Monoalkyl Ester

Earlier discussion established that the carbon-carbon bond cleavage of o-benzoquinone with cupric methoxide/water in pyridine is
followed by a very rapid competing reduction of o-benzoquinone by
copper(I)-species produced in the cleavage reaction which provides
the copper(II)-catecholate complex (Scheme IV). It should also be
remembered that bispyridine cupric chloride alone in pyridine was
inactive in cleaving the carbon-carbon bond in both catechol and
benzoquinone. All experimental observations suggest that the main
role of cupric chloride in the oxidation of o-benzoquinone and catechol with "Cu-Reagent" is to scavenge cuprous hydroxide (or cuprous
oxide) and the copper(I)-salt of the monoalkyl muconate as they are
generated, and to form an apparently innocuous mixture of cuprous
chloride and basic copper(II)-monomethyl muconate. The overall transformation of catechol can then be summarized in Scheme VI (the black
arrows indicate transformations between catechol and o-benzoquinone
with cupric methoxy hydroxide dimer or copper(I)-oxygen species, and
the white ones represent transformations induced by cupric or cuprous
chloride).

Reaction between catechol and cupric methoxy hydroxide dimer as
discussed earlier (eq 23) would provide the intermediate B (see Note
57), which could collapse (see Notes 58 and 59) to either C or D.

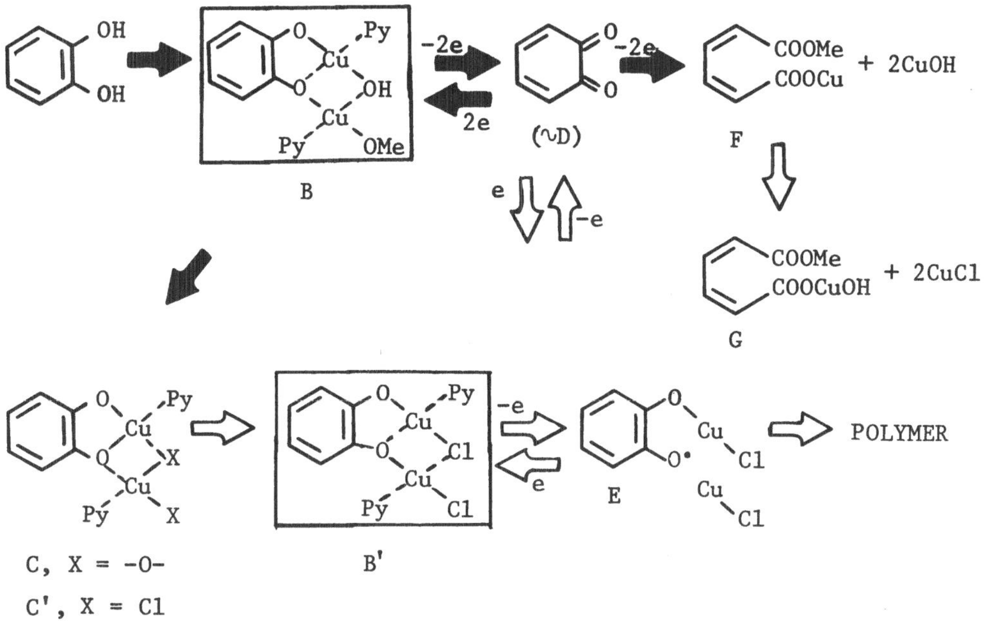

SCHEME VI

Further reaction of *o*-benzoquinone (∿D) with cupric methoxy hydroxide
(Scheme V) provides the copper(I)-salt of monomethyl muconate F and
cuprous hydroxide (or cuprous oxide). However, in the case when the
substrate is catechol, the net forward reaction is not as efficient
as when the substrate was *o*-benzoquinone itself, mainly due to the
parallel, but irreversible, transformation of the intermediate B
into copper(II)-catecholate complex C.

The fact that copper(II)-catecholate reacts with cupric chloride
(presumably *via* initial formation of the complex C') to give, under
thermodynamically controlled conditions, the same polymer that is
formed by reaction of *o*-benzoquinone with cuprous chloride (also
under thermodynamically controlled conditions) was already used as
evidence for the intermediacy of the semiquinone radical anion.
However, since kinetic reaction between *o*-benzoquinone and cuprous
chloride provides catechol after hydrolysis, and since this reaction,
before hydrolysis, does not give the copper(II)-catecholate itself,
it follows that both of these reactions produce the same di-copper
(II)-catecholate intermediate B' (see Note 60). Presumably, a rapid
one electron transfer from the substrate to one of the copper(II)-
centers in the intermediate B' produces the corresponding semiquinone radical anion E, the same intermediate produced by a single
electron reduction of *o*-benzoquinone by cuprous chloride. Under
thermodynamic conditions this radical intermediate undergoes a relatively slow, but irreversible polymerization.

When the reaction of either catechol, o-benzoquinone, or copper (II)-catecholate is carried out with the active "Cu-Reagent", $e.g.$, pyridine cupric methoxy chloride complex A in the presence of water, the carbon-carbon bond cleavage reaction becomes exclusive (eq 22). Clearly, the generated cupric methoxy hydroxide dimer (Scheme II) now effectively converts o-benzoquinone ($\sim$D) into the cleavage product F, regardless of whether the o-benzoquinone was produced directly from the intermediate B (and hence from catechol) or from the intermediate B' [and hence from copper(II)-catecholate and cupric chloride (C')]. In addition to transforming the otherwise stable copper(II)-catecholate into the reaction intermediate B', cupric chloride also acts as an efficient agent for converting the generated copper(II)-species into cuprous chloride. The cuprous chloride remains innocuous in the present system as long as there is any active cupric methoxy hydroxide dimer left to cleave the intermediate o-benzoquinone before the thermodynamically controlled formation of polymer can intervene via E.

The flow of electrons in pairs, or singly in rapid succession, between the substrate and the attacking copper dimer in the intermediate B in which the copper(II)-centers are held together by the bridging oxygen ligands, is essentially simultaneous.[47,48] This would also be true for the related intermediate formed during the carbon-carbon bond cleavage (Scheme V). Thus, the radical anion intermediate E is bypassed and the side reaction leading to polymeric material is suppressed. On the other hand, the electron transfer between the organic substrate and copper(II)-centers in the intermediate B' occurs in a stepwise manner (see Note 55). The radical anion intermediate E, formed by one electron transfer, can then undergo slower but irreversible polymer formation under thermodynamic conditions.

ON THE QUESTION OF ACTIVATION OF MOLECULAR OXYGEN BY MONOOXYGENASES AND DIOXYGENASES

The experiments with $^{18}O_2/H_2O$ and $O_2/H_2{}^{18}O$ unequivocally demonstrated that the oxygen atoms incorporated into the products of oxidation of phenols and catechols, catalyzed by certain iron-containing oxidases, come from the molecular oxygen and not from the water.[2,3] Two possible modes of enzyme-catalyzed activation of molecular oxygen (I and II) that are consistent with these observations were mentioned in the introduction. The salient feature of both of these is the implication that the new carbon-carbon bonds in the product $S(OH)_2$ are formed by transferring the electrons from the substrate directly to the activated molecular oxygen.[4]

Two alternative possibilities are illustrated schematically in III and IV below. The observed incorporation of oxygen atoms in the substrate SH_2 could be achieved by reaction of the substrate with a

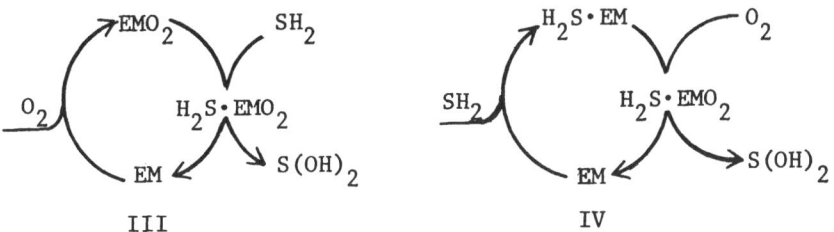

reagent EMO_2 that was generated either by a partial (two electron)
or a complete (four electron) reduction of molecular oxygen by an
independent and spin-allowed reaction between the enzyme metal center,
EM, and molecular oxygen. In either case, the electrons from the
substrate are completely transferred to the active reagent, EMO_2,
with formation of the new carbon-oxygen bonds in $S(OH)_2$ and regener-
ation of the original metal center, EM. In the absence of molecular
oxygen, the flow of electrons from the organic substrate stops at
the metal center, EM, which now contains all of the electrons provided by the substrate. In the presence of molecular oxygen, however, the electrons continue to flow from the metal center, EM, to
the O_2, which now acts as the ultimate acceptor of electrons origi-
nating from the substrate, SH_2. This provides again the active EMO_2
reagent, which is then ready to reenter reaction cycle and accept
another set of electrons from a second molecule of the substrate.

Since the active EMO_2 reagent, rather than molecular oxygen,
acts as the electron acceptor in these transformations, and since
the final transfer of electrons to molecular oxygen occurs from the
generated metal center, EM, a direct reaction between oxygen and
organic substrate does not take place. Consequently, it appears that
if the metal center in the enzyme molecule were indeed capable of
performing the suggested transformations, then the overall oxidation
of an organic substrate would not require a special mechanism for
activation of molecular oxygen. The fact that the experiments with
labeled oxygen and water provided no products that contained the
oxygen atoms from water[2,3] strongly suggests that the effective reaction site was not accessible to external water. This should not
be very surprising since most active sites in enzymes are located in
the internal, hydrophobic region of the enzyme molecules. Moreover,
even if the water is a product of the reaction as, for example, in
monooxygenase oxidations, it will be "squeezed out", away from the
reaction site into the pool of water present as a solvent.

Unfortunately, up to now there were no example of the cleavage
of carbon-carbon bonds in o-benzoquinones and catechols by simple
derivatives of metal oxides that could serve as a model for the transformations outlined in III and IV. For this reason, the possibility
that similar reactions catalyzed by oxygenases may indeed proceed by
some mechanism analogous to those represented schematically in III

and IV, was not considered very seriously.[19] However, our work demonstrates conclusively, for the first time, that the cleavage of the carbon-carbon bond in o-benzoquinones and catechols can be brought about by a particular copper(II)-reagent in the presence, as well as in the absence, of molecular oxygen. The overall transformation can be reasonably explained by the generalized mechanistic schemes III and IV in which "EM" and "EMO$_2$" stand for copper(I)- and the active copper(II)-species, and "SH$_2$" stands for o-benzoquinone or catechol.

Clearly, the existence of the nonenzymatic reaction pathway for the cleavage of the carbon-carbon bonds in the absence of molecular oxygen is not evidence that the enzymatic reaction indeed proceeds by an analogous mechanism. Nevertheless, such a facile transformation must now be taken into consideration as a possible alternative to the widely accepted mechanistic schemes based on the enzymatic activation of molecular oxygen.[2-4]

CONCLUSION

For the first time it was demonstrated conclusively that the cleavage of the carbon-carbon bond in catechol and o-benzoquinone can be brought about by a particular copper(II)-reagent in the absence of as well as in the presence of molecular oxygen. The overall anaerobic transformation of catechol to cis,cis-muconic acid monomethyl ester involves a two electron oxidation of catechol to o-benzoquinone, followed by a second two electron oxidation of o-benzoquinone to the muconic acid ester. The active copper(II)-species can be generated by reaction of cupric methoxide with water in pyridine and are equivalent to a dimeric cupric methoxy hydroxide complexed with pyridine. The oxidation agent ("Cu-Reagent") can also be generated either by reaction of cuprous chloride with oxygen in pyridine in the presence of methanol, by reaction of an alcohol with a product mixture of the oxidation of cuprous chloride in pyridine, or by addition of a molar equivalent of water to any of the following reagents: to cupric methoxy chloride in pyridine, to pyridine cupric methoxy chloride (complex A) in pyridine, to a mixture of cupric methoxide and cupric chloride in pyridine, or to a mixture of cupric methoxide and pyridine hydrochloride in the same solvent. It appears that all methods provide the same active oxidation agent that exists in pyridine as a mixture of bispyridine cupric chloride and dimeric, oligomeric or polymeric cupric methoxy hydroxide in equilibrium with each other. During the reaction the active copper(II)-agent is reduced to copper(I)-muconate ester and cuprous hydroxide which, under the reaction conditions, can efficiently reduce the intermediate o-benzoquinone to the copper(II)-catecholate complex that resists further oxidation. However, this undesirable reduction is prevented by the present cupric chloride, which scavenges $in situ$ cuprous muconate and cuprous hydroxide and converts them into an innocuous mixture of basic cupric muconate and cuprous chloride. In other words, the electrons provided by catechol and *o*-benzoquinone are transferred first to the active copper(II)-agent from where, eventually, they could be transported to their ultimate destination, the molecular oxygen. In the absence of molecular oxygen the flow of electrons from the organic substrates stops at the copper stage, which at the end of the reaction sequence exists in copper(I)-state. In the presence of oxygen the electrons continue to flow from the copper(I)-center to the molecular oxygen thus providing thermodynamically stable copper(II)-oxygen species, which are then again ready to reenter the reaction cycle and accept another set of electrons from the organic substrates.

Since the active copper(II)-species, rather than molecular oxygen, act as electron acceptor in these transformations, and since the final transfer of electrons to molecular oxygen occurs from generated copper(I)-species, a direct reaction between oxygen and organic substrate does not take place. Therefore, in the present system a special mechanism for activation of molecular oxygen is not required.

Clearly, the existence of the nonenzymatic reaction pathway for the cleavage of the carbon-carbon bonds in the absence of molecular oxygen is not evidence that the enzymatic reaction also proceeds by an analogous mechanism. Nevertheless, such a facile transformation must now be taken into consideration as a possible alternative to the widely accepted mechanistic schemes based on the enzymatic activation of molecular oxygen[2,4]

ACKNOWLEDGMENT

The authors wish to thank Dr. Willis Hammond for his contributions to the early stage of this work, and Messrs. Bruce Van Buskirk, William McGrath, Francis Ruggiero, and John Rodler for their valuable assistance in all phases of this work. We are also indebted to the staff of our Chemical Physics Department for their analytical services.

REFERENCES AND NOTES

1. For a general discussion of the metabolic and respiratory mech-
    anisms see A. L. Lehninger, "Biochemistry", Worth, New York,
    NY, 1975.
2. For a general review of monooxygenases and dioxygenases see
    "Molecular Mechanisms of Oxygen Activation", O. Hayaishi, Ed.,
    Academic Press, New York, NY, 1974.
3. (a) G. A. Hamilton, "Molecular Mechanisms of Oxygen Activation",
    O. Hayaishi, Ed., Academic Press, New York, NY, 1974, p. 405;
    (b) G. A. Hamilton, "Advances in Enzymology", F. F. Nord, Ed.,
    John Wiley and Sons, New York, NY, 1969, p. 55; (c) M. M. T.
    Khan and A. E. Martell, "Homogeneous Catalysis by Metal Com-
    plexes", Vol. 1, Academic Press, New York, NY, 1974; (d) W.
    Ullrich, *Angew. Chem. Int. Ed. Engl.*, $\underline{11}$, 701 (1972); (e) J.
    H. Olive and S. Olive, *ibid.*, $\underline{13}$, 29 (1974); (f) J. H. Fuhnhop,
    ibid., $\underline{15}$, 618 (1976); (g) J. H. Wang, *Acc. Chem. Res.*, $\underline{3}$, 90
    (1970).
4. O. Hayaishi in ref. 2, chapter 1.
5. M. M. Rogić, J. Vitrone and M. D. Swerdloff, *J. Am. Chem. Soc.*,
    $\underline{97}$, 3848 (1975); $\underline{99}$, 1156 (1977).
6. M. M. Rogić, K. P. Klein, J. M. Balquist, and B. C. Oxenrider
    J. Org. Chem., $\underline{41}$, 482 (1976).
7. M. M. Rogić, M. T. Tetenbaum and M. D. Swerdloff, *J. Org. Chem.*,
    $\underline{42}$, 2748 (1977).
8. M. M. Rogić, K. P. Klein, T. R. Demmin and B. C. Oxenrider,
    J. Am. Chem. Soc., submitted for publication.
9. For a preliminary report see M. M. Rogić, T. R. Demmin and W.
    B. Hammond, *J. Am. Chem. Soc.*, $\underline{98}$, 7441 (1976).
10. M. M. Rogić and T. R. Demmin, *J. Am. Chem. Soc.*, $\underline{100}$, 0000
    (1978).
11. W. Brackman and E. Havinga, *Rec. Trav. Chim.*, $\underline{74}$, 937, 1021,
    1070, 1100, 1107 (1955).
12. A. P. Terentiev and Ya. D. Magilyanskii, *Doklady Acad. Nauk
    S.S.S.R.*, $\underline{103}$, 91 (1955); *C. A.*, $\underline{50}$, 4807 (1956).
13. K. Kinoshita, *Bull. Chem. Soc.*, *Jpn.;* $\underline{32}$, 777, 780, 783 (1959);
    K. Kinoshita, *J. Chem. Soc. Jpn.*, *Pure Chem. Sect.*, $\underline{75}$, 48
    (1954).
14. A. S. Hay, H. S. Blanchard, G. F. Endres and J. W. Eustance,
    J. Am. Chem. Soc., $\underline{81}$, 6335 (1959).
15. a) A. S. Hay, *J. Polym. Chem.*, $\underline{58}$, 581 (1962); (b) G. F. Endres
    and J. Kwiatek, *ibid.*, $\underline{58}$, 593 (1962); (c) G. F. Endres, A. S.
    Hay and J. W. Eustance, *J. Org. Chem.*, $\underline{28}$, 1300 (1963); (d)
    A. S. Hay and G. F. Endres, *Polym. Lett.*, $\underline{3}$, 887 (1965); (e)
    H. Finkbeiner, A. S. Hay, H. S. Blanchard and G. F. Endres,
    J. Org. Chem., $\underline{31}$, 549 (1966).
16. However, a direct cleavage of phenol to *cis,cis*-muconic acid
    ester aldehyde, followed by rapid oxidation of the latter to
    the acid ester $\underline{1}$, cannot be completely ruled out at this point.
17. J. Tsuji and H. Takayanagi, *J. Am. Chem. Soc.*, $\underline{96}$, 7349 (1974).

18. J. Tsuji, H. Takayanagi and I. Sakai, *Tetrahedron Letters*, 1245 (1975).

19. As early as in 1937, F. Kubowitz proposed [*Biochem. Z.*, **292** 221 (1937); **299**, 32 (1939] that in oxidation of catechol to *o*-benzoquinone, catalyzed by copper oxidases, the function of oxygen was to regenerate copper(I)-enzyme to copper(II)-enzyme, which was the actual oxidizing reagent. Similarly, in their studies of copper(II)-catalyzed polymerization of phenols, Hay, Endres and their co-workers[15] felt that the role of oxygen was to reoxidize the copper(I)-back to the copper(II)-species which was the actual oxidizing reagent.

20. It is assumed at this point, for the sake of simplicity, that the oxidation of cuprous chloride in pyridine in the presence of methanol leads to pyridine cupric methoxy chloride complex A and two equivalents of water, a reagent equivalent to that formed in Method A. However, as it will become apparent later in the text, the actual oxidizing reagent is a mixture of several species (*vide infra*).

21. C. E. Kramer, G. Davies, R. B. Davis and R. W. Slaven, *J. Chem. Soc., Chem. Comm.*, 606 (1975).

22. Y. Ogata and T. Morimoto, *Tetrahedron*, **21**, 2791 (1965).

23. H. Praliand, Y. Kodrafoff, G. Condurier and M. V. Mathieu, *Spectrochimica Acta*, **30A**, 1389 (1974); E. Ochiai, *Tetrahedron*, **20**, 1831 (1964).

24. G. Condurier, H. Praliand and M. V. Mathieu, *ibid.*, **30A**, 1399 (1974).

25. M. Berthelot, *Ann. Chim. Phys.*, [5] **20**, 503 (1880).

26. M. Groger, *Z. Anorg. Chem.*, **28**, 154 (1901).

27. We have also prepared copper-oxygen species by a direct oxidation of copper metal in the presence of catalytic amounts of cuprous chloride.

28. For example see W. E. Hatfield, *ACS Symp. Ser.*, **5**, 108 (1974), and references therein.

29. P. Jeffrey Hay, J. C. Thibeault and R. Hoffman, *J. Am. Chem. Soc.*, **97**, 4884 (1975).

30. The EPR spectra of either of the three reagents in pyridine clearly showed the presence of approximately 50% of the total amount of copper in a paramagnetic state. The spectrum is consistent with that of bispyridine cupric chloride in pyridine and exhibits general features very similar to that described earlier.[23]

31. C. H. Brubaker, Jr. and M. Wicholas, *J. Inorg. Nucl. Chem.*, **27**, 59 (1965).

32. R. W. Adams, E. Bishop, R. L. Martin and G. Winter, *Aust. J. Chem.*, **19**, 207 (1966).

33. It should be remembered, however, that cupric oxide, once isolated and "aged", is insoluble in either of these solvents. Consequently, if present at all in these solutions, cupric oxide must be stabilized in monomeric or oligomeric form by pyridine/methanol, as it is being formed *in situ*.

34. Cupric methoxide was prepared according to the published procedure of ref. 31.

35. T. Tsuda, T. Hashimoto and T. Saegusa, *J. Am. Chem. Soc.*, **94** 658 (1972).

36. A reaction of cuprous *tert*-butoxide with water in pyridine does take place. While a solution of the cuprous *tert*-butoxide in pyridine is stable, the resulting reaction mixture, after treatment with water, does not undergo appreciable disproportionation over several hours. However, an overnight reaction produced appreciable amounts of what appears to be a polymeric cuprous oxide and small amount of copper(0) and cupric oxide. Presumably, after addition of water, the copper(I)-species in pyridine is either cuprous hydroxide or cuprous oxide hydrate, present in oligomeric form.

37. The reaction mixture, before hydrolysis, appears homogeneous and does not deposit cupric oxide on standing. Consequently, it is possible that in solution the copper(II)-catecholate complex and cupric hydroxide exist as a mixture of the isomeric copper (II)-catecholate/cupric hydroxide complexes (*vide infra*).

38. While cuprous hydroxide undergoes ready reaction with cupric chloride, cuprous chloride does not react with cupric methoxide/water in pyridine.

39. D. G. Brown, J. T. Reinprecht and G. C. Vogel, *Inorg. Nucl. Chem. Letters*, **12**, 399 (1976).

40. The molecular weight of the 4-*tert*-butylcatecholate pyridinecopper(II)-complex in benzene solution was found to be 918.0, indicating that it exists as a trimer. We assume that in pyridine solution the trimer either completely dissociates to the corresponding monomer containing two pyridine ligands, or, at least, that it exists under the reaction conditions in equilibrium with such a monomer. We will discuss the structure and EPR spectra of this and other related copper(II)-catecholate complexes elsewhere.

41. Brown and coworkers[42] have shown that 3,5-di-*tert*-butylcatecholate-1,10-phenanthrolinecopper(II), and 3,5-di-*tert*-butylcatecholate-2,2'-dipyridylcopper(II) undergo a reaction with oxygen in solution to give a mixture of a cleavage product and unidentified copper complexes.

42. D. G. Brown, Z. Beckmann, C. H. Ashby, G. C. Vogel and J. T. Reimprecht, *Tetrahedron Lett.*, 1363 (1977).

43. Presumably, all "Cu-Reagents" should behave in the same way.

44. (a) For a review of oxidation by copper(II), see W. G. Nigh, "Oxidation by Cupric Ion," in "Oxidation in Organic Chemistry", W. S. Trahanovsky, Ed., Academic Press, New York, NY, 1973; (b) R. A. Sheldon and J. K. Kochi, *Advances in Catalysis*, **25**, 272 (1976); (c) J. K. Kochi, "Free Radicals", Vol. 1, J. K. Kochi, Ed., John Wiley and Sons, New York, NY, 1973, p. 591 and references therein.

45. F. A. Cotton and G. Wilkinson, "Advanced Inorganic Chemistry", 3rd ed., John Wiley and Sons, New York, NY, 1972, p. 905.

46. H. Hashimoto, T. Noma and T. Kawaki, *Tetrahedron Lett.*, 3411 (1968).

47. Alternatively, the carbon-carbon bond cleavage may involve two simultaneous one-electron oxidations to provide the muconic acid ester and copper(I)-hydroxide and methoxide, followed by reaction between the acid ester and the copper(I)-species to give the copper(I)-muconate and methanol (or water).

48. Recently, Fenton, Schroeder and Lintvedt described a bimolecular copper(II)-system capable of accepting two electrons simultaneously.[49]

49. D. E. Fenton, R. R. Schroeder and R. L. Lintvedt, *J. Am. Chem. Soc.*, **100**, 1931 (1978).

50. We do not have evidence that the copper(II)-catecholate indeed exists in solution as the suggested mixed complex with basic copper(II)-muconate. The structures of inorganic and simple aliphatic basic cupric salts are usually portrayed as combinations of normal salts, cupric hydroxide, cupric oxide and water. However, Kaeding and Shulgin[51] have shown that various basic cupric salts exist as such.

51. W. W. Kaeding and A. T. Shulgin, *J. Org. Chem.*, **27**, 3551 (1962).

52. For a brief review of the copper-containing oxidases, see R. Malkin, in "Inorganic Biochemistry", G. L. Eichhorn, Ed., Elsevier, New York, NY, 1973, Chapter 21.

53. R. Malkin and B. G. Malmstrom, *Adv. Enzymol.*, **33**, 177 (1970).

54. J. A. Fee, R. Malkin, B. G. Malmstrom and T. Vanngard, *J. Biol. Chem.*, **244**, 4200 (1969).

55. It is well known[44] that electron transfer oxidations occur mainly with oxy-salts of copper(II)-, whereas ligand transfer oxidations predominate with copper(II)-halides. Accordingly, the singlet state of copper(II) is unreactive in electron transfer reactions. However, the triplet state of the dimer copper(II)-species is largely unreactive and only paramagnetic monomeric copper(II)-species function as electron transfer oxidizing reagents.

56. E. I. Solomon, D. M. Dooley, R.-H. Wang, H. B. Gray, M. Cerdonio, F. Moguo and G. L. Romani, *J. Am. Chem. Soc.*, **98**, 1029 (1976).

57. The exact nature of bonding in this intermediate is, of course, not known.
58. The catecholate C is present in solution, either as a mixed complex with cupric hydroxide or basic copper(II)-muconate, or as a mixture with these copper(II)-species (see Scheme IV).
59. Once formed, the copper(II)-catecholate complex C is stable because the reverse reaction would require that a relatively soft hydroxide or methoxide anion of cupric methoxy hydroxide replace a much harder catecholate bidentate ligand in the copper(II)-catecholate complex C.
60. Presumably, the harder chloride anion from cupric chloride can replace the relatively softer catecholate anion from the copper catecholate in C', but reverse reaction, the replacement of one of the harder chloride anions in the intermediate B' to produce copper(II)-catecholate and cupric chloride as in C', either does not take place readily, or if it does, the ensuing equilibrium favors the intermediate B'.

STEREOCHEMISTRY OF THE S_E2 BROMINE CLEAVAGE OF TETRAALKYLTIN

COMPOUNDS IN CARBON TETRACHLORIDE AND METHANOL SOLVENTS[1]

Richard L. Chambers and <u>Frederick R. Jensen</u>

Department of Chemistry, University of California

Berkeley, California 94720

[This manuscript is enthusiastically dedicated to Professor H. C. Brown on the occasion of his sixty-sixth birthday.]

It was demonstrated previously that the S_E2 cleavage of *sec*-Bu*Sn(neopentyl)$_3$[†] by bromine in methanol occurs with inversion of configuration; this inversion has now been found to be quantitative. It is shown, and explanations are offered for the phenomenon, that cleavage in carbon tetrachloride as solvent (with necessary inhibitors for the concomitant free radical reactions) occurs with retention of configuration. The reversal of stereochemistry may result, not from change in mechanism, but rather from drastic changes in the rate of competing reactions with change of solvent.

INTRODUCTION

A substitution reaction at carbon is generally one of three types: S_N, S_H or S_E. Of the well-characterized reactions, the S_E2 reactions are least understood, and some uncertainty has arisen as to the stereochemistry and mechanisms of reactions of alkyl transition metal compounds with electrophiles. Formally, these reactions appear to be S_E2 substitution on carbon, but some information indicates that this is not always the case. For example, evidence is available that the reaction of alkyliron(II) compounds with halogens occurs with inversion of configuration,[2] but this reaction was shown to involve oxidation of the iron, followed by nucleophilic displacement on carbon[3] (eqs 1 and 2), rather than direct electrophilic cleavage of the carbon metal bond. The reported S_E2 cleavage of

---

[†]The asterisk indicates that the group is chiral.

$$R*Fe^{II}(CO)_2(n^5C_5H_5) \xrightarrow{X_2} R*Fe^{IV}(CO)_2Cp + 2X^- \qquad (1)$$

$$X^- + R*Fe^{IV}(CO)_2Cp \xrightarrow{S_N2} XR* + Fe(II) \qquad (2)$$

sec-butylcobaloxime to yield inverted sec-butyl bromide[4] may occur by a <u>mixture</u> of S_E2 substitution and oxidation of RCo(III) to RCo(IV) by bromine, followed by S_N2 displacement on carbon by bromide ion.[5] Considering these uncertainties as to mechanisms, it appears at this time prudent to put aside temporarily results with alkyl transition metals in considering the mechanisms and stereochemistry of S_E2 reactions on carbon.

Early work with mercurials[6] uniformly indicated the stereo-chemistry of S_E2 reactions to be retention of configuration on carbon and, because of these results, retention became the accepted stereochemistry. This conclusion was strengthened by work with other elements; for example, Grignard reagents undergo reactions with electrophiles with retention of configuration on carbon.[7]

$$\text{(bicycloalkyl)}-MgBr + E^{\delta+}\!\!-\!\!B^{\delta-} \longrightarrow \text{(bicycloalkyl)}-E + BMgBr \qquad (3)$$

$$endo\text{-}MgBr + E^{\delta+}\!\!-\!\!B^{\delta-} \longrightarrow endo\text{-}E + BMgBr \qquad (4)$$

These retention reactions invariably involve, as leaving groups, metals with low-lying vacant orbitals which, as cations, are very strong acids, e.g., XHg^+ and XMg^+. Thus, there exists a strong demand for complexing of the nucleophilic part of the electrophile (very few "bare" electrophiles are available for study) with the leaving metal and, as a consequence, retention may be forced on the system. Thus, for the simple reaction of bromine with alkylmercuric bromides, a closed transition state (eq 5) is expected to be favorable, forcing retention. With a more open transition state (eq 6), the reaction could proceed with either retention or inversion.

$$RHgBr + Br_2 \longrightarrow R\underset{Br}{\overset{Br}{\underset{\diagdown}{\overset{\diagup}{\mid}}}}\!\!Hg\!\!\cdots Br \longrightarrow RBr + HgBr_2 \qquad (5)$$

<u>Forced Retention</u>

$$R-Hg-Br + Br_2 \longrightarrow R\text{-----}Hg\overset{\delta+}{\text{---}}Br \longrightarrow RBr + \overset{+}{H}gBr + Br^- \quad (6)$$

$$\underset{\underset{\text{-}Br^{\delta-}}{\overset{|}{Br}}}{}$$

Retention or Inversion

In the previous work with tin compounds, where S_E2 inversion was observed, a system and conditions were selected which would promote an open transition state and S_E2 inversion.[9] To obtain with confidence an open transition state, several conditions must be met: (a) there must be no low-lying vacant orbitals (for complexation) on the leaving group, (b) the leaving group must readily support a developing positive charge, and (c) if charged molecules are formed, a polar solvent must be used to stabilize them because they will be separated. In order to "load" the system to obtain preferential inversion, bulky groups which were expected to inhibit retention were used. After considering many systems and after some experimental trials, the system selected for that work was trineo-pentyl-*sec*-butyltin. Tetraalkyltin compounds show (as judged by the available literature) slight, if any, Lewis acid properties with ordinary reagents and, thus, apparently do not have accessible low-lying vacant orbitals. It also appears (but the evidence is not firm) that trialkyltin cations are readily formed when supported by polar (but not necessarily covalent) solvation and, so, have at least moderate stability. Methanol was chosen for the polar solvent. Experience had indicated that neopentyl is the most difficult to cleave of simple alkyl groups. Three neopentyl groups on tin will favor backside attack since direct approach on the carbon-tin bonds will be made difficult by steric hindrance and because, of the various R_3Sn^+ cations possible as leaving groups (see below), that having R = neopentyl is expected to be especially stable due to the favorable inductive effect of the neopentyl groups. The use of the same compounds: R_3SnR' (R = neopentyl, R' = other alkyl groups), in related rate studies is especially useful since various alkyl groups affect the rate at which other alkyl groups are cleaved from tin by electrophiles.

In methanol as solvent, with added bromide ion, the *sec*-butyl group is cleaved by bromine with inversion of configuration (eq 7).

$$(S)-(+)-sec\text{-Bu*Sn(neopentyl)}_3 + Br_3^- \xrightarrow[CH_3OH]{Br^-}$$

$$(R)-(-)-sec\text{BuBr} + (\text{neopentyl})_3\text{SnBr} \quad (7)$$

Kinetic experiments established that the transition state is bimolecular and contains only Br_2 and R_4Sn. In methanol as solvent, the
kinetic expression given in equation 8 is rigorously followed over
a wide range of concentrations. These results unambiguously require
the reactions shown in equations 9 and 10.

$$rate = k_2[Br_2][R_4Sn] \tag{8}$$

$$Br_3^- \xrightleftharpoons{1/K} Br^- + Br_2 \quad (K = 177, \ 25°) \tag{9}$$

$$Br_2 + R_4Sn \xrightarrow[slow]{k_2} RBr + R_3SnBr \tag{10}$$

Taken with the stereochemical results, this indicates that the transition state and initial products are as shown in equation 11. From
this equation it is expected that the solvent will play a very
important role in supporting the charged products (Br^- and $^+SnR_3$),
and that this reaction would be energetically nonfavored in a nonpolar solvent because of lack of solvation of the ions and because

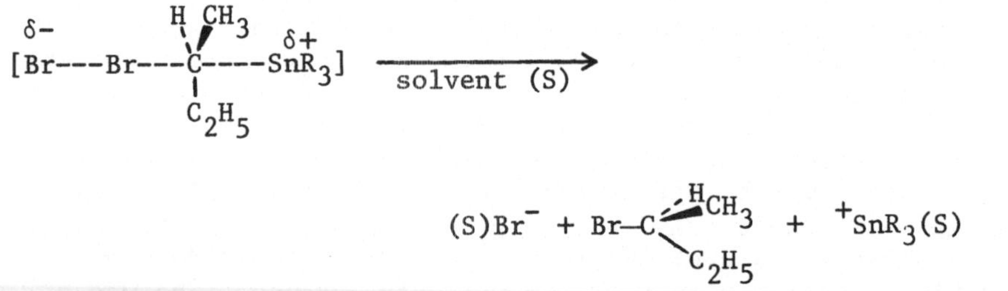

of the large energy necessary to separate the charges. Thus, it
was considered highly desirable to determine the stereochemistry of
this bromodealkylation in a nonpolar solvent as well.

It has recently been reported that *sec*-Bu*Br is formed with
net retention in reactions of *sec*-Bu*Sn(i-Pr)$_3$ and related compounds
with bromine in cyclohexane-methanol.[10] These results are considered
in the discussion section. Retention of configuration has also been
found in the halodemetallation of optically active cyclopropyltin
compounds,[11] but this result is not surprising in view of the high
resistance to inversion on cyclopropane and because of the ease
with which electron transfer or transmetallation may occur in this
system (e.g., eq 12). The remarkable finding that 2,2-diphenyl-1-
methylcyclopropyl is cleaved from tin faster than methyl is readily
understood in terms of either of the above phenomena.

$$+ Br_2 \xrightarrow{slow} + SnMe_3 \xrightarrow{fast} \qquad (12)$$

On the other hand, several cases have been observed in recent years where inversion on carbon occurs in the electrophilic cleavage of alkyl derivatives of main group elements. These include derivatives of a "stacked" mercury system,[12] of thallium,[13] boron[14] and lithium.[15]

In this paper we report and discuss the factors which influence the stereochemistry of the reaction of trineopentyl-*sec*-butyltin with bromine in methanol and in carbon tetrachloride to yield *sec*-butyl bromide.

RESULTS AND DISCUSSION

Samples of (S)-(+)- and (R)-(-)-*sec*-butyltrineopentyltin were synthesized by a previously described indirect procedure[16] and also by direct displacement by trineopentyltin potassium on optically active *sec*-butylmethane-sulfonate. The rotation of enantiomerically pure (S)-(+)-*sec*-butyltrineopentyltin has been observed, by a closed cyclic route, to be 4.19°.[17,18] The substrate was subjected to brominolysis in methanol and in carbon tetrachloride and the resulting *sec*-butyl bromide was purified and analyzed to determine its enantiomeric purity. The rotation of optically pure *sec*-butyl bromide is not known with complete confidence but is probably in the range of 33.5-35°. The value used, 33.7°, is based on unpublished results, and the stereospecificities of the cleavages (Tables 1 and 2) are reported accordingly.

It has previously been demonstrated that the cleavage in methanol is first-order in both organotin and in bromine. Kinetic studies have established that there is no participation by bromide ion, even at high concentrations, and that only bromine reacts with the organotin compound.

In the previous stereochemical study[9] of the cleavage of *sec*-butyltrineopentyltin by bromine in methanol net inversion was observed, but, because the rotations of the pure enantiomers were unknown, it was not possible to determine the degree of inversion. The results of the previous and current work are summarized in Table 1, wherein the rotations for the enantiomers are also listed.

At 45°, the results were not highly reproducible, possibly in part because of the racemization of 2-bromobutane by S$_N$2 reaction

Table 1

Stereochemistry of Bromodemetallation of
sec-Butyltrineopentyltin in Methanol

Conditions[a]	Starting Tin Compound $[\alpha]_D^{23}$ (c 8-10,C_6H_6)	2-Bromobutane $[\alpha]_D$(neat)	% inversion[f,g]
12 hr, 45°[b,c]	-.68°	+1.95°	35.8
12 hr, 45°[b,c]	-.68°	+2.29°	41.4
12 hr, 45°[b,c]	-.61°	+2.44°	49.9
24 hr, 22°[d]	-1.57°	+4.94°	39.1
24 hr, 25°[d]	-1.57°	+8.36°	66.5
24 hr, 25°[e]	-1.57°	+12.54°	99.8

(a) All cleavages done in the absence of light.
(b) Ref. 9.
(c) 0.36 M Br$^-$
(d) No added Br$^-$
(e) 0.122 M Br$^-$
(f) Assuming $[\alpha]_D$ 4.19 for (S)-(+)-*sec*-butyltrineopentyltin; unpublished results.[17,18]
(g) Assuming $[\alpha]_D$ 33.7 for (S)-(+)-*sec*-butyl bromide; unpublished results.

by bromide ion, and possibly in part due to the intervention of a free radical process. The low observed stereospecificity may also result from the intervention of a retention process. These reactions were carried out at 0.2 and 0.4 M in stannane; in the previous kinetic studies, carried out at 10^{-3} M, no evidence was found for the participation of radicals.[9,16] Radical chain processes usually proceed most favorably at high reagent concentrations, whence the reaction is likely to be less stereospecific under such conditions.

In the results at 22° and 25°, the same considerations as above hold, in the absence of added bromide ion. In addition the medium will be less ionic. As has previously been discussed, the inversion mechanism (eq 7) is expected to proceed more favorably in a more ionic medium, but there may be a competing retention mechanism under the above conditions. Thus, many factors could be responsible for these observed varied stereochemistries.

Table 2

Stereochemical Results of Bromine Cleavage of
(S)-(+)-*sec*-Butyltrineopentyltin in Carbon Tetrachloride

	Starting Tin Compound $[\alpha]_D^{23}$ (c 15, C_6H_6)	2-Bromobutane $[\alpha]_D^{23}$ (c 10-20, CCl_4)	% retention
N_2 Atmosphere	+2.07	+.863	5.2
Air Atmosphere	+2.07	+1.511	9.1
O_2 Atmosphere	+2.07	+5.179	31.2
O_2 Saturated	+2.07	+7.553	45.5
5%  (structure) ONO (A)	+2.22	+13.297	74.7
10% 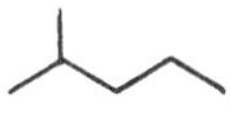 (structure) ONO (A)	+2.22	+16.857	94.7
5% 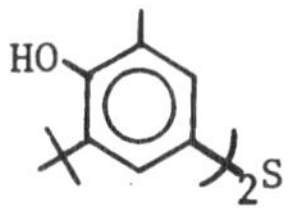 (structure) (B)	+2.22	+15.913	89.4
10% 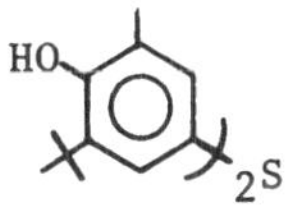 (structure) (B)	+2.22	+16.287	91.5

(a) Temp. 0°, in dark.
(b) See footnotes f and g, Table 1.

    The last entry in Table 1 is, perhaps, most significant. At
high bromide ion concentration, a situation which suppresses radical
components of the reaction and provides a highly ionic medium, the
reaction proceeds, within experimental uncertainty, completely stereospecifically with inversion of configuration (eq 7). To our knowledge, this is the only report of an S$_E$2 reaction with inversion of
configuration, with accompanying kinetic control data, which has
withstood the test of time.

As mentioned above, an S_E2 reaction such as is shown in equation
11 is expected to be unfavorable in a nonpolar solvent because of the
energy necessary to separate the ions. Thus, it was considered crit-
ical to study the reaction in such a solvent; carbon tetrachloride
was chosen for this purpose.

Initial studies in carbon tetrachloride indicated a small net
retention for the reaction of bromine with *sec*-butyl-trineopentyltin
(Table 2). Obviously, we were meeting an old "friend"; the major
problem in S_E2 studies, which, until recently, has been almost
entirely ignored by various workers, is that organometallic compounds
can react at least as well by free radical reactions as by S_E2 processes.[9,19,20] A small net retention was observed in an atmosphere
of nitrogen but, in accord with previous findings with mercurials,
oxygen was found to inhibit the radical component; the largest net
retention was observed with a solution saturated with oxygen (45.5%).
Probably, alkyl and bromine radicals are both involved. To suppress
the free-radical reaction it is necessary to intercept either the
alkyl radical (R·) or the bromine radical (Br·) (eq 13 and 14). It
is difficult to conceive of a better radical trap for R· than Br_2,

$$Br\cdot + R_4Sn \longrightarrow [R_4SnBr] \longrightarrow R_3SnBr + R\cdot \qquad (13)$$

$$R\cdot + Br_2 \longrightarrow RBr + Br\cdot \qquad (14)$$

and difficult to imagine a trap for Br· that is not destroyed by
bromine, but, through perseverance, two were found: isoamyl nitrite
(A) and di-(3-t-butyl-4-hydroxy-5-methyl)sulfide (B), Table 2. The
function of isoamyl nitrite as an inhibitor was discovered in a
screening program, but compounds of this type are now commonly used
as spin-trapping agents for esr studies. Alkyl nitrites are known
to react with radicals to form aldehydes (eq 15). However, hydrogen
abstraction is not expected to compete with the reaction of alkyl
radicals with bromine (albeit the concentration of free bromine is
small). Possibly the inhibition occurs by the simple addition reaction shown in equation 16.

$$R\text{—}CH_2ONO + R'\cdot \longrightarrow RC\overset{\displaystyle{O}}{\underset{\displaystyle{H}}{\diagdown}} + R'H + NO \qquad (15)$$

$$RONO + R'\cdot \longrightarrow RO\text{—}N\underset{\displaystyle{R}}{\overset{\displaystyle{\ddot{O}}}{\diagup}} \qquad (16)$$

Inhibitor B serves by a complex process. It reacts rapidly with 2 moles of bromine and, possibly slower, with still more bromine to form products which serve as inhibitors by trapping bromine or alkyl radicals. At this time the important point is that both A and B are effective in promoting retention in this system; research is underway to attempt to determine how they function.

Addition of oxygen or the inhibitors slows the rate of brominolysis of the tetraalkyltin compounds. These reactions are slow compared to the rate of cleavage in methanol.

While the transition state of the stereospecific inversion mechanism in methanol is easy to specify (eq 11), that for stereospecific retention in CCl_4 cannot be delineated. Kinetic control has not been established for this reaction in nonpolar solvents such as carbon tetrachloride. Although many rate studies have been carried out in many solvents, primarily in the extensive studies of Gielen and Nasielski,[22] the results in nonpolar solvents are often confusing. Many different rate sequences and kinetic expressions have been reported. Our view is that meaningful kinetic information for kinetic control will only be obtained when it can be shown that competing free radical reactions are absent.

However, even when the molecularity and stereochemistry are known, it is expected to be very difficult to determine whether the transition state is open (I) or closed (II). Transition state (I) would produce the two charges in close proximity. Structure (II) shows the tin compound serving as a Lewis acid and, although it is known to be a very weak one, surely it must possess some Lewis acid property. Both types of transition state (I and II) are expected to be more energetically favored in nonpolar solvents than that shown in equation 11.

(Molecularity of 2 assumed)

This dramatic reversal in stereochemistry with change of solvent polarity demonstrates that a delicate balance exists between retention and inversion stereochemistries for concerted electrophilic aliphatic substitution. Studies are underway to determine how other factors, such as alkyl structure, leaving group, and electrophile might affect the stereochemistry.

Recently, Rahm and Pereyre[10] have proposed that the reason that
sec-butyltrineopentyltin reacts with bromine in methanol solution
to produce sec-butyl bromide with inversion instead of retention of
configuration is that the starting tin compound is too hindered for
frontside attack to occur. In support of this proposition, they
found that reaction of bromine with sec-butyl-triisopropyltin yields
sec-butyl bromide with retention of configuration in mixed methanol-
cyclohexane solvent. Also, they observed some net inversion (2%)
for the reaction of bromine with sec-butyltriisopropyltin in chlorobenzene, a solvent in which tin compounds have been reported to
undergo brominolysis by a closed transition state.[23] Their overall
conclusion is that retention of configuration is probably the main
stereochemical course of the bromodemetallation of tetraalkyltin
compounds in the presence of methanol.

Their point regarding the role of the neopentyl groups in
favoring inversion is well taken. However, they have ignored the
role of solvent, and have used methanol-cyclohexane rather than
methanol; the present result demonstrates that solvent can play a
key role in determining stereochemistry. Also, they reported that
bromine cleaves sec-butyltrineopentyltin in chlorobenzene with net
(2%) inversion of configuration and, in the context given, this
result was used to support the view that the trineopentyl compounds
have a strong preference for reaction by an inversion process. They
did not show that radical pathways could not be excluded. General
experience with stereochemical studies has shown that it is not
prudent to make proposals or draw conclusions from a 2% stereochemical result.

The results presented herein demonstrate that, in the absence
of radical reactions in carbon tetrachloride as solvent, the sec-
butyltrineopentyltin compounds are cleaved nearly stereospecifically
with retention of configuration. It is reasonable to expect that a
similar situation holds in chlorobenzene as solvent in the absence
of radical reactions. Although some clarifications of experimental
results may be necessary, the results with sec-butyltrineopentyl-
and sec-butyltriisopropyltin compounds in methanol, carbon tetrachloride and solvent mixtures indicate that there exists a delicate
balance between retention and inversion stereochemistries for S_E2
reactions of organotin compounds, unless a system is "loaded" to
give one or the other.

Factors expected to promote inversion (eq 11) are:

(1) polar solvents; (2) high ionic strength in solvent; (3) R'
to be electron-donating so as to stabilize $R_3'Sn^+$; (4) R' to be very
large sterically so as to promote planar $R_3'Sn^+$ stabilization;
(5) R' to have a structure which suppresses, sterically or electron-
ically, front-side reaction; (6) the absence of low-lying vacant
orbitals on $R_3'SnR$. For retention, the opposite of the above stated
conditions would be expected to be desirable.

In general, for S$_E$2 reactions, the preferred stereochemistry is also expected to be influenced by the leaving group (metal), by the electrophile and by electronic factors in the alkyl group being cleaved.[24]

Acknowledgement is made to the National Institutes of Health (GM15373) and National Science Foundation (GP-33669) for support of this work.

REFERENCES

1. Abstracted in part from the Ph.D. thesis of R. L. Chambers (1974), University of California, Berkeley.
2. G. M. Whitesides and D. J. Boschetto, *J. Am. Chem. Soc.*, 93, 1529 (1971).
3. a) T. G. Attig and A. Wojeicki, *J. Am. Chem. Soc.*, 96, 6460 (1973), but for another point of view, see b) D. A. Stack and M. C. Baird, *J. Am. Chem. Soc.*, 98, 5539 (1976); T. C. Flood and D. L. Miles, *J. Organometal. Chem.*, 127, 33 (1977).
4. F. R. Jensen, V. Madan, and D. H. Buchanan, *J. Am. Chem. Soc.*, 93, 5283 (1971).
5. R. Dreos, A. Tauzher, N. Marisch and G. Costa, *J. Organometal. Chem.*, 108, 235 (1976).
6. F. R. Jensen and B. Rickborn, "Electrophilic Substitution of Organomercurials," McGraw-Hill, New York, 1968.
7. F. R. Jensen and K. L. Nakamaye, *J. Am. Chem. Soc.*, 88, 3437 (1966).
8. The apparent symmetry-forbidden nature of four-centered reductions is easily obviated by timing of reactions, e.g., prior complexation on the bond or the metal.
9. F. R. Jensen and D. D. Davis, *J. Am. Chem. Soc.*, 93, 4048 (1971).
10. A. Rahm and M. Pereyre, *J. Am. Chem. Soc.*, 99, 1672 (1977).
11. K. Sisido, S. Kozima, and K. Takizawa, *Tetrahedron Lett.*, 33, (1967); K. Sisido, T. Miyanisi, and T. T. Isida, *J. Organometal. Chem.*, 23, 117 (1970); P. Backelmans, M. Grelen, and J. Naszielski, *Tetrahedron Lett.*, 1149 (1967).
12. J. J. Miller, Ph.D. Thesis, University of California, Berkeley (1966).
13. D. Heyman, Ph.D. Thesis, University of California, Berkeley (1968).
14. a) H. C. Brown and C. F. Lane, *Chem. Commun.*, 521 (1971); b) M. Gielen and R. Fosty, *Bull. Soc. Chim. Belg.*, 83, 333 (1974); and D. E. Bergbreiter and D. R. Rainville, *J. Organometal. Chem.*, 121, 19 (1976).
15. W. H. Glaze, C. L. Selman, A. L. Ball and L. E. Bray, *J. Org. Chem.*, 34, 641 (1969).
16. F. R. Jensen and D. D. Davis, *J. Am. Chem. Soc.*, 93, 4047 (1971).

17. Unpublished results.
18. Pereyre and co-workers have reported the rotations to be 9.8°, ref. 10, and more recently 3.0°, A. Rahm and M. Pereyre, *J. Organometal. Chem.*, <u>88</u>, 79 (1975). A major obstacle in determining the correct rotation is the difficulty of purifying the pertinent compounds.
19. Ref. 6, p. 67, 69, 76-82, 80, 89, 99, 120, 137-141.
20. F. R. Jensen and D. Heyman, *J. Am. Chem. Soc.*, <u>88</u>, 3438 (1966).
21. F. R. Jensen and L. H. Gale, *J. Am. Chem. Soc.*, <u>81</u>, 1261 (1959); *J. Am. Chem. Soc.*, <u>82</u>, 148 (1960).
22. M. Gielen and J. Nasielski in, "Organotin Compounds," A. K. Sawyer, Ed., Vol. 3, Marcel Dekker, New York, N.Y., 1972, p. 652 and results cited therein.
23. S. Bone, M. Gielen and J. Nasielski, *J. Organometal. Chem.*, <u>9</u>, 443 (1967).
24. Unpublished results.

SELECTIVE REDUCTIONS USING METAL HYDRIDES

Clinton F. Lane

Aldrich-Boranes, Inc., Route 3, Sheboygan Falls

Wisconsin 53085

A discussion of selective reductions is very appropriate at a symposium honoring Professor Herbert C. Brown for his contributions to chemistry. Dr. Brown's Ph.D. thesis (under the direction of Professor H. I. Schlesinger) involved a study of the reduction of aldehydes and ketones by diborane.[1] This was the first application of the use of a boron hydride for the reduction of organic compounds. It was during this period, too, that Professor Schlesinger and his coworkers at the University of Chicago discovered both sodium borohydride[2] and lithium aluminum hydride (LAH).[3]

For over 40 years Dr. Brown has maintained an active interest in selective reductions with boron and aluminum hydrides. He has published over 65 journal articles in this area alone. Interestingly, Dr. Brown's investigation (with B. C. Subba Rao) of the reduction of organic compounds by diborane led to the discovery of the hydroboration reaction.[4] Hydroboration provided, for the first time, a convenient method for the preparation of organoboranes, and this development kindled vigorous activity in the study of organoboranes as intermediates in organic syntheses. This is the point where Aldrich Chemical Company entered the picture.

Professor Brown felt that working synthetic chemists were not adequately utilizing this new chemistry of organoboranes. He reasoned that part of the problem may have been that a major hurdle existed if the chemist had to prepare the starting borane reagent before he could apply a new synthetic reaction based on organoboranes. In the spring of 1972 Dr. Brown was able to persuade Dr. Alfred R. Bader, President of Aldrich Chemical Company, to set up a subsidiary, Aldrich-Boranes, Inc., to make readily available the various boron-based reagents necessary to facilitate application of this new field of synthetic chemistry.

Aldrich-Boranes did the best they could and within a year most of the more important boron reagents were prepared and they were subsequently offered by Aldrich Chemical. It soon became apparent that the boron hydride reagents with the best sales were those which were reported to be useful selective reducing agents. Therefore, for reasons of economic necessity, we at Aldrich-Boranes have, in recent years, directed almost all of our new product development attention to the preparation and utilization of various boron hydride reducing agents. The selectivity and synthetic utility of some of our more useful reducing agents will now be summarized. For more extensive discussions, please consult the review articles which have appeared in *Aldrichimica Acta*.[5-8]

The following table summarizes the observed reactivities of the main functional groups toward the most common hydride reducing agents. The symbol (+) indicates a rapid reaction at 0-25°C, the symbol (-) indicates slow or insignificant reaction at 0-25°C and the symbol (±) indicates a variable rate depending upon structure.

Obviously this summary is a gross oversimplification, but it does illustrate the fact that LAH is an extremely powerful reducing

Reactivity of the Hydride Reducing Agents

	$NaBH_4$	BH_3-THF	$LiAlH_4$
Aldehyde	+	+	+
Ketone	+	+	+
Acid chloride	+	-	+
Lactone	-	+	+
Epoxide	-	+	+
Ester	-	±	+
Carboxylic acid	-	+	+
Carboxylic acid salt	-	-	+
Amide	-	+	+
Nitrile	-	+	+
Nitro	-	-	+
Olefin	-	+	-

agent while NaBH$_4$ is a relatively mild reducing agent. Borane-THF
is intermediate in reactivity. Although it is not apparent from
this summary, NaBH$_4$ and BH$_3$-THF display distinctly different modes
of action.

Sodium borohydride is a nucleophilic, or "basic", reducing
agent and attacks electropositive atoms, i.e., centers of low
electron density. On the other hand, borane is an electrophilic,
or "acidic", reducing agent and attacks electronegative atoms, i.e.,
centers of high electron density. With such different reducing
characteristics, these two reagents tend to complement each other
and they are exceedingly useful for selective reductions. In a
similar manner, LAH and AlH$_3$ are complementary in reducing characteristics.

An even more important property of these metal hydride reducing
agents is their ability to undergo significant change in reactivity
upon modification. One or more of the M-H groups can be changed to
M-Y, where Y represents a wide variety of possible groups. Also,
the cation can be varied in borohydride salts, and the Lewis base
can be varied in borane complexes. Consequently, an almost infinite
variety of new metal hydride reagents can be prepared, some of which
show vastly improved characteristics as selective reducing agents.
This point will become even more apparent later in the present
discussion.

Of the reagents shown in the Table, BH$_3$-THF has seemed most
interesting to us, and most of our activity at Aldrich-Boranes has
been directed toward determining its synthetic utility. We have
found the selective reduction of the carboxylic acid group and the
amide group to be particularly useful. The following are some
examples of selective reductions of carboxylic acids which were
taken from the literature (eqs 1-3). In these and all subsequent
examples the % yield given is for isolated purified product.

$$\text{(1)}$$

82%

N. M. Yoon and H. C. Brown and coworkers[9]

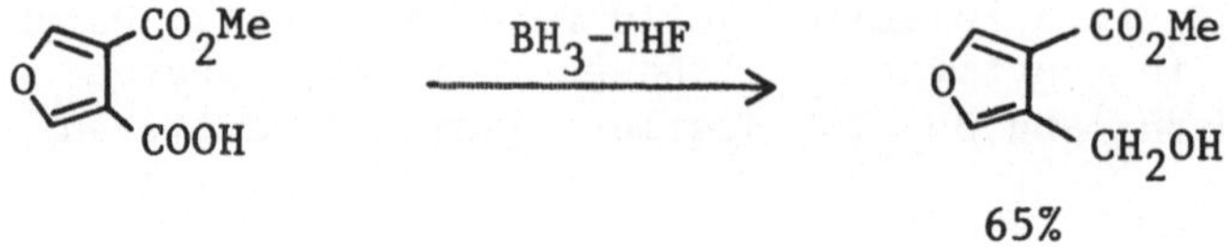

$$(2)$$

65%

E. J. Corey and coworkers[10]

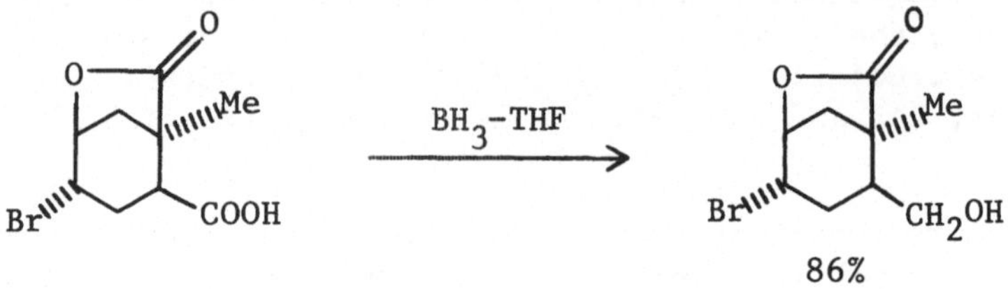

$$(3)$$

86%

E. J. Corey and H. S. Sachdev[11]

Brown and coworkers described the selective reduction of the acid group in the presence of a cyano group, as shown in equation 1. Equations 2 and 3 are particularly interesting and indicate that an acid group can be reduced selectively even when an ester or lactone group is nearby. These two reductions, each being one step in a natural product synthesis, were reported by Professor E. J. Corey and coworkers. A great many additional examples could be cited. Indeed, when selectivity, yield and the mild reaction conditions are all taken into account, borane is probably the reagent of choice for the reduction of carboxylic acids.

Borane-THF is also widely used for the selective reduction of amides. Again, many examples could be cited, but the following will suffice (eqs 4-6).

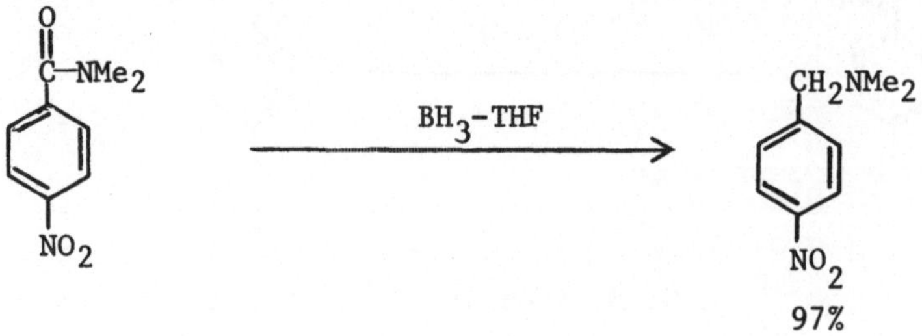

$$(4)$$

97%

H. C. Brown and P. Heim[12]

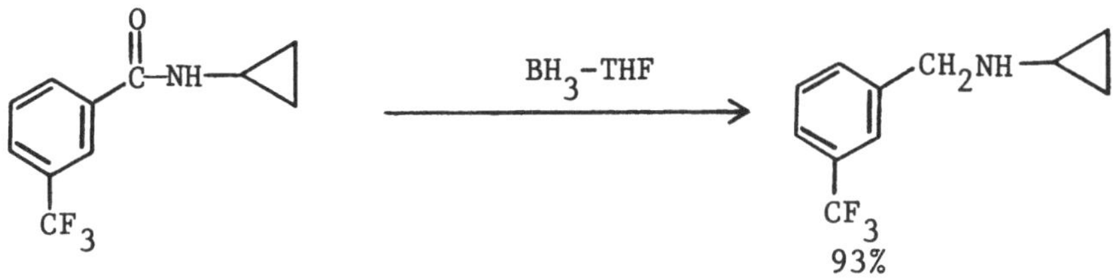

$$(5)$$

H. J. Barbander and W. B. Wright, Jr.[13]

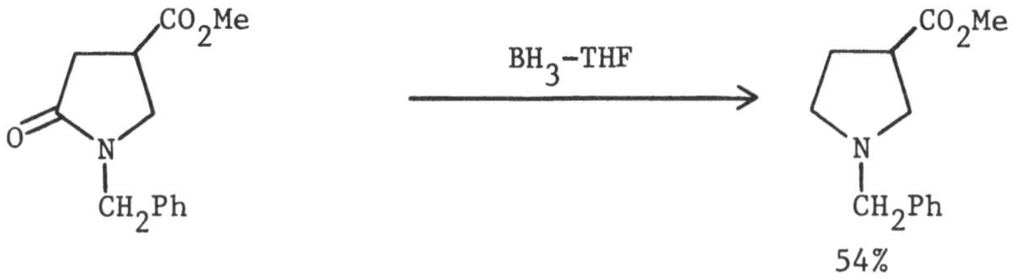

$$(6)$$

M. J. Kornet and coworkers[14]

LAH is a popular reagent for the reduction of amides. However, it cannot be used in the cases cited above; the nitro, N-cyclopropyl, $-CF_3$, N-benzyl and ester group would all be reduced.

Borane-THF is a useful laboratory reagent, but the complex has certain properties which limit its utility as a commercial reagent. It can only be offered as a 1M solution in THF, and the solution must be refrigerated to maintain hydride activity. The commercial reagent contains a small amount of $NaBH_4$ (<5 mole %) to retard cleavage of the THF. All of these disadvantages are eliminated with our borane-methyl sulfide (BMS) reagent. BMS is available as the neat liquid with a BH_3 concentration ten times (~10M) that of BH_3-THF. It can be stored for months at room temperature without loss of hydride activity and is apparently stable indefinitely when refrigerated. Also, BMS is soluble in, and unreactive toward, a wide variety of aprotic solvents. As a convenience, Aldrich also offers BMS as standard solutions in diethyl ether, THF, toluene, and methylene chloride. Finally, our work seems to indicate that the reactivity of BMS parallels that of BH_3-THF[15] though somewhat higher temperatures are usually required.

BMS is an extremely valuable reagent for selective reductions. The following are some examples from our work (eqs 7-9).[15]

$$Br(CH_2)_{10}COOH \xrightarrow{\text{BMS}} \xrightarrow{\text{MeOH}} Br(CH_2)_{10}CH_2OH \qquad (7)$$

$$95\%$$

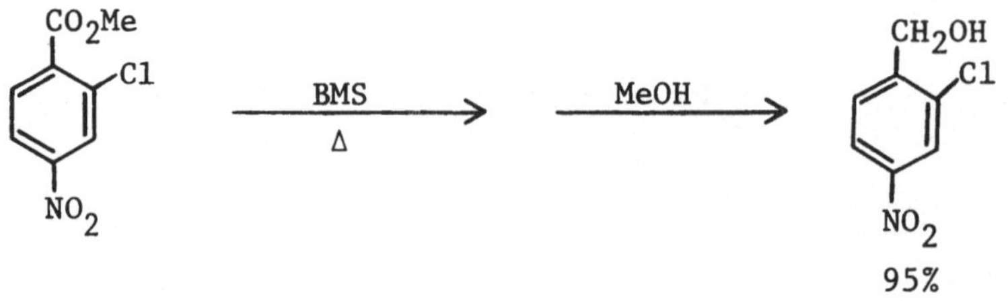

(8)

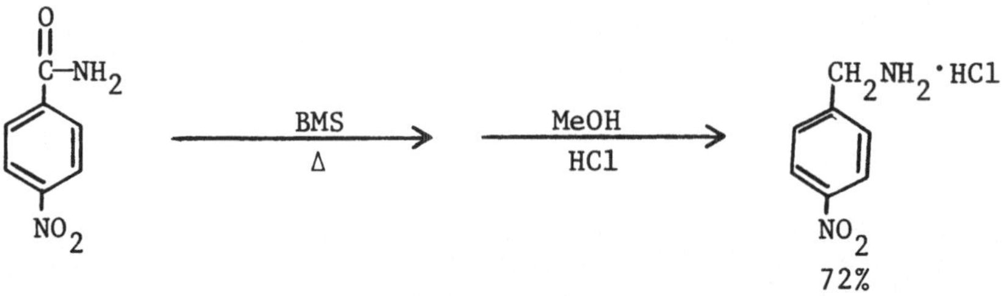

(9)

Reductions of esters and primary amides with a borane reagent normally require heat if complete reduction is to be achieved. With BH_3–THF, a large excess of the reagent is necessary to compensate for loss of hydride activity on heating. Only the stoichiometric quantity of BMS is required, however, because of its improved thermal stability. Side reactions resulting from the use of excess reagent are thereby avoided and vastly improved selectivity is thus possible.

Another important advantage of BMS reductions is the ease of isolation of products. In the reduction of carboxylic acids or esters (e.g., eqs 7 and 8), it is only necessary to pour the reduction mixture into excess methanol or ethanol; removal of the solvent, along with trialkyl borate, on the Roto-vac leaves an alcohol residue that is boron-free and of satisfactory purity for most applications.[15] In the reduction of amides (e.g., eq 9), it is necessary to add anhydrous hydrogen chloride to convert the boron–nitrogen intermediates to trialkyl borate and amine hydrochloride salt. The preparation of 4-nitrobenzylamine·HCl (eq 9) illustrates a special technique where only the stoichiometric amount of methanol is added, followed by treatment with hydrogen chloride gas. Cooling causes crystallization of the amine hydrochloride, which is isolated in 72% yield by simple filtration.[15]

The case of BMS illustrates how the properties of BH_3 can be modified by varying the Lewis base used to form the borane complex. It is also possible to modify the reactivity and selectivity of BH_3 by substituting other groups for hydrogen on boron. A large variety of monosubstituted and disubstituted boranes are known and many of these reagents are useful selective reducing agents.

9-Borabicyclo[3.3.1]nonane (9-BBN) is probably the most widely known disubstituted borane;[6] it is readily prepared by cyclic hydroboration of 1,5-cyclooctadiene (eq 10).

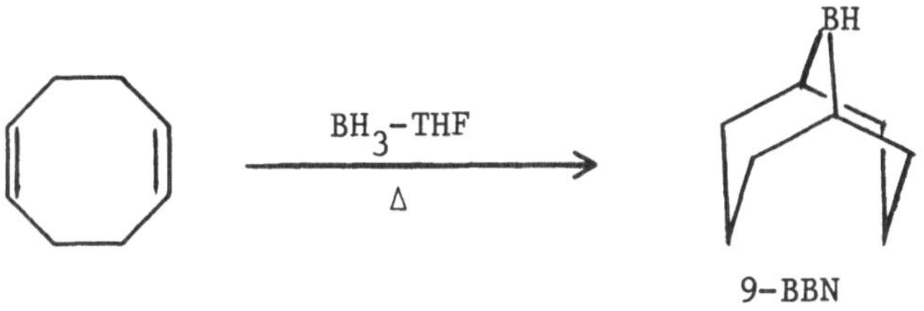

$$\text{(10)}$$

9-BBN exhibits unusual thermal stability for a dialkylborane and this, together with its availability, led Professor Brown and his students to study extensively its hydroboration characteristics. They have also discovered many interesting synthetic reactions of the resulting B-alkyl-9-BBN derivatives. As a selective reducing agent, 9-BBN cleanly converts α, β-unsaturated ketones to allylic alcohols. This reduction can even be carried out selectively in the presence of an ester functionality (eq 11). LAH or diisobutyl-aluminum hydride can also be used to convert α, β-unsaturated ketones to allylic alcohols, but these reagents would also reduce the ester group.

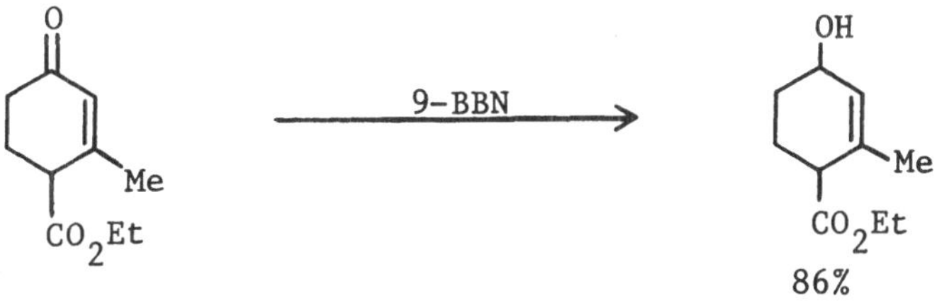

$$\text{(11)}$$

As discussed above, it is apparent that BH_3-THF is a useful selective reducing agent. BMS and 9-BBN illustrate how "BH_3" can be modified to provide even more selective reducing agents. Sodium borohydride can also be modified to give a variety of highly selective reducing agents.

The reaction of liquid HCN with $NaBH_4$ provides a ready synthesis of sodium cyanoborohydride (eq 12).[7] The electron-withdrawing cyano

$$NaBH_4 + HCN \longrightarrow NaBH_3CN + H_2 \qquad (12)$$

group decreases the reactivity of the boron hydride. Thus, $NaBH_3CN$ is hydrolyzed 10^8-fold slower than $NaBH_4$. $NaBH_4$ is stable in aqueous alkali but is rapidly hydrolyzed at pH 7 or lower and so can only be used under neutral or alkaline conditions. On the other hand, $NaBH_3CN$ is stable in aqueous acid to pH 3. It is also soluble in a wider variety of solvents than is $NaBH_4$. All of these properties combine to make $NaBH_3CN$ a milder and more selective reducing agent than $NaBH_4$,[7] as the following interesting examples show (eqs 13-15).

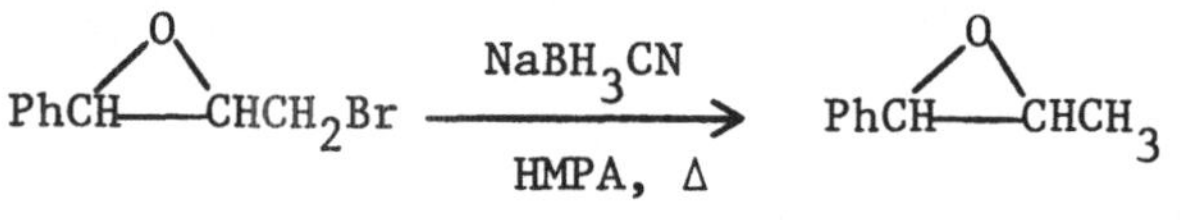

$$PhCH\!\!-\!\!CHCH_2Br \xrightarrow[\text{HMPA, } \Delta]{NaBH_3CN} PhCH\!\!-\!\!CHCH_3 \qquad (13)$$

R. O. Hutchins and coworkers[17]

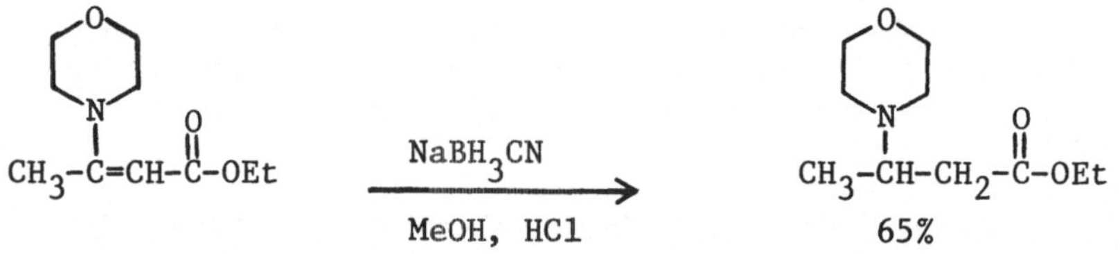

$$CH_3\text{-}C=CH\text{-}\overset{O}{\overset{\|}{C}}\text{-}OEt \xrightarrow[\text{MeOH, HCl}]{NaBH_3CN} CH_3\text{-}CH\text{-}CH_2\text{-}\overset{O}{\overset{\|}{C}}\text{-}OEt \qquad (14)$$

65%

R. F. Borch and coworkers[18]

$$CH_3\text{-}\overset{O}{\overset{\|}{C}}\text{-}(CH_2)_3\text{-}\overset{O}{\overset{\|}{C}}O(CH_2)_6CN \xrightarrow[\text{2) } NaBH_3CN, \Delta]{\text{1) } TsNHNH_2} CH_3CH_2(CH_2)_3\text{-}\overset{O}{\overset{\|}{C}}O(CH_2)_6CN \qquad (15)$$

75%

R. O. Hutchins and coworkers[19]

Equation 13 shows how $NaBH_3CN$ can act as a highly selective simple nucleophile, displacing bromide but leaving the epoxide ring intact. The reaction shown in equation 15 involves the reduction of an intermediate tosylhydrazone and provides a mild, highly selective alternative to the Wolff-Kishner process. An intermediate iminium ion $C=N^+$ is reduced in the reaction illustrated in equation 14. This reduction of iminium ions represents the most important application of NaBH$_3$CN. Even more important, the iminium ion is reduced much faster than a carbonyl group, enabling selective reductive aminations to be carried out without isolation of the intermediate imine. Equations 16–18 provide a few interesting examples.

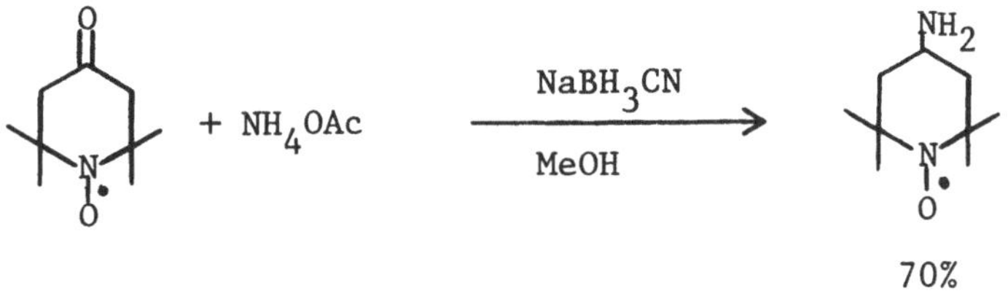

$$ \text{(16)} $$

70%

G. M. Rosen[20]

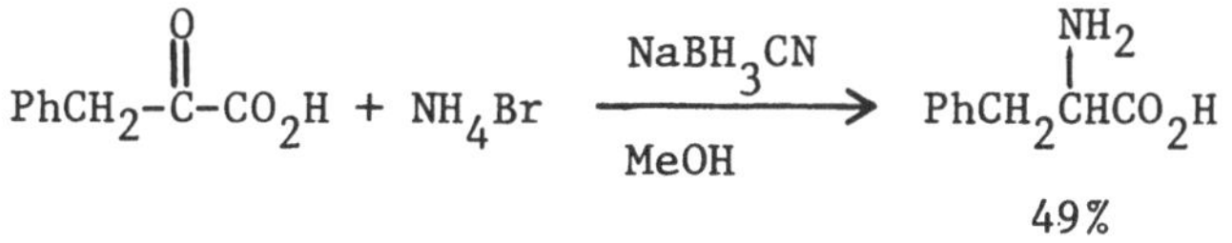

$$ PhCH_2-\overset{O}{\overset{\|}{C}}-CO_2H + NH_4Br \xrightarrow[\text{MeOH}]{\text{NaBH}_3\text{CN}} PhCH_2\overset{NH_2}{\underset{\vert}{CH}}CO_2H \qquad \text{(17)} $$

49%

R. F. Borch and coworkers[18]

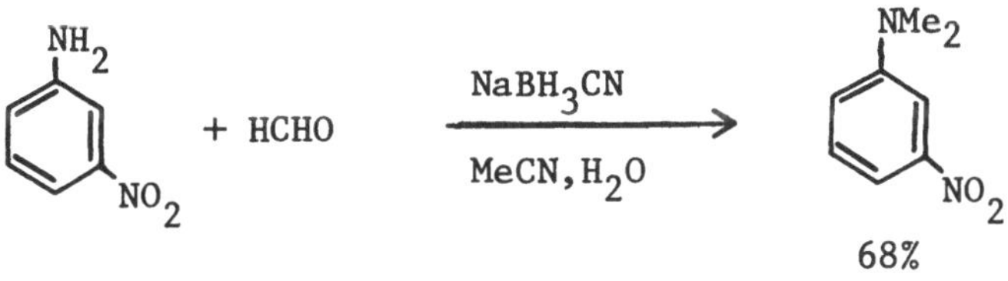

$$ \text{(18)} $$

68%

R. F. Borch and A. I. Hassid[21]

The above examples illustrate how it is possible to carry out reductive aminations in the presence of nitroxide free radical, carboxylic acid, and the nitro group. In addition, amides, esters, lactones, nitriles, and epoxides are also inert toward NaBH$_3$CN under conditions necessary to achieve reductive amination.[7]

Obviously, the -CN group decreases the reactivity of borohydride. It is also possible to increase the reactivity of borohydride with electron-donating alkyl groups.

Lithium hydride reacts readily with triethylborane in THF, giving lithium triethylborohydride (eq 19).[8] This is our

$$Et_3B + LiH \xrightarrow{THF} Li(Et_3BH) \tag{19}$$

Super-Hydride® reagent, and it is an exceptionally powerful nucleo-phile. Alkyl halides react with Super-Hydride to give the hydro-carbon (eq 20). This reaction involves a clean S_N2 displacement;

$$Li(Et_3BH) + RX \xrightarrow{THF} RH + Et_3B + LiX \tag{20}$$

the reactivity decreases in the order 1° > 2° >> 3°. In fact, 1° bromides are reduced in a manner of minutes (eq 21). This dis-placement is completely devoid of any carbonium ion character; thus,

$$RCH_2Br \xrightarrow[25°, \ 2 \ min]{Super-Hydride} RCH_3 \tag{21}$$

exo-norbornyl bromide gives only *endo* deuteride with Super-Deuteride® (eq 22).

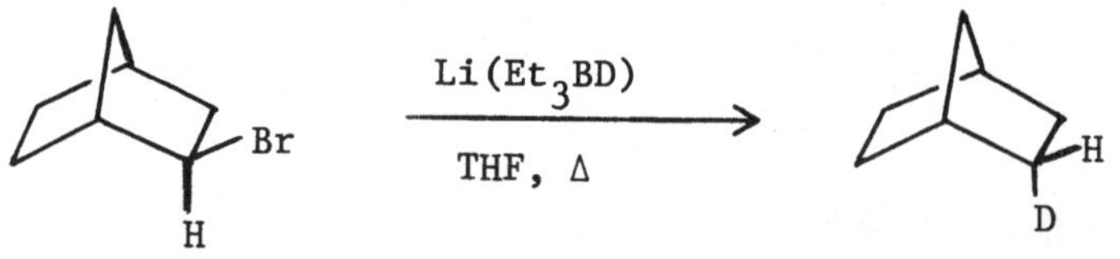

$$\tag{22}$$

This rapid and clean S_N2 displacement is not limited to halides. For example, a trialkylamine can be the leaving group in the selective demethylation of quaternary ammonium salts (eq 23).[22] This reaction is very sensitive to steric effects – only methyl groups are attacked.

$$PhCH_2NMe_3I \xrightarrow{Super-Hydride} PhCH_2NMe_2 \tag{23}$$
$$92\%$$

At Aldrich-Boranes, we have found that Super-Hydride is very reactive toward ester groups.[23] Esters are reduced rapidly even at low temperatures. The following selective reductions were reported (eqs 24 and 25).[23]

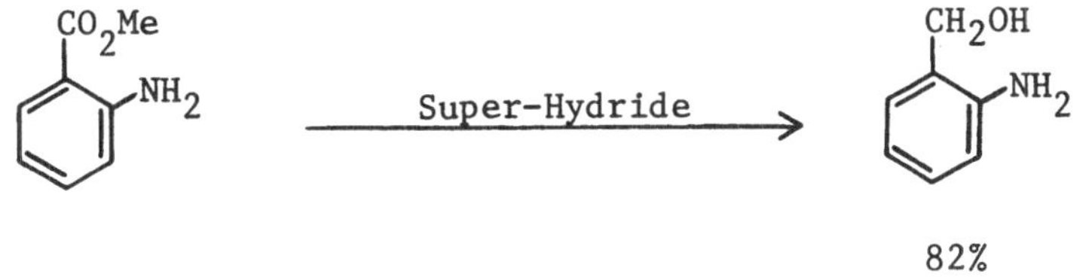

$$\text{(24)}$$

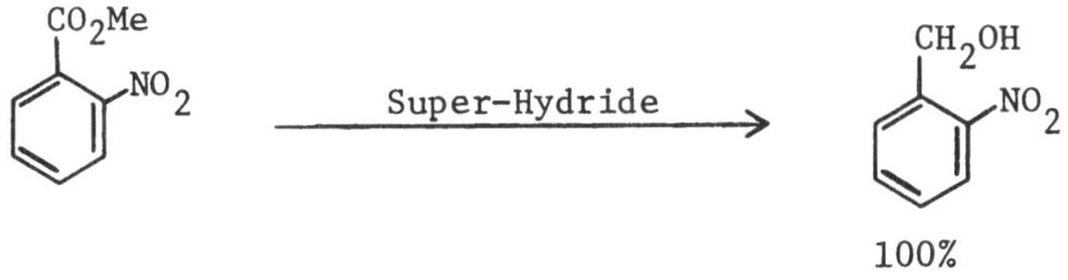

82%

$$\text{(25)}$$

100%

Interestingly, the amino group is completely inert toward Super-Hydride. Thus, for the reaction shown in equation 24, only 2 equivalents of Super-Hydride were required to reduce the ester group. That is, no additional Super-Hydride needed to be added for such side-reactions with the $-NH_2$ group as H_2 evolution or amine-borane formation.

Potentially, the most useful application of Super-Hydride is for the highly selective reductive ring opening of epoxides. Even though Super-Hydride is a very powerful reducing agent, regioselective and stereoselective reduction of epoxides is possible, as illustrated in equations 26 and 27.[24] Reductions of epoxides with the other hydride reducing agents, such as $NaBH_4$, LAH, or BH_3-THF, invariably results in mixtures of products.

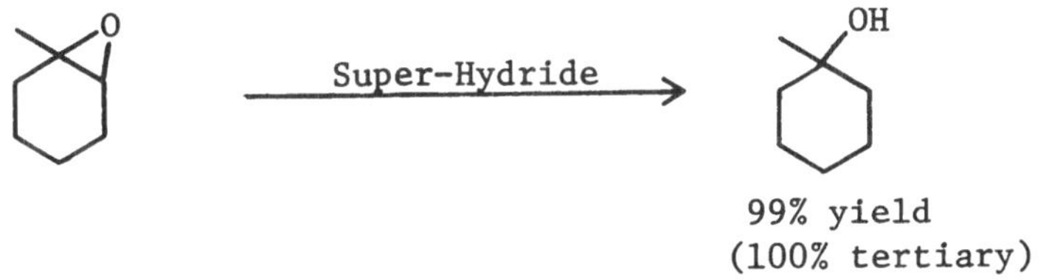

$$\text{(26)}$$

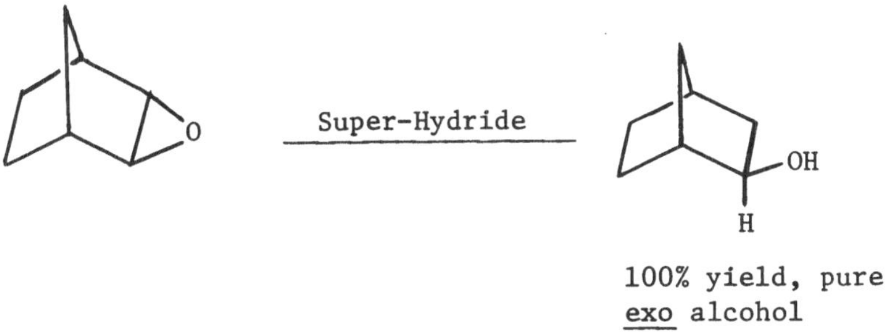

99% yield
(100% tertiary)

$$\text{(27)}$$

100% yield, pure
exo alcohol

    Almost all trialkylboranes can be converted into trialkylboro-
hydrides, and a wide variety of trialkylboranes is readily available
through hydroboration. Consequently, numerous trialkylborohydrides
can be prepared. When the alkyl group is bulky, a highly stereoselective reducing agent can be prepared. Specific examples are
our Selectride reagents.

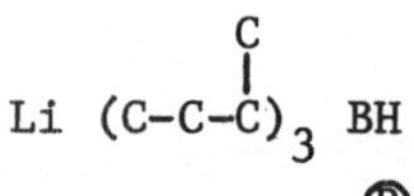

$$\text{Li (C-C-}\overset{\overset{\textstyle C}{|}}{\text{C}}\text{)}_3 \text{ BH}$$

L-Selectride®

(Lithium tri-*sec*-butylborohydride)

H. C. Brown and S. Krishnamurthy[25]

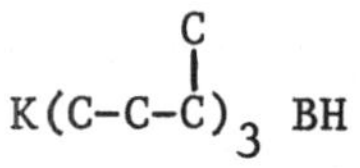

$$\text{K(C-C-}\overset{\overset{\textstyle C}{|}}{\text{C}}\text{)}_3 \text{ BH}$$

K-Selectride®

(Potassium tri-*sec*-butylborohydride)

C. A. Brown[26]

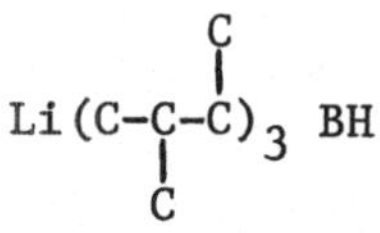

$$\text{Li(C-}\overset{\overset{\textstyle C}{|}}{\underset{\underset{\textstyle C}{|}}{\text{C}}}\text{-C)}_3 \text{ BH}$$

LS-Selectride™

(Lithium trisiamylborohydride)

S. Krishnamurthy and H. C. Brown[27]

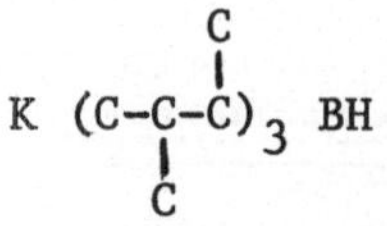

$$\text{K (C-}\overset{\overset{\textstyle C}{|}}{\underset{\underset{\textstyle C}{|}}{\text{C}}}\text{-C)}_3 \text{ BH}$$

KS-Selectride™

(Potassium trisiamylborohydride)

C. A. Brown and S. Krishnamurthy[28]

These reagents were initially developed for the stereoselective reduction of ketones. The results obtained for the reduction of the three methylcyclohexanones are summarized below (eqs 28–30).

(28)

L-Selectride:>99% *cis*

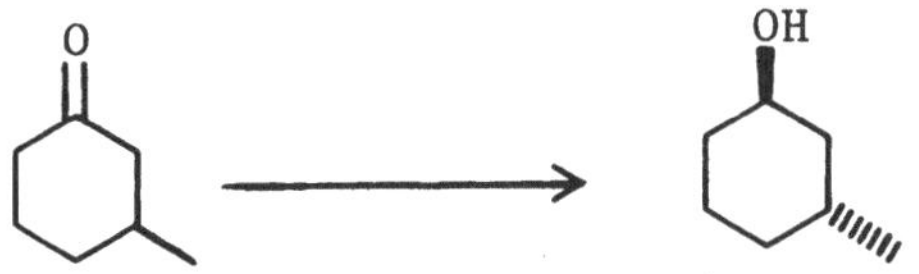

(29)

L-Selectride: 85% *trans*
LS-Selectride: >99% *trans*

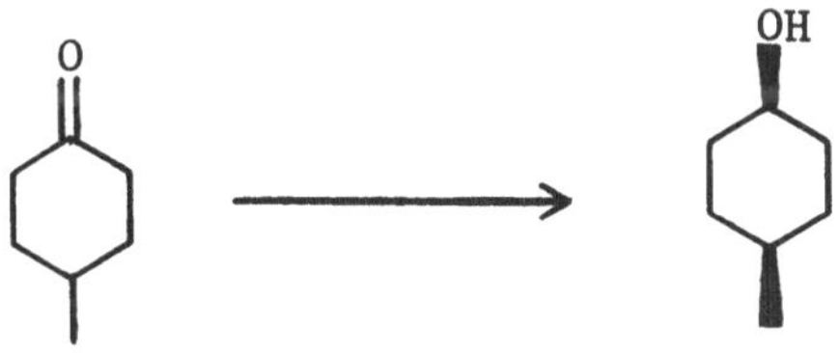

(30)

L-Selectride: 80% *cis*
K-Selectride: 88% *cis*
LS-Selectride: >99% *cis*

In all cases, the reaction involves an equatorial attack by borohydride on the carbonyl group, giving an axial hydroxy group upon hydrolysis. Thus, these highly sterically congested reducing reagents attack the least hindered side of the carbonyl group. This provides a superb example of how Professor Brown's long-standing interest in steric effects has profoundly affected his research in a different area.

The Selectride reagents have found wide application for stereoselective reductions, most notably in prostaglandin synthesis.[8]
However, various workers have discovered other applications for
these versatile reagents, now that they are commercially available.
For example, Professor Bruce Ganem has developed a procedure for
reductive alkylation of enones using our Selectride reagents,[29] and
Professor John Gladysz has utilized Super-Hydride and the Selectride
reagents to synthesize a variety of organometallics.[30]

It should now be apparent that the reactivity and selectivity
of both BH_3 and BH_4^- can be greatly modified by appropriate substitution. Consequently, a large variety of selective boron hydride
reducing agents are now available and many more will undoubtedly be
developed in the future. It is hoped that Aldrich-Boranes, Inc.
will be in a position to make them commercially available.

The reactivity and selectivity of AlH_3 and AlH_4^- can also be
modified by substituting other groups for hydride on aluminum. The
following are a few of the substituted aluminum hydride reagents
which are commercially available.

$$\underset{\text{C}}{\overset{\overset{\displaystyle C}{|}}{(C\text{-}C\text{-}C)_2AlH}}$$

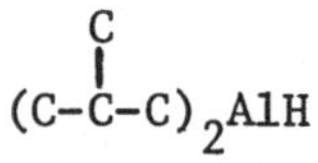

Diisobutylaluminum hydride

(DIBAL-H)

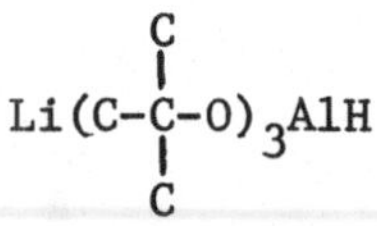

$$Li(C\overset{\overset{\displaystyle C}{|}}{\underset{\underset{\displaystyle C}{|}}{\text{-}C\text{-}O}})_3AlH$$

Lithium tri-*tert*-butoxyaluminum hydride

$$NaAlH_2(OCH_2CH_2OMe)_2$$
Red-al®

(Sodium bis(2-methoxyethoxy)aluminum hydride)

$$NaAlH_2Et_2$$

Sodium diethylaluminum hydride

(OMH-1)

The aluminum hydride reagents are useful selective reducing agents, their properties being, in many cases, complementary to those of the boron hydride reagents. Consequently, Aldrich-Boranes actively uses both the boron hydrides and the aluminum hydrides for the manufacture of fine chemicals and for custom syntheses. The following are some of the more useful applications that various workers have found for these aluminum hydrides as selective reducing agents.

cyclic enone $\xrightarrow{\text{DIBAL-H}}$ unsaturated alcohol[31,32]

nitrile $\xrightarrow{\text{DIBAL-H}}$ aldehyde[32]

ester $\xrightarrow{\text{DIBAL-H}}$ aldehyde[32]

acid chloride $\xrightarrow{\text{Li}(\underline{t}\text{-BuO})_3\text{AlH}}$ aldehyde[33]

unsaturated acid $\xrightarrow{\text{Red-al}}$ unsaturated alcohol[34]

nitrostyrenes $\xrightarrow{\text{Red-al}}$ phenethyl amines[35]

α,β-unsaturated aldehyde $\xrightarrow{\text{OMH-1}}$ allylic alcohol[36]

Other metal hydrides can also be used for selective reductions. Recently, Tom Cole and Professor Roland Pettit described the preparation and utility of a new reducing agent, tetramethylammonium hydridoirontetracarbonyl.[37] This iron hydride is prepared as shown in equation 31 and we now offer this new reagent under the trade name TMA-Ferride.[TM]

$$Fe(CO)_5 + 3KOH + Me_4NBr \xrightarrow{H_2O} \quad (31)$$

$$Me_4N[HFe(CO)_4] + KBr + K_2CO_3 + H_2O$$

TMA-Ferride

The use of TMA-Ferride provides a mild and facile method for the selective reduction of acid chlorides to aldehydes in excellent yields (eq 32). This procedure was applied to several representative acid chlorides (for example, eqs 33 and 34).[37] The ease and

$$2R\overset{O}{\overset{\|}{C}}Cl + 3Me_4N[HFe(CO)_4] \xrightarrow{CH_2Cl_2}$$

$$2R\overset{O}{\overset{\|}{C}}H + 2Me_4NCl + Me_4N[HFe_3(CO)_{11}] + CO$$

(32)

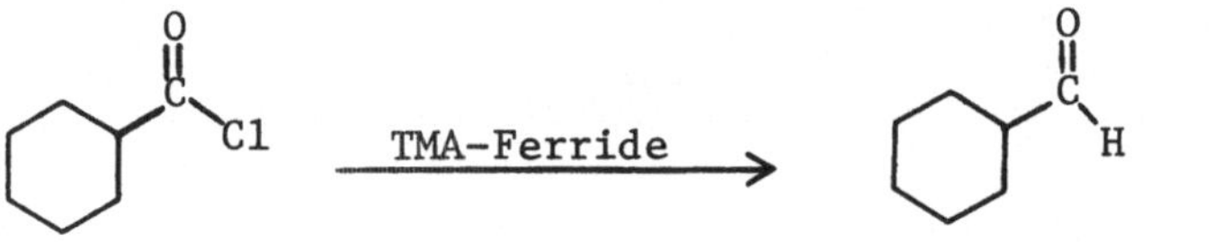

95% gc yield

(33)

75% isolated yield

(34)

simplicity of this conversion of acid chlorides to aldehydes represents an improvement over existing methods.

At alkaline pH, $HFe(CO)_4^-$ is not hydrolyzed in water or in alcohols, but the iron hydride is sensitive to acid. It is also very sensitive to oxygen.

Sodium and potassium hydridoirontetracarbonyl have previously been shown to be useful for reductive alkylations,[38] hydroacylations,[39] dehydrogenations,[40] desulfurizations,[41] and hydrogenations,[42] but the alkali metal hydridoirontetracarbonyl reagents can only be prepared and utilized in concentrated alkaline solutions. TMA-Ferride now provides a source for the interesting hydridoirontetracarbonyl anion in neutral solution.

TMA-Ferride is only the beginning. In the future, there should be available numerous other iron hydride reducing agents. Other boron hydride and aluminum hydride reducing agents should also become available. Eventually, the synthetic chemist should have at his disposal a whole arsenal of selective reducing agents - reagents capable of selectively reducing a given functional group in the presence of all other functional groups, reagents capable of 100% regio- and stereoselectivity, reagents capable of 100% enantiomeric enrichment for an asymmetric reduction, reagents capable of carrying out reductions in water at room temperature at pH 7. Obviously, much work remains to be done, but the future looks very bright for metal hydrides as selective reducing agents.

REFERENCES

1. H. C. Brown, H. I. Schlesinger, and A. B. Burg, *J. Am. Chem. Soc.*, 61, 673 (1939).
2. H. I. Schlesinger and H. C. Brown in collaboration with 18 coworkers, *J. Am. Chem. Soc.*, 75, 186-222 (1953).
3. A. E. Finholt, A. C. Bond, Jr., and H. I. Schlesinger, *J. Am. Chem. Soc.*, 69, 1199 (1947).
4. For an interesting historical account of this work, see H. C. Brown, "Boranes in Organic Chemistry," Cornell University Press, Ithaca, NY, 1972.
5. Borane Complexes: C. F. Lane, *Aldrichimica Acta*, 10, 41 (1977); see also, C. F. Lane, *Chem. Rev.*, 76, 773 (1976).
6. 9-BBN: C. F. Lane, *Aldrichimica Acta*, 9, 31 (1976); see also, H. C. Brown and C. F. Lane, *Heterocycles*, 7, 453 (1977).
7. NaBH$_3$CN: C. F. Lane, *Aldrichimica Acta*, 8, 3 (1975); see also, C. F. Lane, *Synthesis*, 135 (1975).
8. Super-Hydride® and the Selectride reagents: S. Krishnamurthy, *Aldrichimica Acta*, 7, 55 (1974).
9. N. M. Yoon, C. S. Pak, H. C. Brown, S. Krishnamurthy, and T. P. Stocky, *J. Org. Chem.*, 38, 2786 (1973).
10. E. J. Corey, D. N. Crouse, and J. E. Anderson, *J. Org. Chem.*, 40, 2140 (1975).
11. E. J. Corey and H. S. Sachdev, *J. Org. Chem.*, 40, 579 (1975).
12. H. C. Brown and P. Heim, *J. Org. Chem.*, 38, 912 (1973).
13. H. J. Brabander and W. B. Wright, Jr., *J. Org. Chem.*, 32, 4053 (1967).
14. M. J. Kornet, P. A. Thio, and S. I. Tan, *J. Org. Chem.*, 33, 3637 (1968).
15. C. F. Lane, *Aldrichimica Acta*, 8, 20 (1975).
16. S. Krishnamurthy and H. C. Brown, *J. Org. Chem.*, 42, 1197 (1977).
17. R. O. Hutchins, B. E. Maryanoff, and C. A. Milewski, *Chem. Commun.*, 1097 (1971).
18. R. F. Borch, M. D. Bernstein, and H. D. Durst, *J. Am. Chem. Soc.*, 93, 2897 (1971).

19. R. O. Hutchins, C. A. Milewski, and B. E. Maryanoff, *J. Am. Chem. Soc.*, **95**, 3662 (1973).

20. G. M. Rosen, *J. Med. Chem.*, **17**, 358 (1974).

21. R. F. Borch and A. I. Hassid, *J. Org. Chem.*, **37**, 1673 (1972).

22. M. P. Cooke, Jr., and R. M. Parlman, *J. Org. Chem.*, **40**, 531 (1975).

23. C. F. Lane, *Aldrichimica Acta*, **7**, 32 (1974).

24. S. Krishnamurthy, R. M. Schubert, and H. C. Brown, *J. Am. Chem. Soc.*, **95**, 8486 (1973).

25. H. C. Brown and S. Krishnamurthy, *J. Am. Chem. Soc.*, **94**, 7159 (1972).

26. C. A. Brown, *J. Am. Chem. Soc.*, **95**, 4100 (1973).

27. S. Krishnamurthy and H. C. Brown, *J. Am. Chem. Soc.*, **98**, 3383 (1976).

28. C. A. Brown and S. Krishnamurthy, Abstract ORGN 43, 175th National American Chemical Society Meeting, Anaheim, California, March 1978.

29. J. M. Fortunato and B. Ganem, *J. Org. Chem.*, **41**, 2194 (1976).

30. J. A. Gladysz, G. M. Williams, W. Tam, and D. L. Johnson, *J. Organometal. Chem.*, **140**, C1 (1977); J. A. Gladysz and J. C. Selover, *Tetrahedron Lett.*, 319 (1978); J. A. Gladysz and W. Tam, *J. Am. Chem. Soc.*, **100**, 2545 (1978); J. A. Gladysz, J. L. Hornby, and J. E. Garbe, *J. Org. Chem.*, **43**, 1204 (1978).

31. K. E. Wilson, R. T. Seidner, and S. Masamune, *Chem. Commun.*, 213 (1970).

32. E. Winterfeldt, *Synthesis*, 617 (1975).

33. H. C. Brown and B. C. Subba Rao, *J. Am. Chem. Soc.*, **80**, 5377 (1958); H. C. Brown and P. M. Weissman, *Israel J. Chem.*, **1**, 430 (1963).

34. M. Cerny and J. Malek, *Collect. Czech. Chem. Commun.*, **36**, 2394 (1971).

35. J. R. Butterick and A. M. Unrau, *Chem. Commun.*, 307 (1974).

36. C. F. Lane, unpublished results.

37. T. E. Cole and R. Pettit, *Tetrahedron Lett.*, 781 (1977).

38. G. P. Boldrini, M. Panunzio, and A. Umani-Ronchi, *Chem. Commun.*, 359 (1974) and *Synthesis*, 733 (1974); Y. Watanabe, M. Yamashita, T. Mitsudo, M. Tanaka, and Y. Takegami, *Tetrahedron Lett.*, 1879 (1974); Y. Watanabe, T. Mitsudo, S. C. Shim, and Y. Takegami, *Chem. Lett.*, 1265 (1974).

39. T. Mitsudo, Y. Watanabe, M. Yamashita, and Y. Takegami, *Chem. Lett.*, 1385 (1974).

40. H. Alper, *Tetrahedron Lett.*, 2257 (1975).

41. H. Alper, *J. Org. Chem.*, **40**, 2694 (1975); H. Alper and H.-N. Paik, *J. Org. Chem.*, **42**, 3522 (1977).

42. R. Noyori, I. Umeda, and T. Ishigami, *J. Org. Chem.*, **37**, 1542 (1972).

RADIONUCLIDE INCORPORATION *via* ORGANOBORANES

George W. Kabalka

Chemistry Department, University of Tennessee

Knoxville, Tenneessee 37919

The role of organoboranes in organic synthesis has expanded immensely since Professor Brown first reported their simple preparation *via* hydroboration.[1,2] Examination of the current literature leads to the conclusion that the expansion is far from over. Certainly the presentations at this symposium support such a conclusion.

The synthetic utility of the organoboranes is due to the fact that the boron atom is readily replaced by a wide variety of nuclides.[3-5] Generally the reactions are stereospecific with retention of configuration at the carbon from which the boron is displaced; although mercuration,[6] cyclopropane formation,[7] and certain halogenations[8] have been found to occur with inversion of configuration at the carbon originally attached to boron. Representative examples of the boron replacement reactions are presented in equations 1-4.

$$R_3B \xrightarrow{\text{"H"}} RH \tag{1}$$

$$R_3B \xrightarrow{\text{"O"}} ROH \tag{2}$$

$$R_3B \xrightarrow{\text{"M"}} RM \tag{3}$$

$$R_3B \xrightarrow{\text{"X"}} RX \tag{4}$$

This summary is, of course, not exhaustive but it does illustrate the versatility of the organoboranes. Specific examples of the metallation[9] and halogenation[8] reactions are presented in equations 5 and 6.

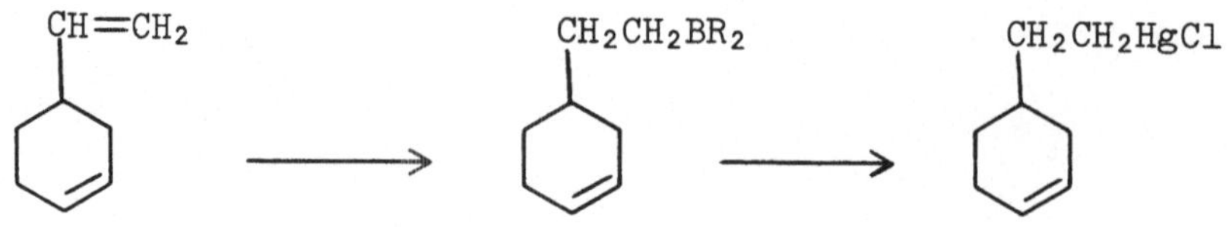

$$\text{(5)}$$

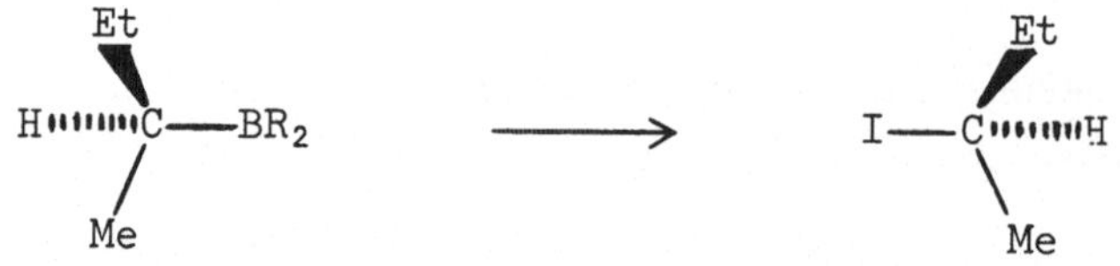

$$\text{(6)}$$

The transmetallation reaction has also been utilized to generate unstable organometallic intermediates[10,11] as illustrated in equation 7.

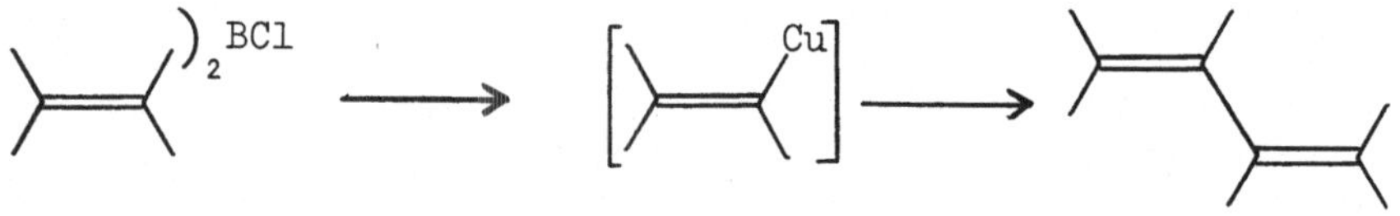

$$\text{(7)}$$

It is somewhat surprising that organoborane technology has not been applied more extensively to incorporation of less abundant nuclides. The stereospecific incorporation of deuterium <u>via</u> deuterioboration and/or solvolysis of organoboranes by acetic acid-<u>d</u> exemplifies the potential of these reactions (eqs 8 and 9).

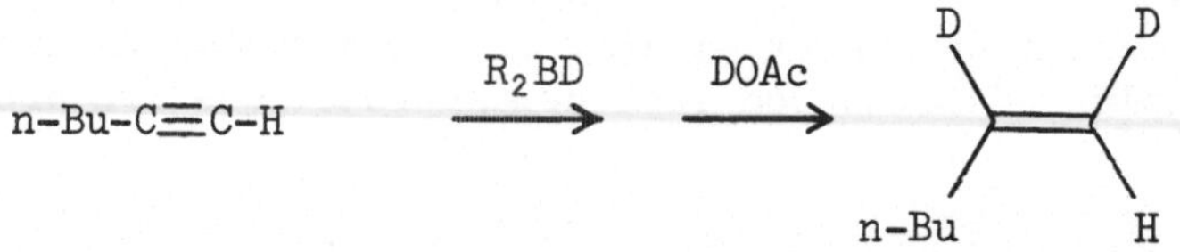

$$n\text{-Bu-C}\equiv\text{C-H} \xrightarrow{R_2BD} \xrightarrow{DOAc} \quad \text{(8)}$$

$$n\text{-Bu-C}\equiv\text{C-D} \xrightarrow{R_2BD} \xrightarrow{HOAc} \quad \text{(9)}$$

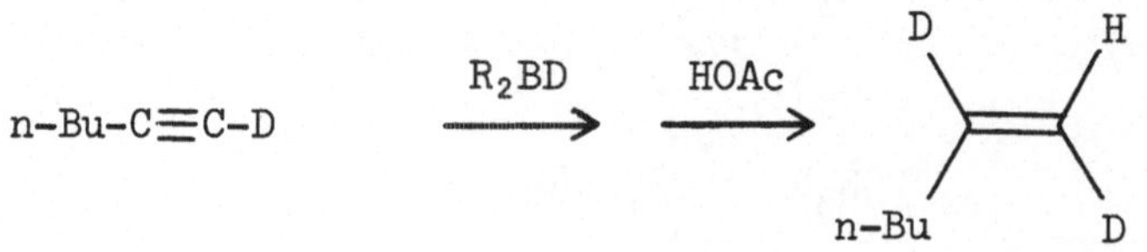

We recently demonstrated that hydroboration involves a *cis* addition of boron hydride (or boron deuteride) to the double bond in simple, acyclic alkenes by utilizing deuterium incorporation reactions.[12] Our results are summarized in equations 10 and 11.

D   D R_2BH H   BR$_2$ D R_2BD H   H
 \   / $\longrightarrow$ H   D $\longleftarrow$ \   /
  C=C   H   D   C=C   (10)
 /   \   n-Bu   /   \
n-Bu   H   D   n-Bu

threo

D   H R_2BH BR$_2$ R_2BD D   H
 \   / $\longrightarrow$ H   D $\longleftarrow$ \   /
  C=C   D   H   C=C   (11)
 /   \   n-Bu   /   \
n-Bu   D   H   n-Bu

erythro

The use of organoborane technology for the incorporation of radionuclides has not been explored. The synthesis of materials containing radionuclides has been of interest since their discovery. Labeled materials have been used extensively as tracers in biological systems[13] and as probes of reaction mechanisms.[14,15]

During the past few years, my group has been investigating the incorporation of radionuclides via organoborane technology. We are fortunate to have as collaborators Professor Clair J. Collins of Oak Ridge National Laboratory's Chemistry Division and Dr. Raymond L. Hayes of Oak Ridge Associated Universities' Medical and Health Sciences Division. Our objective is to demonstrate the utility of organoboranes in the area of radionuclide incorporation. We feel that the organoboranes offer several advantages when compared to traditional incorporation techniques. These include high yields, stereospecificity of the reactions, and the fact that a variety of functional groups are tolerated during the synthetic manipulations.

In the area of radiopharmaceuticals, we have focused on the incorporation of $K^{14}CN$ and ^{14}CO because they readily react with organoboranes. Professor Brown, and others, have developed these reactions such that tertiary, secondary and primary alcohols are readily attainable.[16] [Alternately, ketones and aldehydes can be synthesized.] The reactions are summarized in equations 12-14.

The utility of the carbonylation reaction in the synthesis of biologically interesting molecules has already been demonstrated.[17]

We are utilizing the carbonylation reaction to synthesize carbon-14 labeled 6-hydroxymethylbenzo(a)pyrene[18] for metabolic

$$R_3B \xrightarrow[\text{H}^-]{\text{CO}} \rightarrow \xrightarrow[\text{EtOH}]{\text{KOH}} RCH_2OH \qquad (12)$$

$$R_3B \xrightarrow[100°, \text{H}_2\text{O}]{\text{CO}} \xrightarrow[\text{EtOH}]{\text{KOH}} R_2CHOH \qquad (13)$$

$$R_3B \xrightarrow[100°]{\text{CO}} \xrightarrow[\text{H}_2\text{O}_2]{\text{KOH}} R_3COH \qquad (14)$$

studies (eq 15). This reaction competes effectively with the carbonation reactions of the Grignard and organolithium reagents. However, the carbonylation reaction should be applicable to substrates containing functional groups.

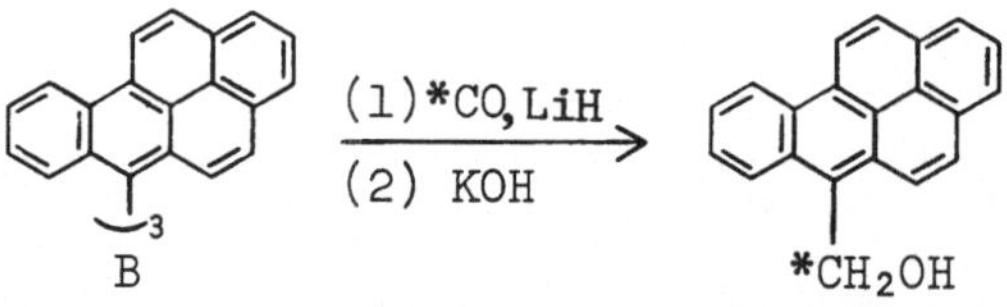

$$(15)$$

We are also examining the reaction of cyanide[19] ion with organoboranes as a route to ^{14}C-labeled estradiol. The material is to be used to test current theories concerning ablative hormone therapy.[20] A potential synthesis is presented below.

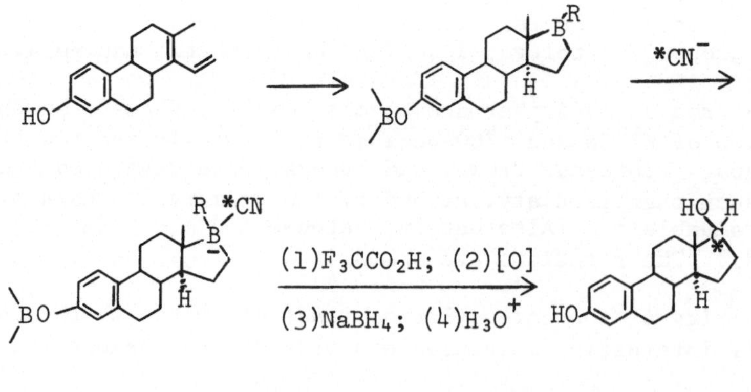

$$(16)$$

We are presently utilizing the iodination reaction to demonstrate that halogen radionuclides can be conveniently incorporated
via organoborane technology. The synthesis of ^{131}I-labeled
16-iodopalmitic acid, an effective myocardial localizer,[21] is illustrated below.

$$CH_2=CH(CH_2)_{13}CO_2K \xrightarrow{BH_3} B[CH_2(CH_2)_{14}CO_2K]_3$$

$$\xrightarrow[\text{(2) } H_3O^+]{\text{(1) } \overset{*}{I}_2, \text{ NaOCH}_3} \overset{*}{I}-CH_2(CH_2)_{14}CO_2H \tag{17}$$

Our investigations are not, however, limited to the synthesis
of radiopharmaceuticals. We are utilizing organoborane technology,
coupled with radionuclide incorporation, to probe molecular structure. For example, in collaboration with Professor Clair J.
Collins, we are developing a method to measure the number of
olefinic linkages in complex and polymeric materials such as coal.
In these materials, the mass of the carbon present in olefinic
linkages is quite small, generally amounting to less than 1% of
the total mass. These small percentages can be measured accurately
using tracer techniques since radionuclides such as ^{14}C can be precisely measured in quantities less than one part per billion.

The sequence we have developed is illustrated schematically:

$$\sim\!\!\sim\!\!CH=CH\!\!\sim\!\!\sim \xrightarrow{BH_3} \sim\!\!\sim\!\!\underset{}{CH_2}\overset{BH_2}{CH}\!\!\sim\!\!\sim$$

$$\xrightarrow{CH_2=CH_2} \sim\!\!\sim\!\!\underset{CH}{\overset{BEt_2}{|}}\!\!\sim\!\! \xrightarrow{*CO} \sim\!\!\sim\!\!\underset{CH}{\overset{Et_2\overset{*}{C}BO}{|}}\!\!\sim\!\! \tag{18}$$

$$Polymer = \sim\!\!\sim\!\!\sim$$

The concept is straightforward. Each alkene will eventually
incorporate one labeled carbon. Determination of the quantity of
labeled carbon yields the quantity of olefinic linkages originally
present. The scheme will, of course, yield the correct answer only
if few olefin linkages are present since the presence of a large
number of olefinic linkages would lead to dialkylborane formation
which would produce erroneous results. The study is still in

progress but our control experiments indicate that it is possible
to determine the quantity of olefinic linkages present in structures such as coal.

In the near future we will be applying these new techniques to
a variety of related systems.

REFERENCES

1. H. C. Brown and B. C. Subba Rao, *J. Am. Chem. Soc.*, 78, 5694
    (1956).
2. H. C. Brown, "Hydroboration," W. A. Benjamin, New York, N.Y.,
    (1962).
3. H. C. Brown, "Boranes in Organic Chemistry," Cornell University
    Press, Ithaca, N.Y. (1972).
4. H. C. Brown, "Organic Syntheses Via Boranes," Wiley-Interscience,
    New York, N.Y. (1975).
5. G. M. Cragg, "Organoboranes in Organic Synthesis," Dekker, New
    York, N.Y. (1973).
6. D. E. Bergbreiter and D. P. Rainville, *J. Am. Chem. Soc.*, 98,
    1290 (1976).
7. H. L. Goering and S. L. Trenbeath, *J. Am. Chem. Soc.*, 98, 5016
    (1976).
8. H. C. Brown, N. R. DeLue, G. W. Kabalka, and H. C. Hedgecock,
    Jr., *J. Am. Chem. Soc.*, 98, 1290 (1976).
9. R. C. Larock and H. C. Brown, *J. Am. Chem. Soc.*, 92, 2467
    (1970).
10. Y. Yamomoto, H. Yatagai, K. Maruyama, A. Sonoda, and S.
    Murahashi, *J. Am. Chem. Soc.*, 99, 5652 (1977).
11. H. C. Brown, N. C. Hébert, and C. H. Snyder, *J. Am. Chem. Soc.*,
    83, 1001 (1961).
12. G. W. Kabalka, R. J. Newton, and J. Jacobus, *J. Org. Chem.*,
    43, 1567 (1978).
13. A. P. Wolf, D. R. Christman, J. S. Fowler, and R. M. Lambrecht,
    "Radiopharmaceuticals and Labelled Compounds," Vol. 1, IAEA,
    Vienna (1973).
14. C. J. Collins and N. S. Bowman, "Isotope Effects in Chemical
    Reactions," ACS Monograph 167, Van Nostrand Reinhold Co.,
    New York, N.Y. (1970).
15. E. Buncel and C. C. Lee, "Isotopes in Organic Chemistry,"
    Vol. I, Elsevier, Amsterdam, Netherlands (1975).
16. (a) H. C. Brown, *Acc. Chem. Res.*, 2, 65 (1969); (b) A. Pelter,
    M. G. Hutchings, and K. Smith, *Chem. Commun.*, 1048 (1971).
17. E. Negishi, M. Sebanski, J. J. Katz, and H. C. Brown,
    Tetrahedron, 32, 925 (1976).
18. R. Freudenthal and P. W. Jones, "Carcinogenesis," Vol. 1,
    Raven Press, New York, N.Y. (1971).
19. A. Pelter, M. G. Hutchings, and K. Smith, *Chem. Commun.*, 1529
    (1970).

20. K. M. J. Menon and J. R. Reel, "Steroid Hormone Action and
     Cancer," Chapters 1 and 9, Plenum Press, New York, N.Y.
     (1976).
21. G. D. Robinson and F. W. Zielinski, *J. Labelled Compd. Radio-
     pharm.*, XIII, 220 (1977).

ASYMMETRIC SYNTHESIS *via* BORANES: CHIRAL ALLENIC BORANES AND

TRIALKYLBORANE REDUCING AGENTS

M. Mark Midland

Department of Chemistry, University of California

Riverside, CA 92521

Our research has been involved in the development and application of reactions of organoboranes, especially those which lead to chiral products. Organoboranes are well known for their ability to transform the alkyl group into various products with complete retention of configuration of the boron-carbon bond (eq 1).[1] Likewise, chiral boranes, such as diisopinocampheylborane, may be used to induce optical activity during the hydroboration process (eq 2).[2]

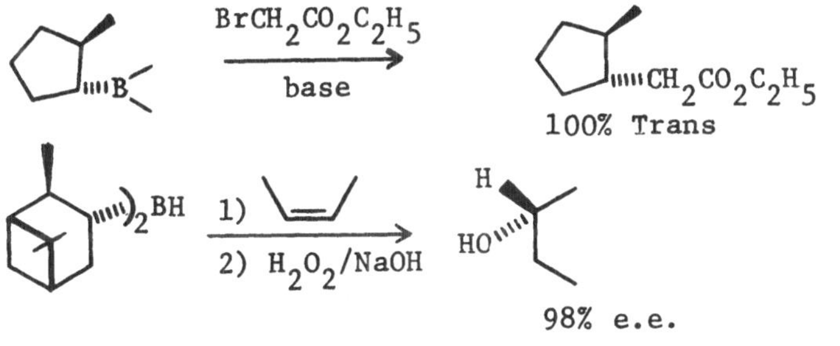

In this article we will discuss two other methods of introducing chirality into organic compounds. The first involves the tranformation of propargyl acetates into allenes or acetylenes. The second is the use of trialkylboranes as chemo- and enantio-selective reducing agents.

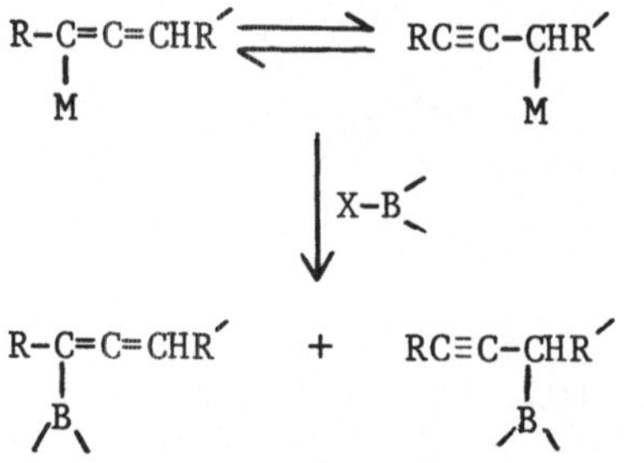

$$R-C=C=CHR' \rightleftharpoons RC\equiv C-CHR'$$
$$\underset{M}{|} \qquad\qquad \underset{M}{|}$$

$$\downarrow X-B\diagdown$$

$$R-C=C=CHR' \quad + \quad RC\equiv C-CHR'$$
$$\underset{B\diagdown}{|} \qquad\qquad\qquad \underset{B\diagdown}{|}$$

Preparation of allenic boranes *via* reactive organometallics

SCHEME I

THE CONVERSION OF PROPARGYL ACETATES INTO ALLENES AND ACETYLENES

Allenic organoboranes are potentially valuable intermediates in organic synthesis. They possess both a vinylic and an allylic portion and, thus, could react as either kind of organoborane. Conventional methods for making allenic boranes have used allenic lithium or Grignard reagents (Scheme I).[3] This route has the disadvantage that reactive functional groups must not be present. Furthermore the allenic lithium or Grignard reagent may undergo an allenic-propargylic rearrangement and thus give a mixture of products.

Zweifel has developed a route to allenic boranes which circumvents the problem of the reactive organometallics (Scheme II).[4] He treats propargyl chloride with methyllithium and then adds a trialkylborane to the resulting lithium propargyl chloride. The resulting "ate" complex rearranges to an allenic borane which may be protonated with acetic acid to give a terminal allene. The alkyl group migrates with retention of configuration.

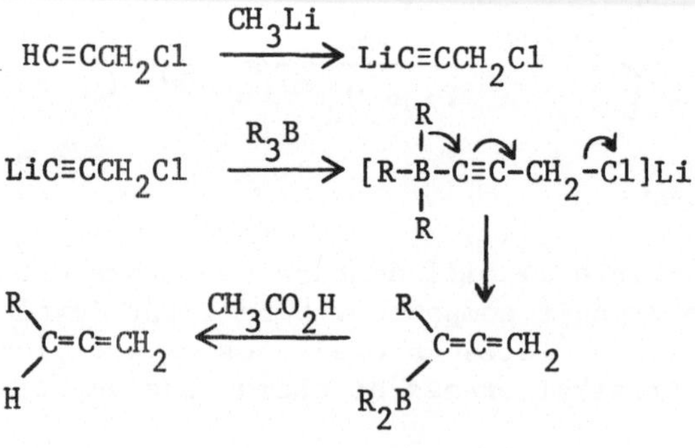

$$HC\equiv CCH_2Cl \xrightarrow{CH_3Li} LiC\equiv CCH_2Cl$$

$$LiC\equiv CCH_2Cl \xrightarrow{R_3B} [R-B-C\equiv C-CH_2-Cl]Li$$

$$\underset{H}{\overset{R}{\diagdown}}C=C=CH_2 \xleftarrow{CH_3CO_2H} \underset{R_2B}{\overset{R}{\diagdown}}C=C=CH_2$$

Zweifel's route to allenes

SCHEME II

The Zweifel method is an attractive route to allenes but is limited in scope by the availability of a variety of the required propargyl chlorides.

A. Preparation of Allenes from Propargyl Acetates

A few years ago we were studying the chemistry of lithium alkynyltrialkylborates. We felt that if we could indirectly make an alkynyl "ate" complex containing an appropriate γ-leaving group, we could develop a general route to allenic boranes. Our approach was to prepare the parent ethynyl "ate" complex[5] (Scheme III), convert it into the dilithium species,[6] add the ethynyl group to a ketone and then convert the resulting alkoxide into an acetate.

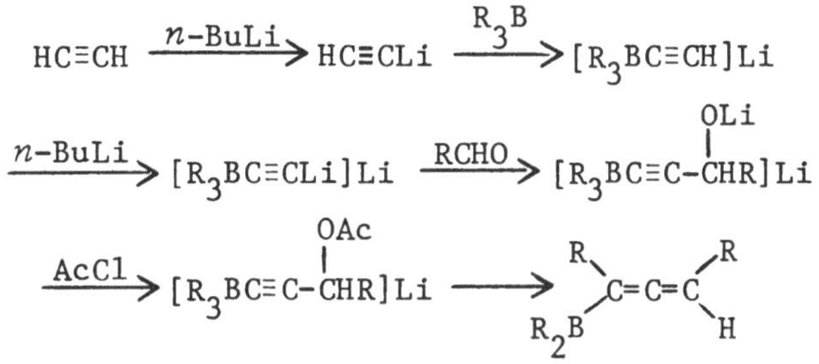

$$HC\equiv CH \xrightarrow{\ n\text{-BuLi}\ } HC\equiv CLi \xrightarrow{\ R_3B\ } [R_3BC\equiv CH]Li$$

$$\xrightarrow{\ n\text{-BuLi}\ } [R_3BC\equiv CLi]Li \xrightarrow{\ RCHO\ } [R_3BC\equiv C-\overset{\overset{\displaystyle OLi}{|}}{C}HR]Li$$

$$\xrightarrow{\ AcCl\ } [R_3BC\equiv C-\overset{\overset{\displaystyle OAc}{|}}{C}HR]Li \longrightarrow \underset{R_2B}{\overset{R}{>}}C=C=\underset{H}{\overset{R}{<}}$$

Ethynyl "ate" complex route to allenic boranes

SCHEME III

This approach worked quite well for aldehydes and gave the allene after acetic acid protonation in 80% yield.[7] However, the reaction was not general in that ketones failed to give allenic product. We suspect that the acetyl chloride was reacting with the acetylenic group in preference to the tertiary alkoxide in a manner analogous to that observed in the parent trialkylalkynylborate system (eq 3).[8] We next decided to attempt a direct route for the preparation of the required propargyl acetate-borane "ate" complex. A variety of propargyl acetates is readily available from the reaction of monolithium acetylide with ketones or aldehydes[9] followed by acetylation. We soon found that the desired lithium salt was readily formed without interference by the acetate group by reaction of the

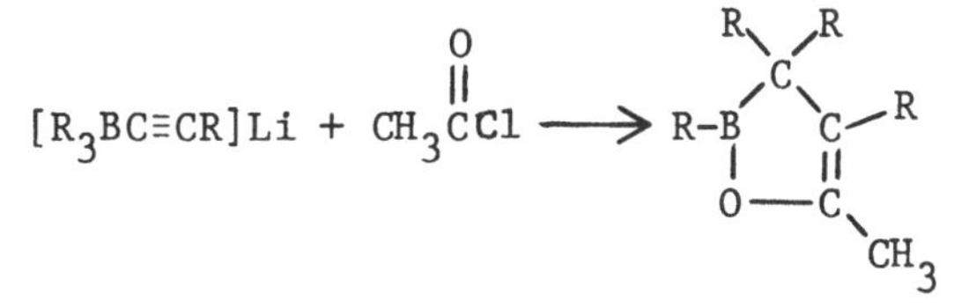

$$[R_3BC\equiv CR]Li + CH_3\overset{\overset{\displaystyle O}{\|}}{C}Cl \longrightarrow R-B\overset{\displaystyle \underset{R}{\overset{R}{>}}C}{\underset{O-C}{\overset{|}{}}}\overset{C-R}{\underset{\diagdown CH_3}{\overset{\|}{}}} \qquad (3)$$

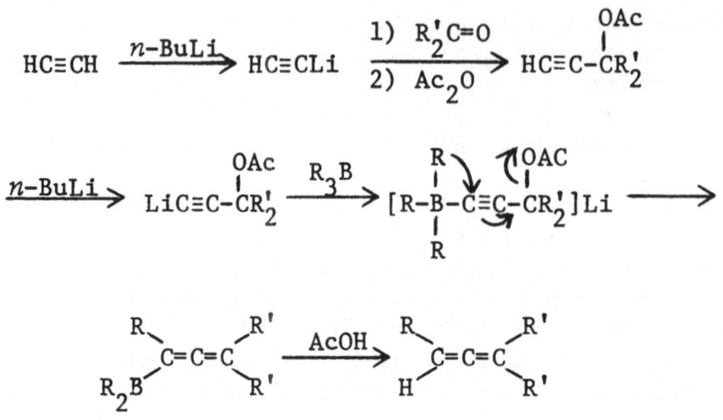

$$HC\equiv CH \xrightarrow{\textit{n}-BuLi} HC\equiv CLi \xrightarrow[\text{2) } Ac_2O]{\text{1) } R_2'C=O} HC\equiv C-\overset{\overset{\textstyle OAc}{|}}{C}R_2'$$

$$\xrightarrow{\textit{n}-BuLi} LiC\equiv C-\overset{\overset{\textstyle OAc}{|}}{C}R_2' \xrightarrow{R_3B} [R-\overset{\overset{\textstyle R}{|}}{\underset{\underset{\textstyle R}{|}}{B}}-C\equiv C-\overset{\overset{\textstyle OAC}{|}}{C}R_2']Li \longrightarrow$$

$$\underset{R_2B}{\overset{R}{\diagdown}}C=C=C\underset{R'}{\overset{R'}{\diagup}} \xrightarrow{AcOH} \underset{H}{\overset{R}{\diagdown}}C=C=C\underset{R'}{\overset{R'}{\diagup}}$$

The propargyl acetate route to allenes

SCHEME IV

propargyl acetate with $\textit{n}$-butyllithium at -78° or -120°. Addition of a trialkylborane, followed by warming to room temperature and then protonation with acetic acid gives an allene in high yield[10] (Scheme IV).

The reaction is quite general both with respect to the organo-borane and the propargyl acetate (eq 4); it is equivalent to Crabbé's organocuprate route to allenes.[11] However, the cuprate must be pre-pared from a reactive organometallic, which precludes the use of many functional groups. The borane route easily accommodates func-tional groups in the organoborane (*e.g.*, eq 4).

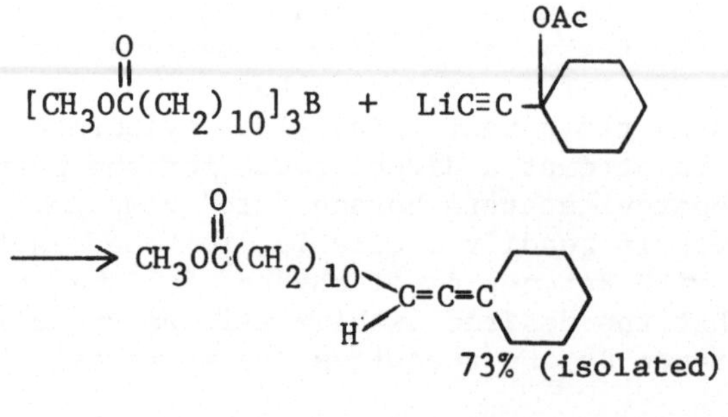

$$[CH_3\overset{\overset{\textstyle O}{\|}}{O}C(CH_2)_{10}]_3B \;+\; LiC\equiv C-\underset{\text{(cyclohexyl with OAc)}}{}$$

$$\longrightarrow CH_3\overset{\overset{\textstyle O}{\|}}{O}C(CH_2)_{10}\underset{H}{\overset{}{\diagdown}}C=C=C\text{(cyclohexyl)}$$

$$\text{73\% (isolated)}$$

(4)

The alkyl group presumably migrates with retention of configur-ation, as Zweifel has observed.[4] Trialkylboranes are well-known for the transformation of the alkyl group into a variety of products with retention of configuration.[1] These reactions are believed to occur through a rearrangement of an organoborate anion in which one of the groups contains a suitable leaving group.

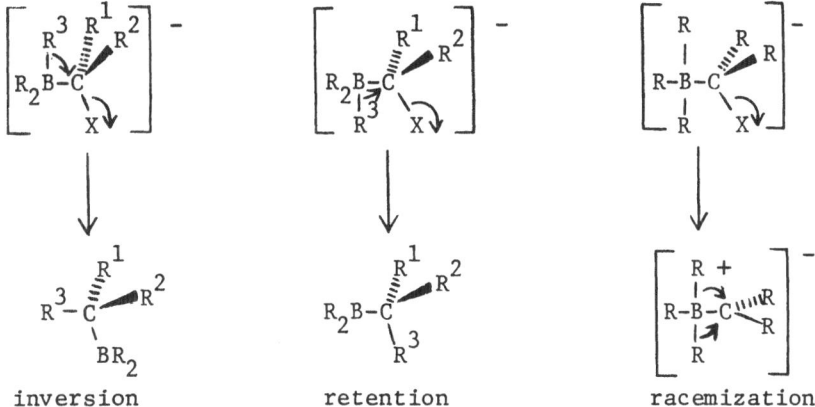

Stereochemistry of the boron ⟶ carbon rearrangement

SCHEME V

Very little is known about the stereochemistry of the carbon containing the leaving group. The rearrangement could occur simultaneously with the X group leaving, giving inversion (S_N2 type) or
retention of configuration. Alternatively, the rearrangement could
occur with prior ionization of the leaving group (S_N1 type) and
subsequent loss of stereochemistry (Scheme V). Pasto has reported
that hydride migration from boron proceeds with inversion of configuration,[12] while Brown and Yamamoto have found that methyl migration
occurs with retention of configuration.[13]

The mechanism of the allenic borane-forming reactions looks
like an S_N2' reaction. Since optically active propargyl acetates
are readily available, the transformation represents a good method
for monitoring the stereochemistry of the rearrangement with respect
to the leaving group.[14] Treatment of (R)-(+)-1-octyn-3-ol acetate[15]
in tetrahydrofuran with n-butyllithium, followed by tri-n-butylbor-
ane, warming to room temperature and protonation with acetic acid
gave the allene. The 5,6-dodecadiene had a positive rotation and
was assigned the S configuration on the basis of the Brewster-Lowe
rule.[16] The alkyl group thus migrates preferentially *anti* to the
acetate leaving group (Scheme VI). Of course, we must assume that
the allenic borane resembles a vinyl-borane and that, as such, it
would be protonated with retention of configuration.[17]

We estimate that the optical purity of the allene is 23%.[18]
Crabbé has obtained similar optical purities *via* the organocuprate
route.[19] Attempts to increase the rotation by changing the leaving
group have so far failed. However, the optical purity of the product is affected by the solvent. For example 50/50 dimethoxyethane/
tetrahydrofuran gives only 15% e.e., while ethyl ether gives 40%

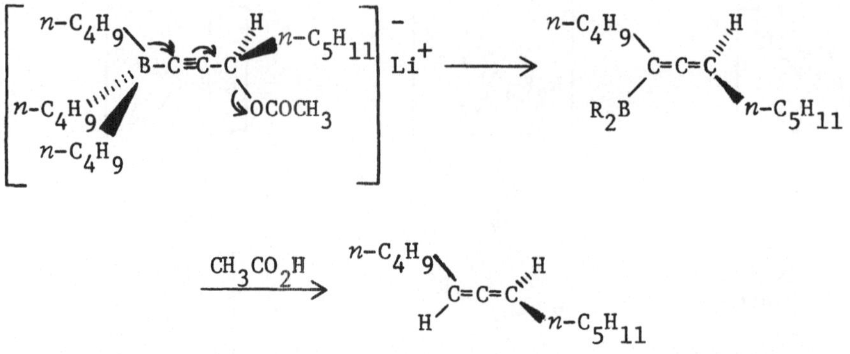

Configuration of the allene formed from an organoborane
and 1-octyn-3-ol acetate

SCHEME VI

e.e., and toluene gives 45% e.e. In the less-coordinating solvent
the lithium ion may be coordinating with the acetate and assisting
in its leaving. Presumably the rearrangement can then occur at a
lower temperature and with a higher selectivity.

The allenic borane appears to be configurationally unstable at
room temperature in tetrahydrofuran. A sample stirred overnight
gave allene of only 3% optical purity. Allenic boranes resemble
allylic boranes and, as such, could undergo a 1,3-boron shift.[20]
However, this shift would not account for the loss of configuration
of the allene (see Scheme VII). Another possibility for loss of
configuration may be that some impurity is causing ionic isomerization or that the boron is lowering the energy barrier for rotation
around the double bonds of the allene.

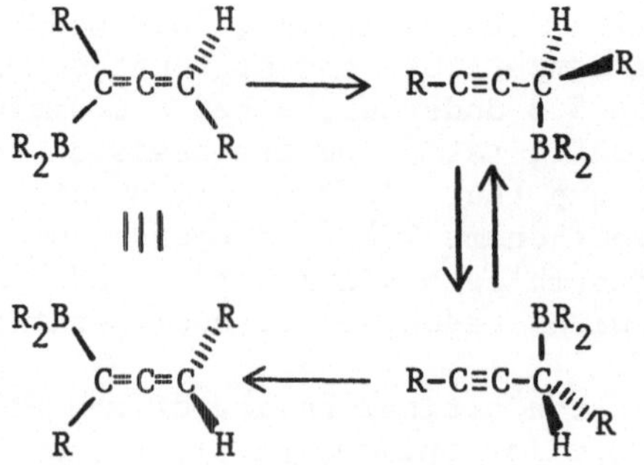

1,3-Shifts of allenic boranes with retention of chirality

SCHEME VII

Chiral allenes are rather hard to obtain by synthesis or resolution. We hope to develop this method into a general procedure
for making optically active allenes.

B. Preparation of Alkynes from Propargyl Acetates

As mentioned above, the allenic boranes may resemble allylic
boranes and, as such, would be expected to undergo reactions typical
of allylic boranes. For example, allylic boranes are readily protonated with water to give rearranged alkenes. Addition of water
to an allenic borane results in rapid protonation with exclusive
formation of an acetylene (eq 5).[10] As with the allylic boranes,
protonation occurs with rearrangement. We can account for the di-
chotomy in protonation with acetic acid and water by looking at the
proposed transition states for these two reactions (Scheme VIII).
With acetic acid, the carbonyl oxygen may participate and give a

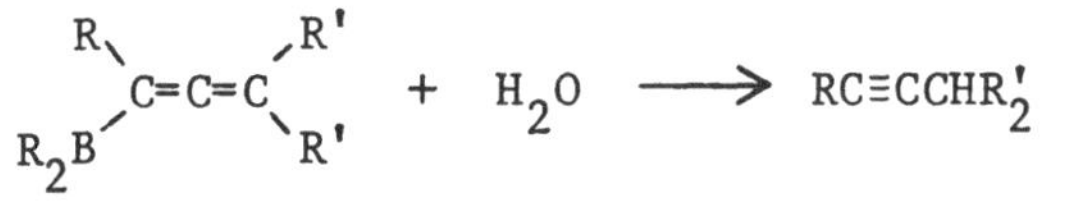

$$
\begin{array}{c}
R \\
\diagdown \\
R_2B
\end{array}
C{=}C{=}C
\begin{array}{c}
R' \\
\diagup \\
\diagdown R'
\end{array}
\quad + \quad H_2O \quad \longrightarrow \quad RC{\equiv}CCHR'_2 \qquad (5)
$$

six-centered transition state as the proton adds to the α-carbon.[22]
With water, a six-centered transition state gives addition of the
proton to the γ-carbon.

Once again, the reaction is quite general with respect to the

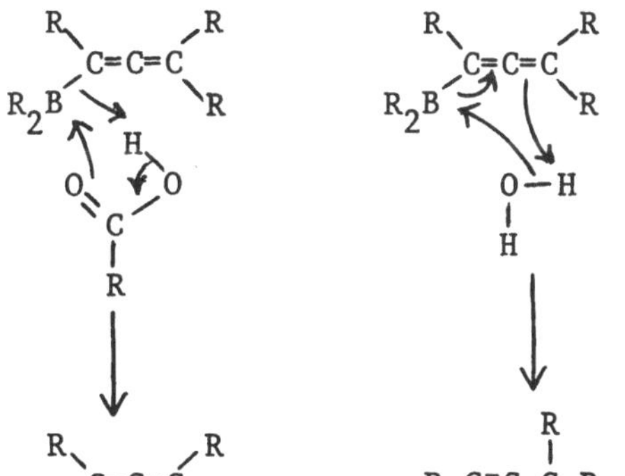

Protonation of allenic boranes with acetic acid and water

SCHEME VIII

trialkylborane and ethynylalkanol acetate. The net transformation is quite remarkable in that an alkyne is added to a carbonyl compound and a hydrocarbon product is produced (eq 6).

$$R_3B \ + \ HC{\equiv}CH \ + \ O{=}CR_2' \longrightarrow RC{\equiv}CCR_2' \qquad (6)$$
$$\underset{H}{\mid}$$

The reaction is useful in making 1,4-enynes (eq 7).

$$\overset{\displaystyle OAc}{\underset{\displaystyle \mid}{}}$$
$$RCH{=}CHCHO \longrightarrow RCH{=}CHCHC{\equiv}CH \longrightarrow RCH{=}CHCH_2C{\equiv}CR' \qquad (7)$$

These in turn could be converted into 1,4-dienes, such as **are** found in fatty acids. The stereochemistry of each double bond could be determined by the choice of the starting α,β-unsaturated aldehyde and the mode of reduction of the alkyne.

A closer look at the proposed transition state for protonation by water shows that a chiral allenic borane should give an alkyne with a chiral hydrocarbon center, since the proton should enter the allene on the same side as the borane (Scheme IX). In order to test

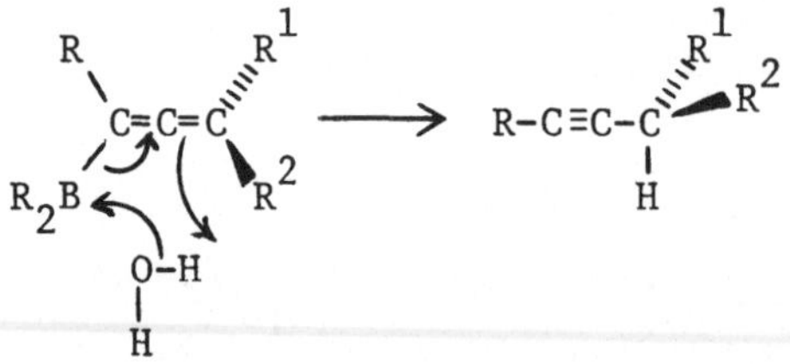

Proposed stereochemistry of the water protonation of allenic borane

SCHEME IX

this idea, we have investigated the possibility of attaching the cholesterol side chain to a steroid nucleus. Pregnenolone acetate was converted to the 20-ethynyl diacetate. Addition of acetylene to pregnenolone has been shown to obey Cram's rule and give the (20S)-ethynylpregn-5-ene-3β,20-diol.[23] The stereochemistry of the propargyl alcohol is such that if the rearrangement of the organoborate were to occur in an *anti* manner, as observed with 1-octyn-3-ol, then the boron should end up on the top side of the molecule. Boron in this arrangement would then be expected to direct protonation at

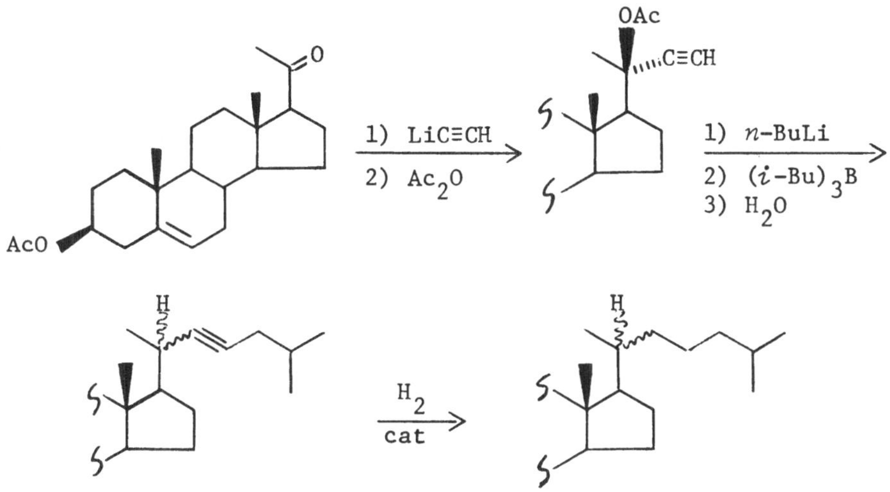

Stereochemistry of the allenic borane ⟶ acetylene transformation.

SCHEME X

C-20 from the top side and this would result in the natural configuration at C-20.

Sequential treatment of the ethynyl diacetate with n-butyllith-
ium, trisobutylborane and water gave an acetylene. (Treatment with
acetic acid produced an allenic steroid.) The acetylene appears to
be a 2:1 mixture of epimers at C-20 (Scheme X). Reduction of the
acetylene was found to give a 2:1 mixture of cholesterol and iso-
cholesterol acetate.

The pregnenolone → cholesterol route does not appear to be an
efficient method for controlling the stereochemistry of the side-
chain. However, the acetylene may be put together in two ways.
$\Delta^{17(20)}$-Pregnene-3β-ol may be hydroborated, to produce the correct
geometry at C-20,[24] and the borane converted into the acetylene with
retention of configuration. We are currently exploring this route
for building up the side chain.

TRIALKYLBORANE REDUCING AGENTS

Trialkylboranes are noted for their tolerance of a variety of
functional groups. It is thus very surprising that trialkylboranes
can be made to react with aldehydes at room temperatures.[25] Mikhailov has reported that trialkylboranes will reduce benzaldehyde at
elevated temperatures (eq 8).[26] Jerry Buhler studied the reaction
in refluxing THF for his Ph.D. thesis.[27] The reaction was slow,

$$(n-C_4H_9)_3B \ + \ C_6H_5CHO \ \xrightarrow[\text{neat}]{100-150°}$$

(8)

$$(n-C_4H_9)_2BOCH_2C_6H_5 \ + \ CH_2=CHC_2H_5$$

requiring 24 hours for completion with tri-n-butylborane. The use of a less sterically demanding borane, B-n-butyl-9-BBN, still gave a slow reaction and required 20 hours for completion. The reaction seemed to have promise as a very chemoselective reducing agent since ketones were found to be reduced even more slowly. However, reduction of aldehydes at 100–150°, or over a 24 hour period, is not very elegant! We felt that if the reaction could be carried out at room temperature in a few minutes then it would be synthetically useful.

A. Reduction of Aldehydes

Since we have recently published a full paper describing factors which contribute to a high rate of reduction,[28] I will summarize our results, and concentrate on the use of trialkylboranes for asymmetric reductions.

The rate of reduction is dramatically affected by substituents in the β- position of the B-alkyl group of 9-BBN. A tertiary β-hydrogen is required for fast reaction (Table I). The ability to form a *syn*-planar B-C-C-H configuration is also very important. Thus, cyclopentyl derivatives are reduced much faster than are cyclohexyl derivatives. The necessity of a planar B-C-C-H arrangement also accounts for the lack of reactivity of the cyclooctyl portion of 9-BBN.

The high degree of reactivity of tertiary β-hydrogens, as compared with secondary or primary, is also reflected in the regiochemistry of the displaced olefin. Thus 1,2-dimethylcyclopentyl-9-BBN gives only 1,2-dimethylcyclopentene, even though a highly reactive secondary cyclopentyl hydrogen atom is present (eq 9).

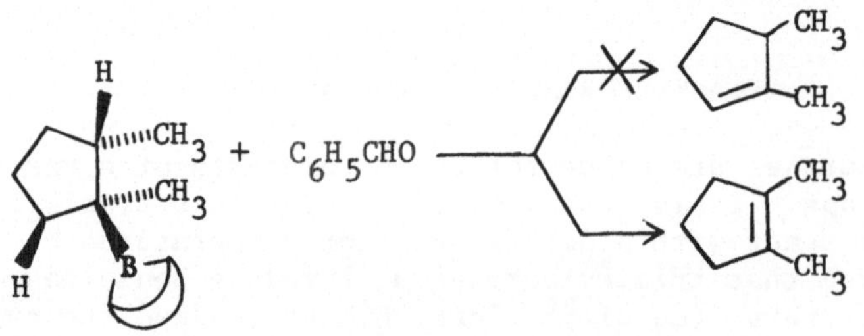

(9)

TABLE I

The Reduction of Benzaldehyde by
Various B-Alkyl-9-BBN Compounds

Alkyl group[a]	$t_{\frac{1}{2}}$(min)[b]
Ethyl	5500
n-Octyl	116
2-phenylethyl	98
sec-Butyl	80
Isobutyl	21
3-Methyl-2-butyl	11
2,3-Dimethyl-2-butyl	7000
Cyclopentyl	15
exo-Norbornyl	20
Cyclohexyl	420
Cyclooctyl	_c
trans-2-Methylcyclopentyl	_c
trans-2-Methylcyclohexyl	67
4-Isocaranyl	_c
1,2-Dimethylcyclopentyl	1000

[a] 0.5 *M* B-Alkyl-9-BBN in THF at 65°
[b] Time for 50% completion of the reaction
[c] The reaction was too fast to measure accurately.

TABLE II

The Reduction of Ketones with B-Alkyl-9-BBN

Alkyl-9-BBN[a]	Ketone[a]	$t_{\frac{1}{2}}$(min)[b]
3-Methyl-2-butyl	Cyclohexanone	1800
3-Methyl-2-butyl	Cyclopentanone	1200
3-Methyl-2-butyl	Acetophenone	1275
Isobutyl	Acetophenone	4000
trans-2-Methylcyclopentyl	Acetophenone	4180
Cyclopentyl	Acetophenone	_c
n-Butyl	Acetophenone	_c

[a] 0.5 M in THF at 65°
[b] Time for 50% completion of the reaction.
[c] The reaction was very slow.

The rate of reduction is also affected by substituents on benz-aldehyde. The general order of reactivity is p-NO$_2$>p-Cl>p-H>p-OCH$_3$>p-N(CH$_3$)$_2$. Ketones are reduced much more slowly than are aldehydes (Table II). For example, 3-methyl-2-butyl-9-BBN reduces benzalde-hyde to the extent of 50% in 11 <u>minutes</u> but requires nearly 22 <u>hours</u> to give 50% reduction of acetophenone. In fact, all ketones tested, including unhindered cyclohexanones, are reduced at least 100-200 times slower than are aldehydes.

The 3-methyl-2-butyl-9-BBN is a very effective reagent for selective reduction of aldehydes in the presence of ketones.[29] For example, competition between benzaldehyde and acetophenone for a single equivalent of 3-methyl-2-butyl-9-BBN resulted in greater than 95% reduction of the aldehyde in 2 hours, with no detectable reduc-tion of the ketone. There have been numerous reports of reagents which will selectively reduce aldehydes in the presence of ketones.[30] However, only two of these, diisopropyl-carbinol on alumina[30g] and 9-BBN-pyridine[30h] are able to reduce an aldehyde in the presence of an unhindered cyclohexanone.

Mikhailov proposed[26] that the reduction occurs *via* a cyclic transition state similar to that for the Meerwein-Pondorf-Verley reduction (eq 10). Our observations support Mikhailov's mechanism,

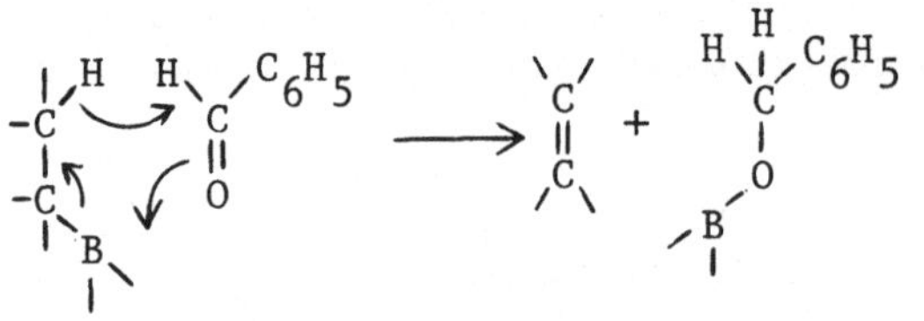

$$\tag{10}$$

as opposed to a dehydroboration-reduction process (eq 11). Thus the rate is dependent on the structure of the carbonyl compound and is

$$R_3B \xrightarrow[\text{slow}]{\Delta} R_2BH + \text{olefin} \xrightarrow[\text{fast}]{C_6H_5CHO} R_2BOCH_2C_6H_5 \tag{11}$$

affected by substituents on benzaldehyde. The reduction can be carried out under conditions in which dehydroboration cannot be important, *i.e.*, room temperature for a few minutes. Finally, the reaction obeys second-order kinetics, as required for the bimolecular process.

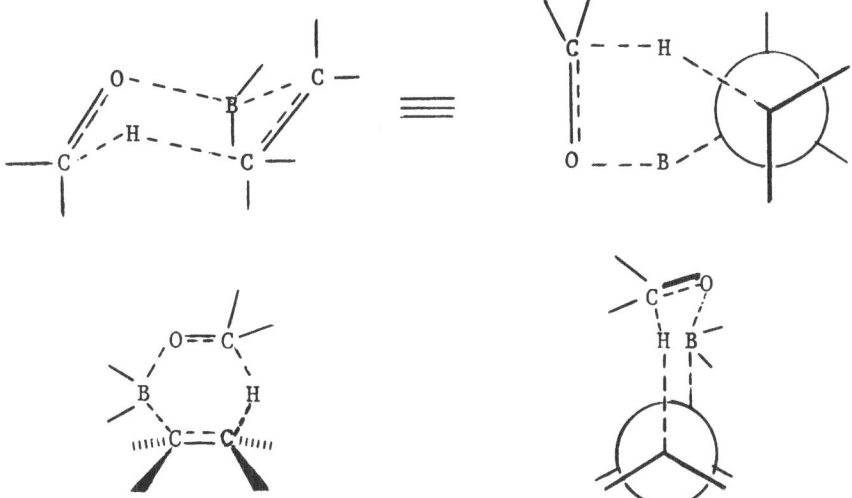

Figure 1. Transition state extremes for the organoborane reduction
of aldehydes

One may envision two extremes for the transition state of the
reduction. The first is chair-like. Such a transition state has
been proposed to account for certain stereoselectivities in the
reduction of alkyl phenyl ketones by Grignard and other reagents.[31]
The other extreme would have a planar arrangement (Fig. 1).

Models indicate that for the rigid bicyclo[3.3.1]nonane ring
to react, only the chair-like transition state is possible. The
B-alkyl group, on the other hand, is free to rotate and assume the
planar structure. Such a situation would allow maximum overlap of
orbitals in the developing π system of the displaced olefin. The
ability to form a planar arrangement of the B-C-C-H bonds plays an
important role, at least in the initial stages, in determining the
rate of reduction.

B. Chiral Organoborane Reducing Agents for Aldehydes

Optically-active terpenes, such as β-pinene, α-pinene, camphene or carene, may be hydroborated with 9-BBN to produce chiral
organoboranes of well-defined structure. These reagents (Fig. 2)
have the reducing "hydride" incorporated into a chiral center and
thus should be effective in inducing optical activity into the reduced product.

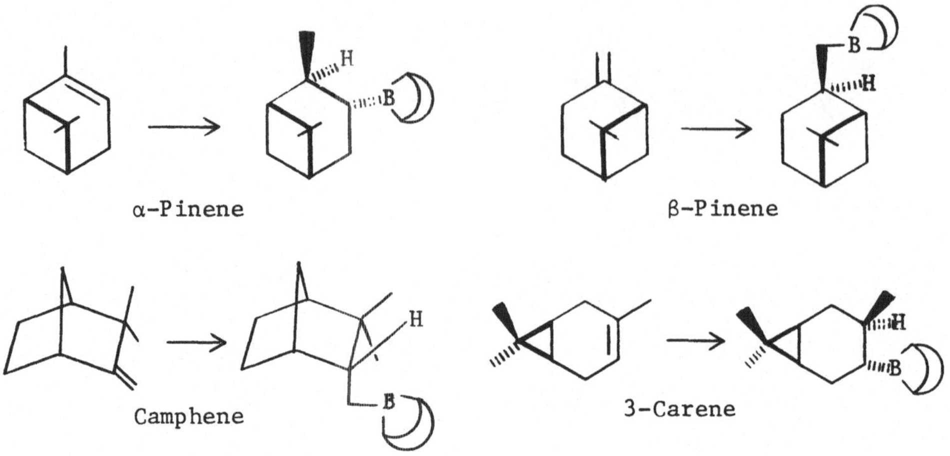

α-Pinene   β-Pinene

Camphene   3-Carene

Figure 2. Chiral organoborane reducing agents

The organoboranes are effective reducing agents for aldehydes so we decided to test the feasibility of preparing optically active primary alcohols using these reagents. The optically active α-deuterio primary alcohols are important substrates for studying the mechanism of chemical and biochemical systems.[32] Because they have low rotations one would like to obtain these compounds in high optical purity. The problem is usually solved by resorting to enzyme systems.[32d,33] Although high optical purity is obtained, the process is tedious at best and not amenable to large-scale preparation.

Treatment of benzaldehyde-α-d with the chiral organoborane reagents results in rapid reduction (usually complete within 30 minutes at room temperature!) and the formation of chiral benzyl α-d alcohol. The borane from β-pinene gives benzyl alcohol with 47% enantiomeric excess (eq 12). This result is quite good in comparison to other chiral reducing agents. For example, the excellent asymmetric reducing agent, $LiAlH_4$/4-dimethylamino-3-methyl-1,2-diphenyl-2-butanol, developed by Mosher gives only 40% e.e.[34]

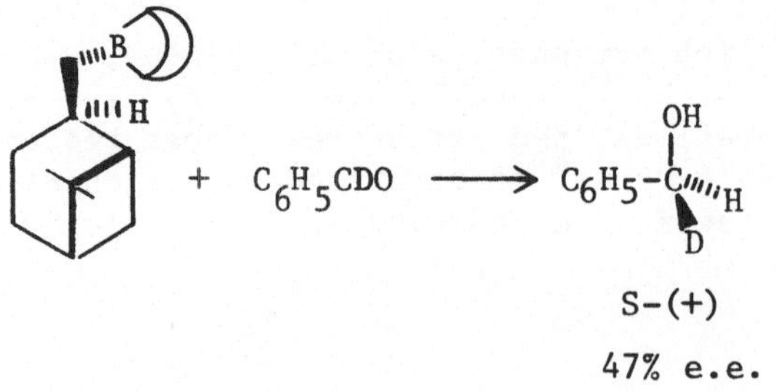

$$+ \quad C_6H_5CDO \longrightarrow C_6H_5-\overset{OH}{\underset{D}{\overset{|}{C}}}{}_{\prime\prime\prime\prime}H \qquad (12)$$

S-(+)

47% e.e.

$$C_6H_5CDO \longrightarrow C_6H_5-\overset{\overset{\displaystyle OH}{|}}{\underset{\underset{\displaystyle D}{|}}{C}}{}_{''''}H \qquad (13)$$

S-(+)
100% e.e.

Using the organoborane from (+)-α-pinene we have achieved asymmetric reductions which approach, or are as good as those obtained with the enzyme systems (eq 13). In fact, using optically pure α-pinene, we cannot detect the minor component of benzyl-α-<u>d</u> alcohol using chiral NMR shift reagents.[34]

The reaction produces S-(+)-benzyl-α-<u>d</u> alcohol. The enantiomer could, in principle, be prepared from (-)-α-pinene; it is conveniently prepared using a deuterated reducing agent prepared from (+)-α-pinene and <u>B</u>-deuterio-9-BBN. This route not only gives a benzyl-α-<u>d</u> alcohol in high enantiomeric purity (81%; 90% when corrected for deuterium incorporation, and 98% when corrected for pinene purity) but also eliminates the need to prepare benzaldehyde-α-<u>d</u>.

The deuterated reducing agents allow one conveniently to reduce a variety of aldehydes (Table III). The reaction gives high enantiomeric purity in all cases studied. Steric effects seem to have little effect; pivalaldehyde gives very good results. Electronic effects do seem to be important, and will be discussed later.

The absolute configuration, as determined by use of NMR shift reagents,[34] is consistently R. Normally one thinks of transition states for asymmetric reductions as involving a dovetailing of a small group and a large group on the reducing agent with a large group and a small group on the ketone[35] (Fig. 3). Steric interactions

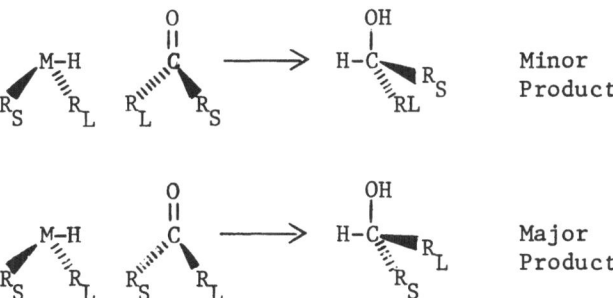

Figure 3. Asymmetric reduction of a ketone by a chiral metal hydride.

TABLE III

The Reduction of Aldehydes with $\underline{B}$-3α-Pinanyl-9-BBN

Product	% e.e.[a]	Corrected for %D[b]	Corrected for % e.e. of α-pinene[c]
$CH_3CH_2CH_2CHDOH$	83	101	101[d]
$CH_3(CH_2)_4CHDOH$	64	83	89
$(CH_3)_3CCHDOH$	70	91	98
$C_6H_5CH=CHCHDOH$	60	78	84
$(CH_3)_2C=CHCH_2CH_2CCH_3=CHCHDOH$	58	75	81
C_6H_5CHDOH	66	86	93
$p-ClC_6H_4CHDOH$	87	94	101
$p-O_2NC_6H_4CHDOH$	86	93	100
$p-CH_3C_6H_4CHDOH$	76	83	89
$p-CH_3OC_6H_4CHDOH$	70	76	82
$p-(CH_3)_2NC_6H_4CHDOH$	61	66	71

[a]As observed by measuring the area of the carbinyl proton signals in the presence of $(Eu(hfc)_3$. [b]Entry 1 is corrected for 90% deuteration, entries 2-5 for 87% deuteration, and entries 7-11 for 96% deuteration. The difference in % deuteration represents different samples of 9-BBN-9-d. [c]The (+)-$\underline{\alpha}$-pinene was 93% e.e.. [d]The pinene was 100% e.e..

are presumably minimized by the interaction of the large groups with the small groups. This leads to a lower energy pathway than when large groups must interact.

The stereochemical approach observed in the 3-pinanyl-9-BBN reduction seems, at first, to contradict this simple explanation for the direction of asymmetric reduction. The large R group approaches over the pinanyl ring and the hydrogen is over the methyl (Fig. 4). However, models suggest that the R group is situated over a well in the pinane framework which resembles the pocket of an enzyme. If the R group were to approach from the opposite side, then the freely rotating methyl group would interfere because the methyl hydrogens would interact with the R group. However, it is hard to believe that even a *tert*-butyl group will approach the borane in this manner. An alternative explanation may be needed.

Earlier we discussed the importance of the ability of the organoborane to form a planar B-C-C-H arrangement. The pinanylborane readily forms such a planar structure and the rate of benzaldehyde reduction is very rapid. Perhaps this requirement is important in

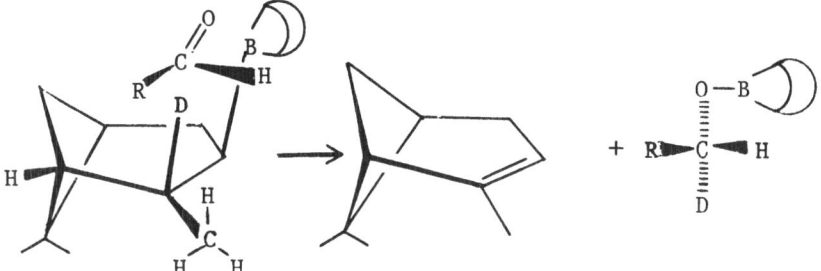

Figure 4. Asymmetric reduction of aldehydes by 3-pinanyl-9-BBN

the early stages of the reaction. As the reaction proceeds, the
carbonyl carbon would begin to take on a tetrahedral hybridization
and the arrangement of atoms may begin to look like the chair
transition state. If the aldehyde has approached the pinanylborane
in the "correct" manner, the R group will assume an equatorial position (Fig. 5). A chair-chair interconversion would give an axial
R group and should not be favored. If the R group approaches the
pinanylborane from the "wrong" side, the structure on the left in
Figure 5 would have an axial R group. Chair-chair interconversion
would give an equatorial group as depicted on the right. However,
this chair form has severe interactions between the pinanyl ring
and the cyclooctyl ring of 9-BBN.

Our thoughts on the transition state are mere speculation at
this time and may be modified with additional data. However, regardless of the precise structure for the transition state, one thing is
clear. The optical purities of the product are consistently high.
Furthermore, the aldehyde consistently approaches the pinanylborane

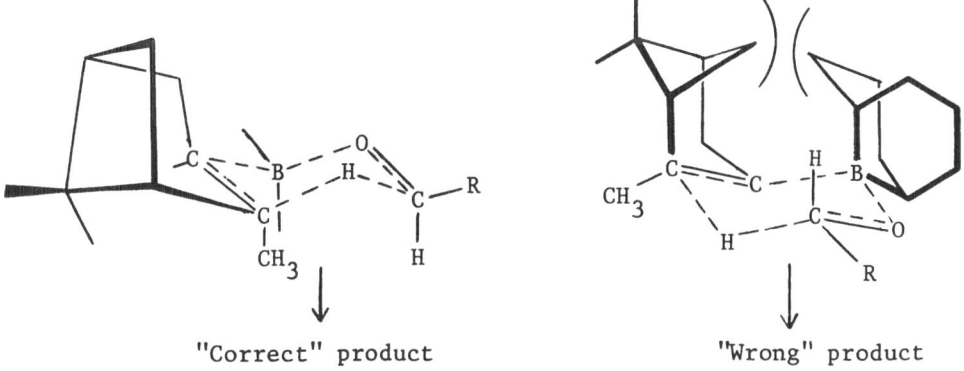

"Correct" product   "Wrong" product

Figure 5. Chair forms for aldehyde reduction by pinanyl-9-BBN

in the same manner. This fact, combined with use of optically active shift reagents,[34] could be very useful in assigning absolute
configuration to optically active primary alcohols.

We have also investigated the reduction of benzaldehyde-α-d
with other chiral 9-BBN derivatives. The borane from (+)-3-carene
gives 61% e.e. (S) (77% when corrected for the optical purity of
the carene). The approach of the aldehyde to the borane is analogous to that in the 3-pinanyl system - the phenyl group is over the
carane ring and the hydrogen atom over the methyl. (-)Camphene gives
75% e.e. (R) (84% when corrected for the optical purity of the camphene). The approach of the reagents in this case is that predicted
by the large-small interaction model, *i.e.* the phenyl approaches so
as to avoid the gem-dimethyl groups.

As mentioned earlier, the enantiomeric purity of the product
is not greatly affected by the steric requirements of the aldehyde.
Thus a *tert*-butyl group has about the same effect as a straight-
chain alkyl group. However, it was noted that electron-donating
groups on benzaldehyde tended to decrease the enantiomeric purity
of the benzyl alcohol product (Table III, entries 6-11). Interestingly, this decrease in selectivity is accompanied by a decrease in
the rate of reduction. Normally one would feel that the faster a
reaction proceeds the less selective it would become. This unusual
behavior, of decreased selectivity with decreased rate, may reflect
the tightness of the transition state. With the fast reactions, the
borane and aldehyde approach one another very closely so that steric
effects are maximized. With the slower reactions, the reactants
may be further apart and the transition state more sloppy. We are
currently examining kinetics and deuterium isotope effects to get
a handle on this problem.

C. Ketone Reductions

The 3-pinanyl-9-BBN reagent is perhaps the most selective chiral reducing agent yet devised for aldehydes. As mentioned earlier,
ketones are reduced only very slowly by the 9-BBN reagents. However, by using excess reducing agent and refluxing the solution
overnight, high-yield reductions of ketones may be obtained. An
effective chiral reducing agent for ketones would be most welcome.
Unfortunately the 3-pinanyl-9-BBN is not an effective chiral reducing agent for acetophenone. We obtain an enantiomeric excess of
only about 10%. Changing the optically-active group does not improve the results significantly.

The ketones are reduced rather slowly. In an attempt to speed
the reaction, we investigated boranes which should be less sterically demanding than 9-BBN, such as the boracyclohexane (borinane)

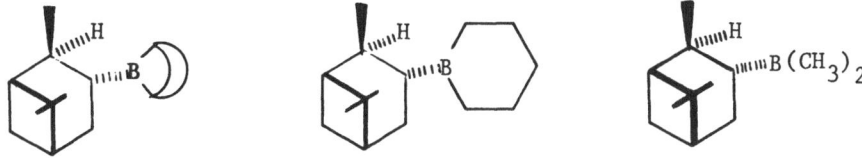

Figure 6. Chiral organoborane reducing agents for ketones .

and dimethylborane. Surprisingly, the 3-pinanyl derivatives of
these boranes are much better chiral reducing agents than is the
9-BBN compound (Fig. 6).

The effectiveness of chiral reductions is increased with de-
creasing size of the "inert" group on boron. Thus, the borocyclo-
hexane gives 35% e.e. of 1-phenyl-ethanol (39% when corrected for
pinene purity). The dimethylpinanylborane looks very promising. It
gives approximately 50% e.e. in reductions of acetophenone. However
the reaction is not behaving properly. The rate is very rapid at
the beginning and then becomes very slow. It is possible that the
organoborane is not pure, but since we have just begun this study
it is really too early to say what is happening.

With the sterically more crowded ketones it is conceivable that
the secondary hydrogen of pinane competes favorably with the tertiary hydrogen (eq 14). However, we have shown by deuterium labeling
that only the tertiary hydrogen does the reduction. Furthermore,

$$\text{(14)}$$

100%   0%

the recovered pinene is identical to the starting α-pinene.

The acetophenone approaches the pinanylborane in the same sense
as does the benzaldehyde (eq 15). The benzene ring is towards the

$$\text{(15)}$$

pinanyl ring and the methyl of the ketone is towards the methyl of the pinane. However, the aldehyde results show that there is a clear preference for all alkyl or aryl groups to approach towards the pinanyl ring. With a ketone one group must obviously approach from the unfavorable side. With this knowledge we have investigated the reduction of α,β-alkynyl ketones. The alkynyl group being cylindrical resembles a hydrogen and fits over the methyl group with ease. A variety of alkynyl ketones gives fair to excellent results (eq 16).[36] Best results are obtained with a large R and small R'. The alkynyl

$$\text{(pinanylborane)} + \underset{\substack{O \\ \parallel}}{R}CC\equiv CR' \longrightarrow \underset{H}{R}-\overset{H}{\underset{}{C}}\overset{OH}{\underset{}{—}}C\equiv CR' \qquad (16)$$

$$45\text{-}90\% \text{ e.e.}$$

group serves as a functionality which may be manipulated into other functionalities.

ACKNOWLEDGMENT

I thank my co-workers, whose names appear on the references. Financial support for this work from the National Institutes of Health (GM 24517), the Research Corporation, the Petroleum Research Fund (administered by the American Chemical Society), the UCR Intramural Research Fund, and the Alfred P. Sloan Foundation is gratefully acknowledged. I also wish to thank Syntex for a gift of steroid starting materials.

REFERENCES

1. H. C. Brown, "Hydroboration," W. A. Benjamin, New York, NY, 1962; H. C. Brown, G. W. Kramer, A. B. Levy, and M. M. Midland, "Organic Syntheses *via* Boranes," Wiley, New York, NY, 1975; H. C. Brown, M. M. Rogić, M. W. Rathke and G. W. Kabalka, *J. Am. Chem. Soc.*, **91**, 2150 (1969).
2. H. C. Brown, N. R. Ayyangar, and G. Zweifel, *J. Am. Chem. Soc.*, **86**, 397 (1964).
3. J. Blais, A. L'Honore, J. Soulie, and P. Cadiot, *J. Organometal. Chem.*, **78**, 323 (1974).
4. T. Leung and G. Zweifel, *J. Am. Chem. Soc.*, **96**, 5620 (1974).
5. M. M. Midland, J. A. Sinclair, and H. C. Brown, *J. Org. Chem.*, **39**, 731 (1974); H. C. Brown, A. B. Levy, and M. M. Midland, *J. Am. Chem. Soc.*, **97**, 5017 (1975).

6. K. Utimoto, Y. Yabuki, K. Okada, and H. Nozaki, *Tetrahedron Lett.*, 3969 (1976).

7. M. M. Midland, the 172nd National Meeting of the American Chemical Society, San Francisco, August 29 - September 3, 1976.

8. M. Naruse, T. Tomita, K. Utimoto and H. Nozaki, *Tetrahedron* 30, 835 (1974); P. Binger, *Angew. Chem. Internat. Edit.*, 6, 84 (1967).

9. M. M. Midland, *J. Org. Chem.*, 40, 2250 (1975).

10. M. M. Midland, *J. Org. Chem.*, 42, 2650 (1977).

11. P. Rona and P. Crabbé, *J. Am. Chem. Soc.*, 91, 3289 (1969).

12. D. J. Pasto and J. Hickman, *J. Am. Chem. Soc.*, 89, 5608 (1967).

13. H. C. Brown, private communication.

14. M. M. Midland and D. C. McDowell, *J. Organometal. Chem.*, in press.

15. J. Fried, C. H. Lin, M. M. Mehra, W. L. Kao, and P. Dalven, *Ann. N. Y. Acad. Sci.*, 180, 38 (1971).

16. J. H. Brewster, "Topics in Stereochemistry," Inter-Science, New York, NY, 1967, Vol. 2; G. Lowe, *Chem. Commun.*, 411 (1965).

17. H. C. Brown and G. Zweifel, *J. Am. Chem. Soc.*, 83, 3834 (1961).

18. W. H. Pirkle and C. W. Boeder, *J. Org. Chem.*, 42, 3697 (1977).

19. J. L. Luche, E. Barreiro, J. M. Dollat, and P. Crabbé, *Tetrahedron Lett.*, 4615 (1975).

20. G. W. Kramer and H. C. Brown, *J. Organometal. Chem.*, 132, 9 (1977) and ref. therein.

21. B. M. Mikhailov, *Organomet. Chem. Rev. A.*, 1 (1972).

22. H. C. Brown, "Hydroboration," W. A. Benjamin, New York, NY, 1962, p. 66.

23. N. K. Chaudhuri, J. A. Williams, R. Nickolson, and M. Gut, *J. Org. Chem.*, 34, 3759 (1969).

24. A. M. Krubiner and E. P. Oliveto, *J. Org. Chem.*, 31, 24 (1966).

25. M. M. Midland, A. Tramontano, and S. A. Zderic, *J. Organometal. Chem.*, 134, C17 (1977).

26. B. M. Mikhailov, Yu. N. Bubnov, and V. G. Kiselev, *J. Gen. Chem. USSR*, 36, 65 (1966).

27. J. D. Buhler, Ph.D. Thesis, Purdue University, 1973.

28. M. M. Midland, A. Tramontano, and S. A. Zderic, *J. Organometal. Chem.*, in press.

29. M. M. Midland, A. Tramontano, *J. Org. Chem.*, 43, 1470 (1978).

30. (a) H. C. Brown, "Boranes in Organic Chemistry," Cornell University Press, Ithaca, NY, 1972, p. 215; (b) R. O. Hutchins and D. Kandasamy, *J. Am. Chem. Soc.*, 95, 6131 (1973); (b) G. W. Gribble and D. C. Ferguson, *J. Chem. Soc. Chem. Commun.*, 535 (1975); (d) C. S. Sell, *Aust. J. Chem.*, 28, 1383 (1975); (e) Y. Yamamoto, H. Toi, A. Sonoda, and S-I. Murahashi, *J. Am. Chem. Soc.*, 98, 1965 (1976); (f) H. C. Brown, S. Krishnamurthy and N. M. Yoon, *J. Org. Chem.*, 41, 1778 (1976); (g) G. H. Posner, A. W. Runquist, and M. H. Chapdelaine, *J. Org. Chem.*, 42, 1202 (1977); (h) H. C. Brown and S. U. Kulkarni, *J. Org. Chem.*, 42, 4169 (1977).

31. D. Nasipuri, C. K. Ghosh, P. R. Mukherjee, and S. Venkataraman,

Tetrahedron Lett., 1587 (1971).

32. (a) D. Arigoni and E. L. Eliel in "Topics in Stereochemistry,"
    Vol. 4, E. L. Eliel and N. L. Allinger, Ed., Wiley-Interscience,
    New York, NY, 1969, pp. 127-244; (b) L. Verbit in "Progress in
    Physical Organic Chemistry," Vol. 7, R. Streitwieser and R. W.
    Taft, Ed., Wiley-Interscience, New York, NY, 1970, p. 51-127;
    (c) K. S.Y. Lau, P. K. Wong, and J. K. Stille, *J. Am. Chem.
    Soc.*, 98, 5832 (1976) (d) E. Caspi and C. R. Eck, *J. Org. Chem.*,
    42, 767 (1977), and references therein.

33. V. E. Althouse, D. M. Feigl, W. A. Sanderson, and H. S. Mosher,
    J. Am. Chem. Soc., 88, 3595 (1966).

34. L. J. Reich, G. R. Sullivan, and H. S. Mosher, *Tetrahedron
    Lett.*, 1505 (1973).

35. J. D. Morrison and H. S. Mosher, "Asymmetric Organic Reactions,"
    Prentice-Hall, Inc., Engelwood Cliffs, NJ, 1970.

36. Research in progress with A. Tramontano and D. C. McDowell.

THE VERSATILE ALKENYLALANES AND ALKENYLBORANES

George Zweifel

Department of Chemistry, University of California, Davis

Davis, California 95616

The discoveries that alkenylalanes[1] and alkenylboranes[2] of predictable stereochemistry can be directly synthesized *via* the monohydroalumination and monohydroboration, respectively, of alkynes have opened a number of exciting new frontiers for chemical research. Of special importance have been the studies directed toward the utilization of these organometallics as intermediates in organic synthesis. These efforts have resulted in the development of many new synthetic procedures of interest to the organic chemist.

An important difference in the chemistry of alkenylaluminum and alkenylboron compounds lies in their modes of reaction with inorganic and organic electrophiles.[3] Thus, the reactions of vinylaluminum compounds with electrophilic reagents generally proceed with stereospecific intermolecular transfers of the alkenyl moiety to afford functionally substituted olefins. On the other hand, alkenylboron compounds exhibit a pronounced tendency to undergo intramolecular transfer reactions. These are usually followed by oxidation, protonolysis, or β-elimination to give the final desired reaction products.

In recent years, we have been investigating the reactions of alkenylaluminum and alkenylboron compounds with a variety of inorganic and organic substrates. Our specific aim has been to develop novel synthetic transformations. Following a brief review on the preparation of these organometallics, the discussion will be focused on a number of selected reactions that in particular are suited for contrasting the chemical behaviors of vinylaluminum and vinylboron compounds toward a given electrophilic reagent.

In 1956, Wilke and Müller[1] reported that the reactions of dialkylaluminum hydrides both with mono- and with disubstituted alkynes can be controlled to give alkenyldialkylalanes. Thus, monohydroalumination of aliphatic terminal alkynes with dialkyl-aluminum hydrides, either neat or in hydrocarbon solvents, proceeds through a regio- and stereo-selective cis-addition of Al-H to the triple bond to give (E)-alkenylalanes. However, it should be noted that the hydroalumination of 1-alkynes with diisobutyl-aluminum hydride is accompanied by small amounts of metallation and bis-hydroalumination (eq 1).[4] With more acidic acetylenes,

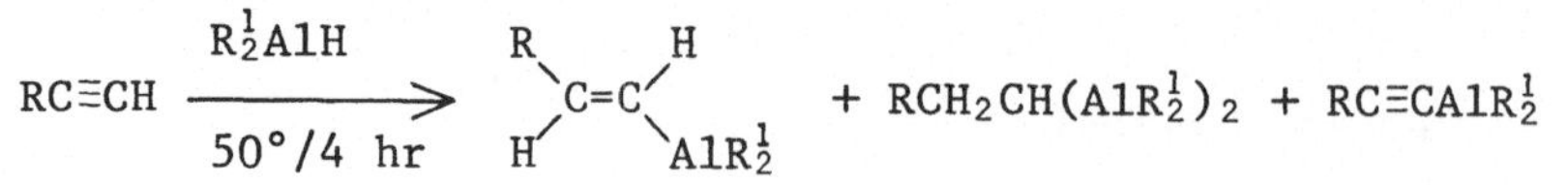

$$RC{\equiv}CH \xrightarrow[50°/4\ hr]{R_2^1AlH} \underset{H}{\overset{R}{>}}C{=}C\underset{AlR_2^1}{\overset{H}{<}} + RCH_2CH(AlR_2^1)_2 + RC{\equiv}CAlR_2^1 \quad (1)$$

R^1 = n-C_4H_9:	90	4	6%
c-C_6H_{11}:	94	4	2%
C_6H_5:	69	2	29%

such as phenylacetylene and conjugated enynes, the amount of metallation increases significantly.

The choice of solvent in hydroaluminations of 1-alkynes markedly influences the course of the Al-H additions. Thus, mono- and dihydroalumination[4,5] or metallation products[6] (eqs 2-4) may be obtained by carrying out the reaction in hydrocarbon, ether, or tertiary amine solvents, respectively. The reaction of di-isobutylaluminum hydride with terminal alkynes in a 3 to 1 ratio

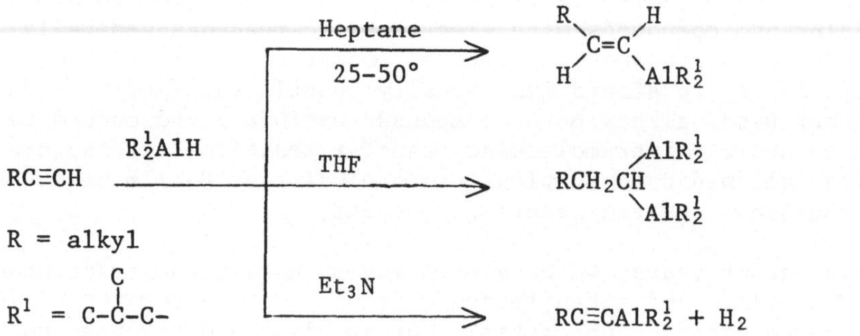

$$\xrightarrow[25-50°]{Heptane} \underset{H}{\overset{R}{>}}C{=}C\underset{AlR_2^1}{\overset{H}{<}} \quad (2)$$

$$RC{\equiv}CH \xrightarrow[THF]{R_2^1AlH} RCH_2CH\overset{AlR_2^1}{\underset{AlR_2^1}{<}} \quad (3)$$

R = alkyl

$$R^1 = C{-}\overset{C}{\underset{|}{C}}{-}C{-} \quad \xrightarrow{Et_3N} \quad RC{\equiv}CAlR_2^1 + H_2 \quad (4)$$

$$RC{\equiv}CH \xrightarrow[Et_3N]{3\ R_2^1AlH} RCH_2C\overset{AlR_2^1}{\underset{AlR_2^1}{\diagup\diagdown}}AlR_2^1 + H_2 \quad (5)$$

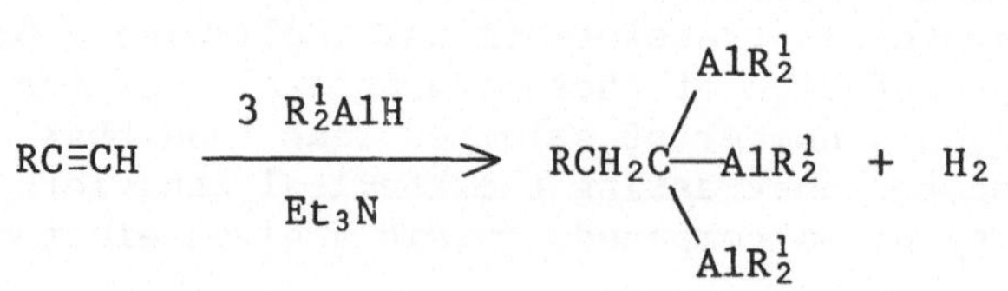

in tertiary amine solvents results in the formation of tris-aluminoalkanes (eq 5).[4,5]

The direct synthesis of *trans*-1-alkenylboranes from 1-alkynes was achieved by Brown in 1959.[2] Treatment of terminal acetylenes with di- and monoalkylboranes afforded the corresponding mono- and dialkenylboranes, respectively. The reaction involves a regioselective *cis*-addition of the B-H bond to the triple bond (eqs 6,7).[2,7]

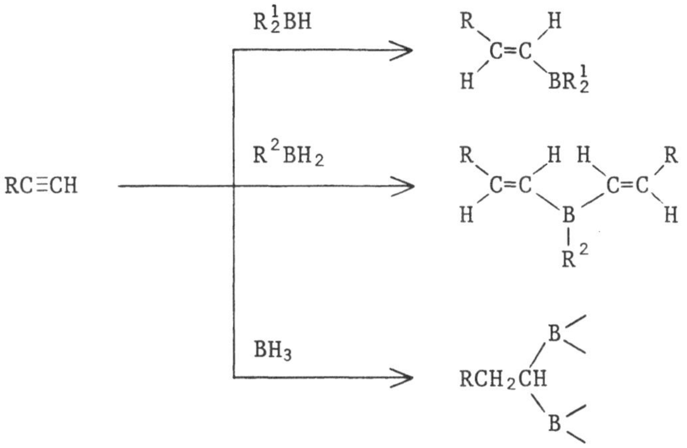

$$RC\equiv CH \quad \xrightarrow{\quad\quad} \quad$$

$$R_2^1BH \longrightarrow \begin{array}{c} R \\ \diagdown \\ \quad C=C \\ \diagup \\ H \end{array} \begin{array}{c} H \\ \diagup \\ \diagdown \\ BR_2^1 \end{array} \qquad (6)$$

$$R^2BH_2 \longrightarrow \begin{array}{c} R \\ \diagdown \\ C=C \\ \diagup \\ H \end{array} \begin{array}{c} H \quad H \\ \diagdown B \diagup \\ | \\ R^2 \end{array} \begin{array}{c} R \\ C=C \\ \diagdown \\ H \end{array} \qquad (7)$$

$$BH_3 \longrightarrow RCH_2CH \begin{array}{c} \diagup B\diagup \\ \diagdown B\diagup \end{array} \qquad (8)$$

R_2^1BH: dialkyl- and dicycloalkylboranes

R^2BH_2:   thexylborane

In contrast to the behavior observed in the monohydroalumination of 1-alkynes, formation of 1-alkynylboranes is negligible even during hydroboration of acidic acetylenes. Attempts to convert 1-alkynes into alkenylboranes using borane (BH_3) resulted in the formation of dihydroborated products (eq 8).[2]

Major differences in regioselectivity are encountered in the additions of Al-H and B-H to triple bonds of disubstituted alkynes as exemplified by the data in Fig. 1.[4,7] Thus, the directions of

	BuC≡CH	BuC≡CCH$_3$	c-C$_6$H$_{11}$C≡CCH$_3$	t-BuC≡CCH$_3$
	↑	↑	↑	↑
R$_2$AlH	98	67	75	85%
R$_2$BH	99	60	92	99%

Figure 1. Regiochemistry of Additions of Diisobutylaluminum Hydride and Dicyclohexylborane to Disubstituted Alkynes

addition of dicyclohexylborane or disiamylborane to unsymmetrically
disubstituted alkynes are markedly affected by the size of the
substituents attached to the triple bond. On the other hand,
diisobutylaluminum hydride is considerably less regioselective.

As pointed out earlier, both the alkenylaluminum and alkenylboron compounds have proved to be useful and versatile reagents
or intermediates for synthetic methodology. For example, sequential treatment of mono- or disubstituted alkynes with diisobutylaluminum hydride or dicyclohexylborane, followed by protonolysis
of the resultant alkenylalanes or alkenylboranes, provides a non-
catalytic method for reduction of triple bonds to give mono- or
cis-disubstituted alkenes, respectively (eq 9,10).[2,7] The fact
that the protonolysis reaction proceeds with retention of con-

$$C_6H_{13}C{\equiv}CH \quad
\begin{array}{l}
1. \quad R_2AlH \\
2. \quad H_3O^+
\end{array}
\longrightarrow C_6H_{13}CH{=}CH_2 \; (+ \; alkane \; + \; alkyne)$$

$$
\begin{array}{l}
1. \quad R_2BH \\
2. \quad AcOH
\end{array}
\longrightarrow C_6H_{13}CH{=}CH_2
$$

$$(9)$$

$$\bigcirc\!\!-C{\equiv}CEt \quad
\begin{array}{l}
1. \quad R_2AlH \\
2. \quad H_3O^+ \\[4pt]
1. \quad R_2BH \\
2. \quad AcOH
\end{array}
\longrightarrow (Z)-\bigcirc\!\!-CH{=}CHEt \qquad (10)$$

figuration provides a convenient method for synthesis of deuterated
olefins of predictable stereochemistry as exemplified in equations
11-14.

$$RC{\equiv}CH \quad
\begin{array}{l}
{>}B{-}H \quad \xrightarrow{AcOD} \quad \begin{array}{c}R\\H\end{array}{>}C{=}C{<}\begin{array}{c}H\\D\end{array} \qquad (11) \\[10pt]
{>}B{-}D \quad \xrightarrow{AcOH} \quad \begin{array}{c}R\\D\end{array}{>}C{=}C{<}\begin{array}{c}H\\H\end{array} \qquad (12) \\[10pt]
{>}B{-}D \quad \xrightarrow{AcOD} \quad \begin{array}{c}R\\D\end{array}{>}C{=}C{<}\begin{array}{c}H\\D\end{array} \qquad (13)
\end{array}$$

$$\text{--}{C{\equiv}CBu} \quad \xrightarrow[\text{2. AcOD}]{\text{1. } R_2BH} \quad \begin{array}{c}X\\H\end{array}{>}C{=}C{<}\begin{array}{c}Bu\\D\end{array} \qquad (14)$$

The conversion of alkenylalanes and alkenylboranes into the corresponding *ate* complexes with methyllithium results in their exhibiting differential behavior toward protonation. Thus addition of hydrogen chloride to alkenylalanates still involves cleavage of the vinylcarbon-aluminum bond to give the corresponding alkenes. On the other hand, conversion of alkenyldialkyl-boranes by methyllithium into the *ate* complexes, followed by addition of a dry ethereal solution of hydrogen chloride, results in protonation of the double bonds and concurrent migration of one alkyl group from boron to the α-carbon (eq 15). Oxidation of the intermediate organoboranes with alkaline hydrogen peroxide produces the corresponding secondary or tertiary alcohols as illustrated in equations 16-18.[8] An interesting sidelight to these reactions is the low migratory aptitude of the methyl group as compared to secondary alkyl groups in the protonation of lithium alkenyldialkylmethylborates.

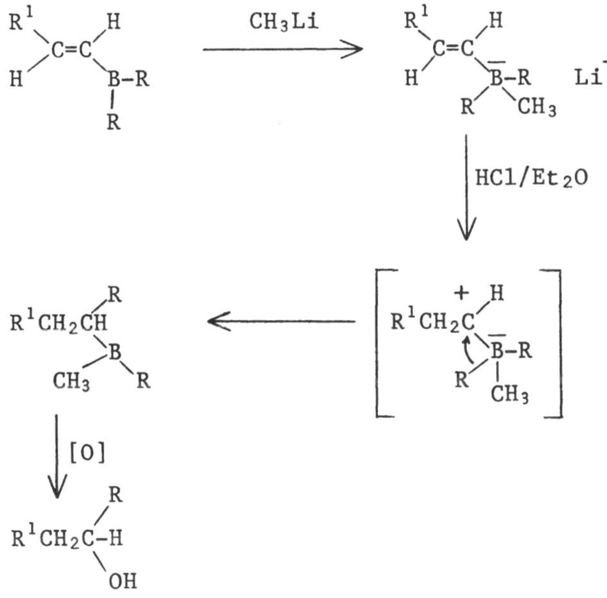

(15)

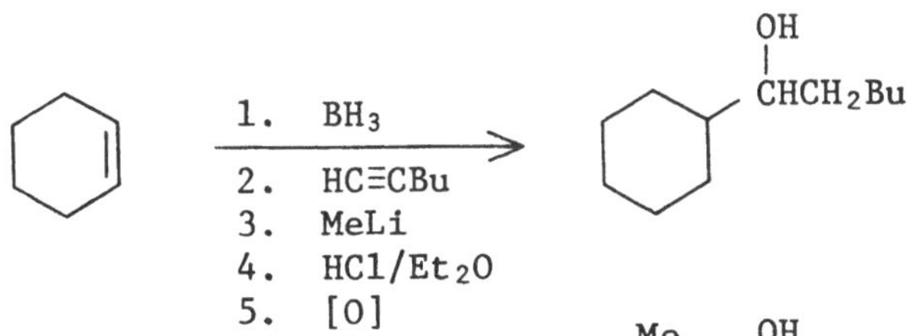

(16)

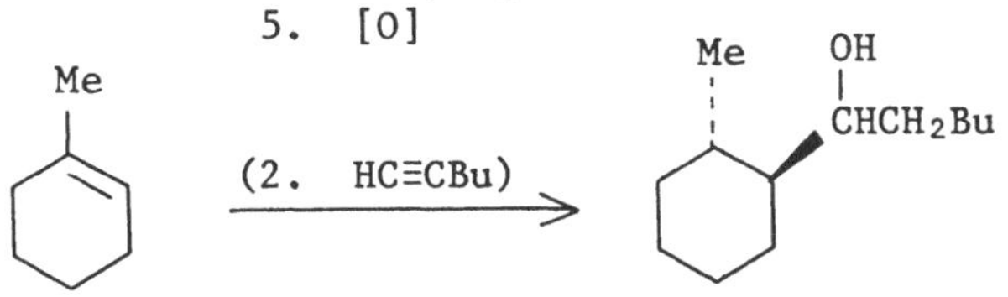

(17)

$$(2. \quad CH_3C{\equiv}C\text{---}) \qquad (18)$$

The contrasting modes of reaction of alkenylaluminum and alkenylboron compounds with electrophiles are particularly evident in their halogenation and cyanohalogenation reactions. Thus, we have shown that treatment of *trans*-1-alkenyldiisobutyl-alanes with N-chlorosuccinimide, bromine, or iodine results in a highly chemoselective cleavage of the vinylcarbon-aluminum bond to afford excellent yields of the corresponding isomerically pure (E)-vinyl halides (eqs 19-21).[9]

$$\begin{array}{ll} 1) & NCS, \ Et_2O, \ -30° \\ 2) & H_3O^+ \end{array} \longrightarrow (E)\text{-RCH=CHCl} \qquad (19)$$

$$\begin{array}{ll} 1) & Br_2, \ Et_2O\text{-pyr.}, \ -78° \\ 2) & H_3O^+ \end{array} \longrightarrow (E)\text{-RCH=CHBr} \qquad (20)$$

$$\begin{array}{ll} 1) & I_2, \ THF, \ -78° \\ 2) & H_3O^+ \end{array} \longrightarrow (E)\text{-RCH=CHI} \qquad (21)$$

On the other hand, halogenations of alkenyldialkylboranes proceed *via* initial addition of the electrophilic halogen to the double bond.

Recently, we have extended the halodealumination reaction to ((Z)-1-alumino-1-alkenyl)silanes derived *via* the hydroalumination of 1-alkynylsilanes[10] with diisobutylaluminum hydride in ether solvent.[11,12] Treatment of the α-silylalkenylalanes with N-chloro-succinimide, bromine, or iodine produces the corresponding ((E)-1-halo-1-alkenyl)silanes in high isomeric purities and yields (eqs 22-24).[12]

$$RC{\equiv}CSiMe_3 \quad \xLeftarrow[\text{2. } Me_3SiCl]{\text{1. } RLi} \quad RC{\equiv}CH$$

$$\begin{array}{ll} 1. & NCS \\ 2. & H_3O^+ \end{array} \longrightarrow \begin{array}{c} R \quad SiMe_3 \\ C{=}C \\ H \quad Cl \end{array} \qquad (22)$$

$$\begin{array}{ll} 1. & Br_2, \ pyr. \\ 2. & H_2O^+ \end{array} \longrightarrow (E)\text{-RCH=C} \begin{array}{c} SiMe_3 \\ Br \end{array} \qquad (23)$$

$$\begin{array}{ll} 1. & I_2 \\ 2. & H_3O^+ \end{array} \longrightarrow (E)\text{-RCH=C} \begin{array}{c} SiMe_3 \\ I \end{array} \qquad (24)$$

Extension of the bromination reaction to the dienylsilyl-alane derived through the chemo- and regioselective hydroalumination of 2-cyclohexenylethynyl(trimethyl)silane produced a mixture of products. However, the dienylsilylalane was successfully converted into the desired bromide by treatment with a predried solution of cyanogen bromide (eq 25).[12]

$$
\text{(cyclohexenyl)}-C{\equiv}C-SiMe_3 \quad \xrightarrow[\text{2. BrCN}]{\text{1.   R}_2\text{AlH}} \quad (25)
$$

(96%)

It is gratifying that the ((Z)-1-chloro- and (Z)-1-bromo-1-alkenyl) silanes are readily accessible through bromine-catalyzed isomerization of the corresponding ((E)-1-chloro- and (E)-1-bromoalkenyl) silanes (eq 26) produced as described above.[12] This

$$
\begin{array}{c} R \\ \diagdown \\ H \end{array} C{=}C \begin{array}{c} SiMe_3 \\ \diagup \diagdown \\ X \end{array} \quad \xrightarrow[\text{h}\nu]{\text{Br}_2} \quad \begin{array}{c} R \\ \diagdown \\ H \end{array} C{=}C \begin{array}{c} X \\ \diagup \diagdown \\ SiMe_3 \end{array} \quad (26)
$$

X = Cl, Br

isomerization of ((E)-1-halo-1-alkenyl)silanes is reminiscent of the bromine-catalyzed *cis,trans* isomerization of 1,2-dibromo-ethylene, which has been subjected to a detailed mechanistic study.[13]

The availability of an operationally simple synthesis of α-haloalkenylsilanes is of considerable interest in connection with synthetic methodology. Thus, it provides a valuable starting point for stereo-specific syntheses of substituted dialkyl-alkenylsilanes, alkenyl halides, and trisubstituted olefins as exemplified in equation 27.[14]

$$
\begin{array}{c} R \\ \diagdown \\ H \end{array} C{=}C \begin{array}{c} SiMe_3 \\ \diagup \diagdown \\ Br \end{array} \quad \xrightarrow[\text{2. EtI}]{\begin{array}{c}\text{1.   } sec\text{-BuLi}\\ \text{THF, }-78°\end{array}} \quad \begin{array}{c} R \\ \diagdown \\ H \end{array} C{=}C \begin{array}{c} SiMe_3 \\ \diagup \diagdown \\ Et \end{array}
$$

$$
\Big\downarrow \begin{array}{l}\text{1.   Br}_2/\text{CH}_2\text{Cl}_2 \\ \\ \text{2.   MeONa}\end{array} \quad (27)
$$

$$
\begin{array}{c} R \\ \diagdown \\ H \end{array} C{=}C \begin{array}{c} Et \\ \diagup \diagdown \\ Me \end{array} \quad \xleftarrow[\text{CuI (0.2 eq.)}]{\text{MeLi}} \quad \begin{array}{c} R \\ \diagdown \\ H \end{array} C{=}C \begin{array}{c} Et \\ \diagup \diagdown \\ Br \end{array}
$$

80%   85%

As mentioned earlier, in contrast to the behavior observed with the alkenylalanes, halogenation of alkenyldialkylboranes proceeds *via* initial addition of the electrophilic halogen to the double bond. For example, the formation of (Z)-1-bromo-1-alkenes resulting from bromination of *trans*-1-alkenyldisiamylboranes, followed by hydrolytic work-up, has been proposed to involve an initial *trans*-addition of bromine to the double bond. This is then followed during the hydrolytic work-up by *trans*-elimination of the elements of disiamylboron bromide from the dibromide intermediate (eq 28).[15]

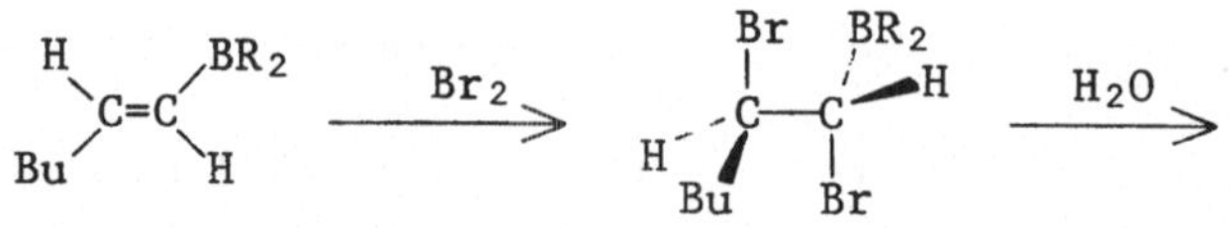

$$
\underset{Bu}{\overset{H}{\diagdown}} C=C \overset{BR_2}{\underset{H}{\diagup}} \xrightarrow{Br_2} \quad H-\overset{\overset{Br}{|}}{\underset{\underset{Bu}{|}}{C}}-\overset{\overset{BR_2}{|}}{\underset{\underset{Br}{|}}{C}}-H \xrightarrow{H_2O} \underset{Bu}{\overset{H}{\diagdown}} C=C \overset{H}{\underset{Br}{\diagup}} \tag{28}
$$

The halogenation of alkenyldialkylboranes takes a completely different course when the halogenating agent is iodine in the presence of a base. Thus, we have shown that such treatment of vinylboranes derived from hydroboration of 1-alkynes with dialkylboranes results in the transfer of one alkyl group from boron to the adjacent carbon. This is then followed by deborono-iodination to give isomerically pure *cis* olefins. Similarly, iodination of vinyldialkylboranes derived from the hydroboration of disubstituted alkynes provides a tool for preparing trisub-stituted olefins.[16] It has been suggested that migration of the alkyl group proceeds with inversion at the migration terminus, and that deboronoiodination occurs in a *trans* manner (eq 29).[16]

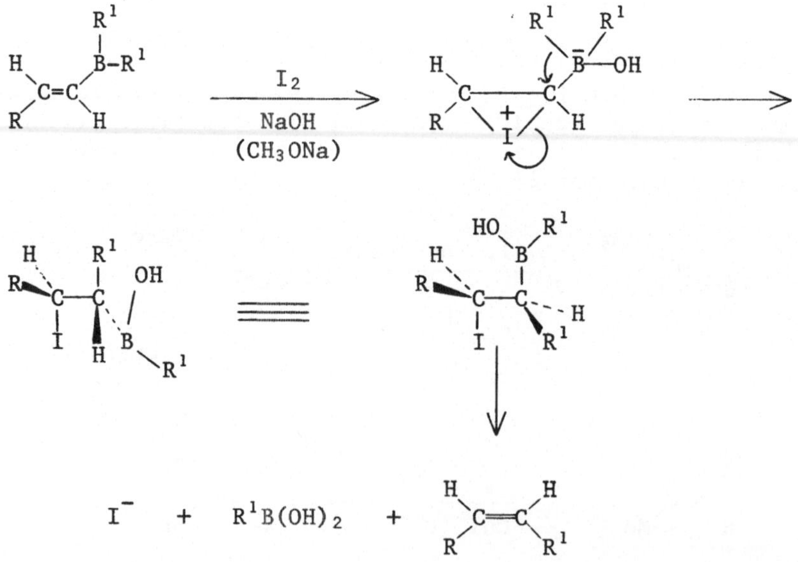

$$
\tag{29}
$$

To the above remarkable sequence of stereoselective steps leading to the formation of olefins must finally be added an additional chemical feature. It has been observed that the migrating group retains its configuration. Thus, it is evident that the iodination reaction provides an efficient route to di- and trisubstituted olefins of defined stereochemistries which are not readily obtainable by current methodology. Several varied examples of such conversions are shown in equations 30-32.[17]

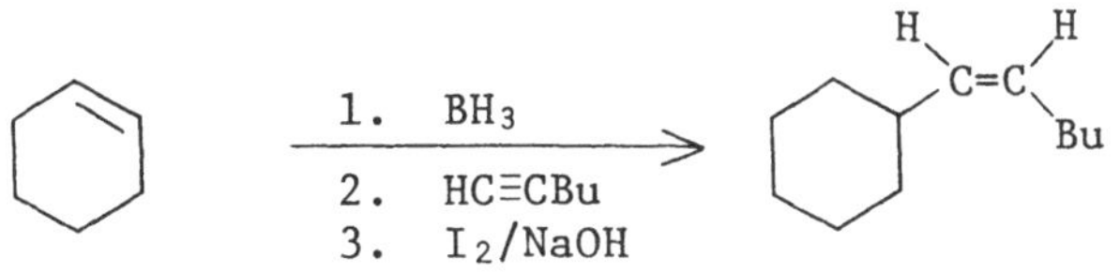
1. BH3
2. HC≡CBu
3. I2/NaOH
H
H
C=C
Bu

(30)

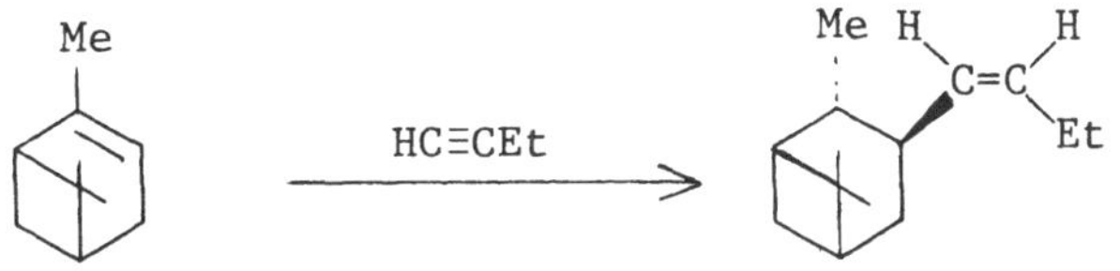
Me
HC≡CEt
Me H
H
C=C
Et

(31)

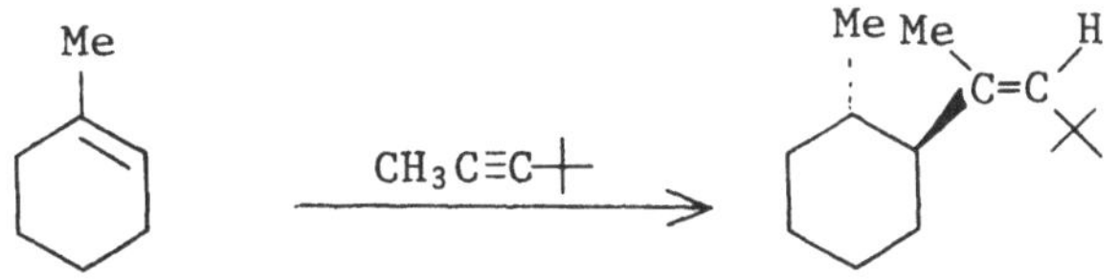
Me
CH3C≡C
Me Me
H
C=C

(32)

Entirely different behaviors are also encountered in the reactions of alkenylaluminum and alkenylboron compounds with cyanogen halides. Although the reactions of alkenylalanes with cyanogen bromide or cyanogen chloride in ether solvents proceed only sluggishly, conversion of the organoalanes into the *ate* complexes with alkyllithium reagents enhances their reactivity. Thus, sequential treatment of alkenylalanes with methyllithium and cyanogen bromide, followed by a hydrolytic work-up, affords the corresponding *trans*-alkenyl bromides (eq 33).[9] By using a cyanogen halide with inverted polarity, as in cyanogen chloride, alkenylalanates may be converted into the corresponding *trans*-α,β-unsaturated nitriles (eq 34).[18]

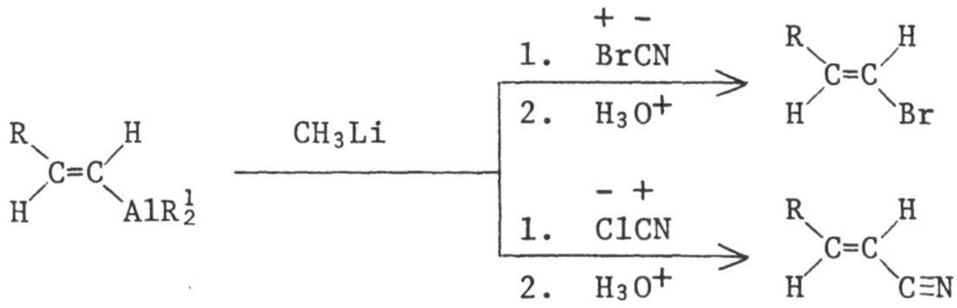
+ -
1. BrCN
2. H3O+
R
H
C=C
H
Br
R
H
C=C
H
AlR12
CH3Li
- +
1. ClCN
2. H3O+
R
H
C=C
H
C≡N

(33)

(34)

In contrast to the substitutions observed with alkenylalanates, the reactions of alkenyldialkylboranes with cyanogen bromide in methylene chloride solvent results in the transfer of one alkyl group from boron to the adjacent carbon atom. Subsequent *syn*-elimination of boron and bromide leads to the observed di- and trisubstituted olefins (eq 35).[19]

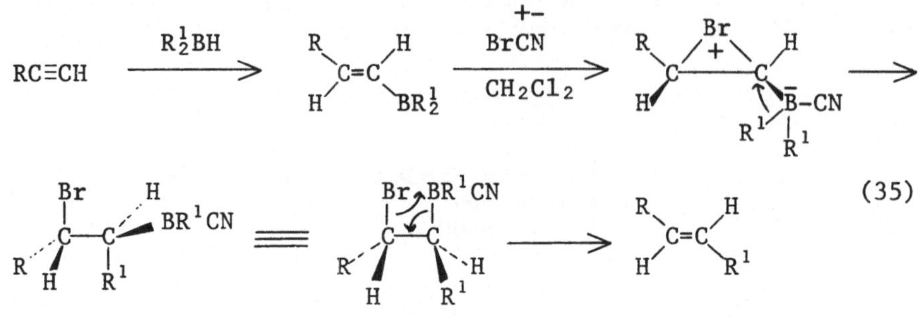

(35)

It should be noted that the olefins produced by the hydroboration-cyanohalogenation of alkynes possess opposite stereochemistries (eq 36) from those obtained in the hydroboration-iodination reaction of the corresponding acetylenes (eq 32).

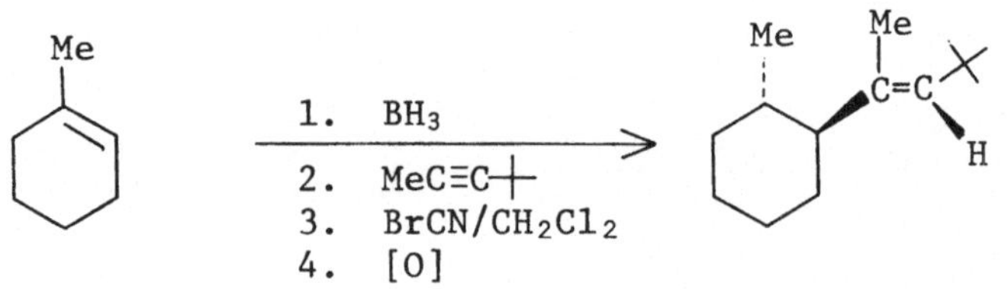

$$
\begin{array}{l}
1. \quad BH_3 \\
2. \quad MeC{\equiv}C{+} \\
3. \quad BrCN/CH_2Cl_2 \\
4. \quad [O]
\end{array}
$$

(36)

Addition of cyanogen chloride to *trans*-1-hexenyldicyclohexylborane in THF followed by oxidation of the intermediate organoborane yielded 1-cyclohexyl-2-cyanohexan-1-ol (eq 37).[20] In this

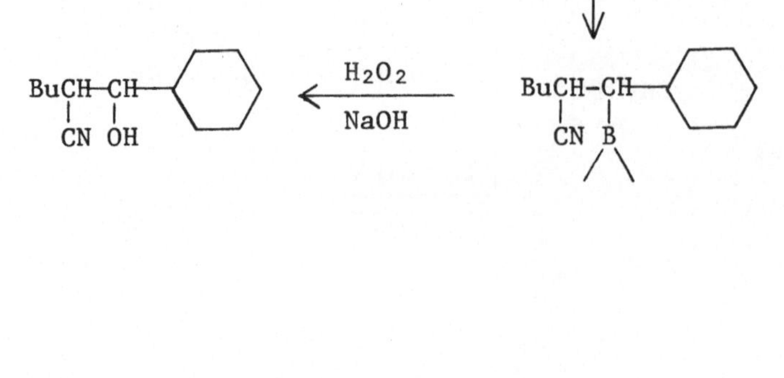

(37)

case, the electrophilic cyanide moiety initiated the migration of the cyclohexyl group. This remarkable reaction represents a method for introducing three different substituents onto the triple bond of an alkyne.

We have shown that hydroboration of 1-halo-alkynes with dialkylboranes affords the synthetically valuable α-haloalkenyl-boranes (eq 38).[21] Unfortunately, attempts to synthesize the

$$RC\equiv CX \qquad X = Cl, Br, I$$

$$\xrightarrow{R_2^1BH} \qquad \begin{matrix} R \\ \diagdown \\ H \end{matrix} C = C \begin{matrix} X \\ \diagup \\ BR_2^1 \end{matrix} \qquad (38)$$

$$\xrightarrow[\text{2. }H_3O^+]{\text{1. }R_2^1AlH} \qquad RC\equiv CH \qquad (39)$$

corresponding α-haloalkenylalanes via the reactions of 1-halo-1-alkynes with diisobutylaluminum hydride have resulted in reduction of the carbon-halogen bonds to give 1-alkynes (eq 39).[22] Recently, however, we have found that 1-chloro-1-alkynes containing primary and secondary alkyl substituents attached at the triple bond undergo $trans$-hydroalumination when reacted with an equimolar amount of lithium aluminum hydride.[23] The resultant α-halo-alkenylalanates (eq 40) are only moderately stable even when

$$RC\equiv CCl \xrightarrow[\text{THF, }0°]{LiAlH_4} \begin{matrix} R \\ \diagdown \\ H \end{matrix} C = C \begin{matrix} \overset{-}{A}lH_3 \ \overset{+}{L}i \\ \diagup \\ Cl \end{matrix} \qquad (40)$$

maintained at 0°. Although we have not yet explored the full potential of these novel organometallic reagents as intermediates for organic synthesis, preliminary results indicated that they may be suitable as precursors for preparing (Z)-1-bromo-1-chloro-1-alkenes. For example, treatment of the α-chlorovinylalanates with three molar equivalents of acetone results in the formation of the corresponding isopropoxy compounds (eq 41). These then react readily with bromine to produce (Z)-1-bromo-1-chloro-alkenes

$$\begin{matrix} R \\ \diagdown \\ H \end{matrix} C = C \begin{matrix} \overset{-}{A}lH_3 \ \overset{+}{L}i \\ \diagup \\ Cl \end{matrix} \xrightarrow[\text{3 equiv.}]{(CH_3)_2CO} \begin{matrix} R \\ \diagdown \\ H \end{matrix} C = C \begin{matrix} \overset{-}{A}l-(OC_3H_7-i)_3 \ \overset{+}{L}i \\ \diagup \\ Cl \end{matrix} \qquad (41)$$

(eq 42).[23] It should be noted that the 1,1-dihaloalkenes formed are very susceptible to isomerization at ambient temperatures. It is evident that the availability of dihaloolefins of defined stereochemistry open new vistas into the methodology for syntheses of substituted olefins.

$$\underset{H}{\overset{R}{}}C=C\underset{Cl}{\overset{Al(OC_3H_7\text{-}i)_3\ \overset{+}{Li}}{}} \xrightarrow{\quad Br_2 \quad} \underset{H}{\overset{R}{}}C=C\underset{Cl}{\overset{Br}{}} \qquad (42)$$

In exploring the synthetic utilities of α-haloalkenylboranes, we have observed that they undergo anionotropic rearrangement when treated with sodium methoxide. The rearrangement occurs with inversion of configuration at the migration terminus bearing the halogen substituent (eq 43). Protonolysis of the intermediate

$$\underset{H}{\overset{R}{}}C=C\underset{\underset{R^1}{\overset{|}{B}}\,{}^{R^1}}{\overset{X}{}} \xrightarrow{\quad NaOMe \quad} \underset{H}{\overset{R}{}}C=C\underset{\underset{OMe}{\overset{|}{\overset{R^1}{}\overset{|}{\bar{B}}\text{-}R^1}}}{\overset{\uparrow X}{}} \longrightarrow \underset{H}{\overset{R}{}}C=C\underset{R^1}{\overset{\overset{MeO}{}\overset{B}{}{}^{R^1}}{}} \qquad (43)$$

vinylborane with carboxylic acids proceeds with retention of configuration to afford *trans* olefins (eq 44),[17,21] whereas oxidation produces the corresponding ketones (eq 45).[21,24] Representative examples of an olefin and of a ketone derived from α-haloalkenyl-

$$\underset{H}{\overset{R}{}}C=C\underset{R^1}{\overset{\overset{MeO}{}\overset{B}{}{}^{R^1}}{}} \xrightarrow[\Delta]{\quad RCO_2H \quad} \underset{H}{\overset{R}{}}C=C\underset{R^1}{\overset{H}{}} \qquad (44)$$

$$\xrightarrow{\quad [O] \quad} RCH_2\underset{\underset{O}{\parallel}}{C}R^1 \qquad (45)$$

dialkylboranes are shown in eqs 46 and 47, respectively.

$$\qquad (46)$$

$$\underset{H_3C}{\overset{CH_3}{\underset{|}{}}}C=CHCH_3 \quad \xrightarrow[\text{2. } BrC\equiv C\text{—}\hspace{-2pt}\bigcirc]{\text{1. } BH_3} \quad$$

3. NaOH–H₂O₂

$$\qquad (47)$$

The anionotropic rearrangement of α-haloalkenylboranes with bases coupled with the iodination reaction provides trialkyl-substituted olefins (eq 48, 49).[25] Extension of the sequential alkyl

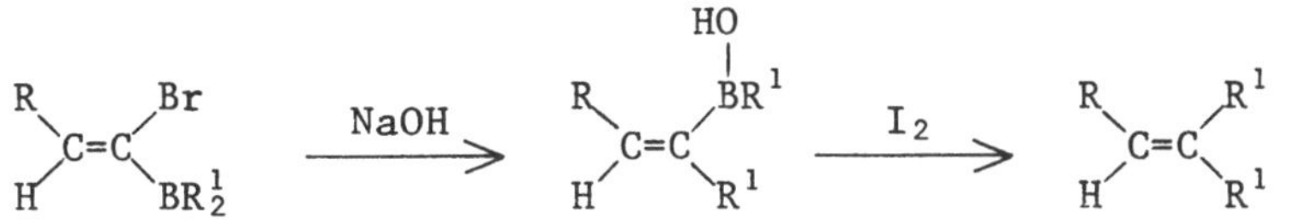

$$(48)$$

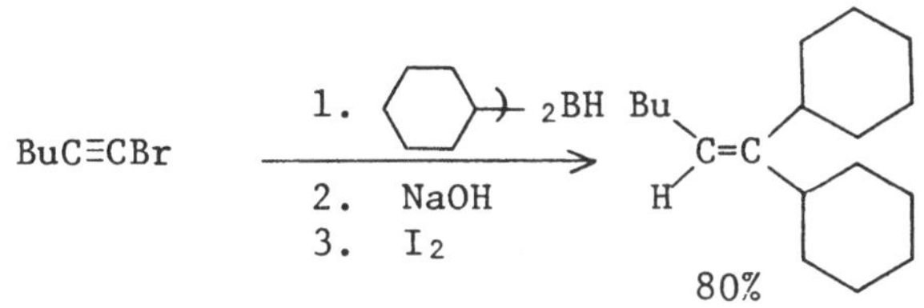

$$(49)$$

group migration to B-(1-bromo-1-alkenyl)boracycloalkanes (eq 50)

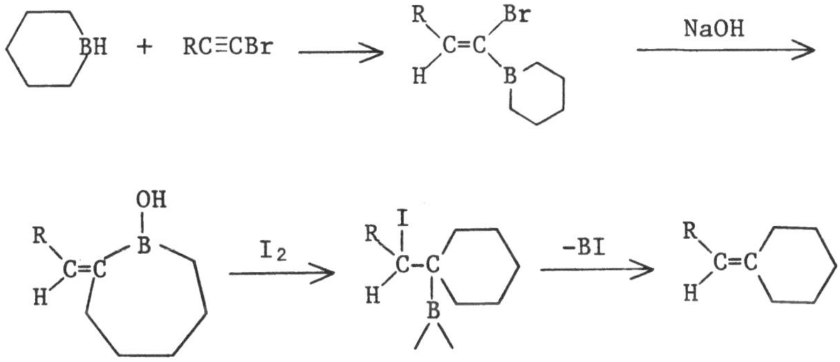

$$(50)$$

derived from the hydroboration of 1-bromo-1-alkynes with borinanes[26] (eq 51) or borepanes,[26] leads to efficient syntheses of

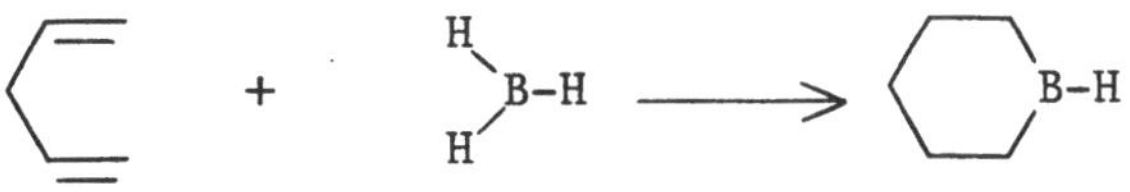

$$(51)$$

alkylidenecycloalkanes (eqs 52-54).[27]

$$(52)$$

$$(53)$$

$$\text{(structure)} + BrC{\equiv}CBu \longrightarrow \text{(ring structure)}{=}CHBu \qquad (54)$$

It appeared to us that α-haloalkenylboranes might serve as precursors for the preparation of 1,1-dihaloalkenes of defined stereochemistry by subjecting them to the bromination-deborono-bromination reactions. However, treatment with bromine of the α-halovinylborane derived from the hydroboration of 1-chloro-1-hexyne with disiamylborane, followed by treatment with sodium hydroxide, produced a mixture of 1-bromo-1-chloro-1-hexene and 2,3-dimethyl-4-chloro-4-nonene (eq 55). On the other hand,

$$RC{\equiv}CCl \xrightarrow{R_2^1BH} \underset{H}{\overset{R}{>}}C{=}C\underset{BR_2^1}{\overset{Cl}{<}} \xrightarrow{Br_2} \cdots$$

$$\xrightarrow{NaOH} \underset{H}{\overset{R}{>}}C{=}C\underset{Cl}{\overset{Br}{<}} \quad + \quad RCH{=}C\underset{R^1}{\overset{Cl}{<}} \qquad (55)$$

$$R = C_4H_9-; \quad R^1 = CH_3CHCH- $$
$$\qquad\qquad\qquad\qquad\quad H_3C\;\; CH_3$$

conversion of the α-chloroalkenyldisiamylborane into the borinic ester derivative by treatment with one equivalent of trimethyl-amine oxide, followed by addition of bromine and a basic work-up, yielded the (Z)-1-bromo-1-chloro-1-alkene (eq 56).[25] This pro-

$$\underset{H}{\overset{R}{>}}C{=}C\underset{BR_2^1}{\overset{Cl}{<}} \xrightarrow{Me_3NO} \underset{H}{\overset{R}{>}}C{=}C\underset{B}{\overset{Cl}{<}} \xrightarrow[\quad 2.\;\; OH^-\quad]{1.\;\; Br_2} \underset{H}{\overset{R}{>}}C{=}C\underset{Cl}{\overset{Br}{<}} \qquad (56)$$

cedure thus offers an alternative to the α-chloroalkenylalanate route (eq 42) for the preparation of 1,1-dihaloalkenes.

As pointed out earlier, there are few intramolecular transfer reactions involving organoaluminum compounds which have been docu- mented. However, it appeared to us that an *ate* complex derived from the reaction of lithium chloropropargylide with a trialkyl-alane might readily undergo such a process. Migration of an alkyl group from aluminum of the *ate* complex to the adjacent carbon, concomitant with an electron pair shift and loss of chloride, would be expected to produce the allenic alane (eq 57). That this does indeed take place was demonstrated by the formation of a 65 to 35 mixture of hexylallene and hexylmethylacetylene on

$$\text{LiC}\equiv\text{CCH}_2\text{Cl} \xleftarrow[\text{-78°}]{\text{RLi}} \text{HC}\equiv\text{CCH}_2\text{Cl}$$

$$\Big\downarrow\begin{array}{c}R_3^1\text{Al}\\-78°\end{array}$$

(57)

$$\underset{\text{Li}}{+}\; R_2^1\text{-Al-C}\!\equiv\!\overset{R^1}{\overset{|}{\text{C}}}\!-\text{CH}_2\!-\!\text{Cl} \longrightarrow \overset{R^1}{\underset{R_2^1\text{Al}}{>}}\text{C}\!=\!\text{C}\!=\!\text{CH}_2$$

treatment of lithium chloropropargylide with tri-n-hexylalane at -78° followed by addition of acetic acid at room temperature (eq 58).[28] Note, however, that we have not yet established whether

$$\overset{+}{\text{Li}}[(\text{C}_6\text{H}_{13})_3\overset{-}{\text{Al}}\text{-C}\!\equiv\!\text{CCH}_2\text{Cl}]\; \xrightarrow[\text{2.}\quad\text{CH}_3\text{CO}_2\text{H}]{\text{1.}\quad-78°\;\rightarrow\;25°}\; \begin{array}{c}\text{C}_6\text{H}_{13}\text{CH}\!=\!\text{C}\!=\!\text{CH}_2\\+\\\text{C}_6\text{H}_{13}\text{C}\!\equiv\!\text{CCH}_3\end{array}$$

the allenic alane initially formed is the actual precursor for the alkylallene.

The above transfer reaction has its counterpart in organoboron chemistry. Thus, addition of trialkylboranes to lithium chloropropargylide at -78°, followed by protonation of the resultant organoboranes at 25°, affords the corresponding alkylallenes containing only small amounts (~3%) of the alkyne byproducts (eqs 59, 60).[29]

1. BH$_3$
2. LiC$\equiv$CCH$_2$Cl (-78°)   (59)
3. CH$_3$CO$_2$H (25°)

(60)

In the course of exploring the chemistry of the organoboranes leading to alkylallenes we have uncovered operationally simple, high yield syntheses of homopropargylic (eq 61) and α-allenic alcohols (eq 62) via sequential treatment of lithium chloropropargylide with trialkylboranes and aldehydes.[30] The most remarkable feature of these synthetic transformations lies in the fact that which alcohol is specifically formed depends solely on the temperature at which the organoborane precursor is maintained prior to its reaction with the aldehyde.

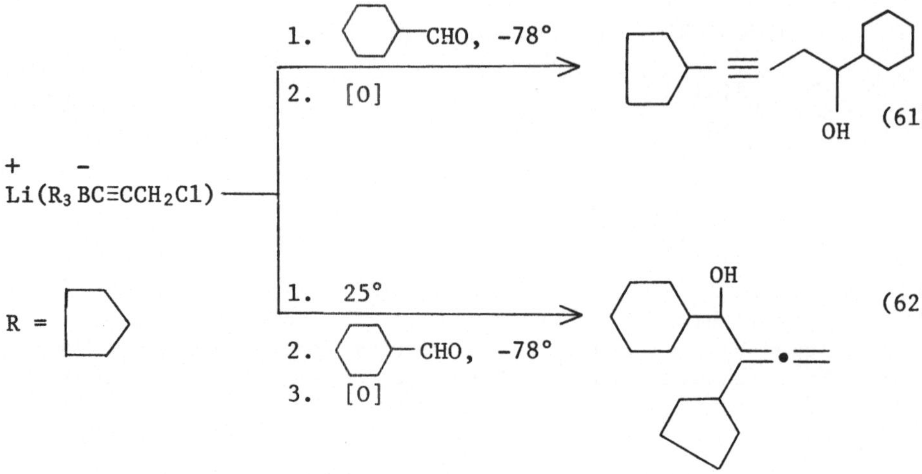

$$
\begin{array}{l}
\text{1.} \quad \text{(cyclohexyl)—CHO, } -78° \\
\text{2.} \quad [O]
\end{array}
\longrightarrow
\quad (61)
$$

$$
\underset{\text{Li}(R_3BC\equiv CCH_2Cl)}{\overset{+\qquad -}{}} \text{—}
$$

$$
R = \text{(cyclopentyl)}
$$

$$
\begin{array}{l}
\text{1.} \quad 25° \\
\text{2.} \quad \text{(cyclohexyl)—CHO, } -78° \\
\text{3.} \quad [O]
\end{array}
\longrightarrow
\quad (62)
$$

Accordingly, reactions leading to the homopropargylic- and α-allenic alcohols may be depicted as follows in equation 63. The *ate* complex formed by the reaction of the trialkylborane with

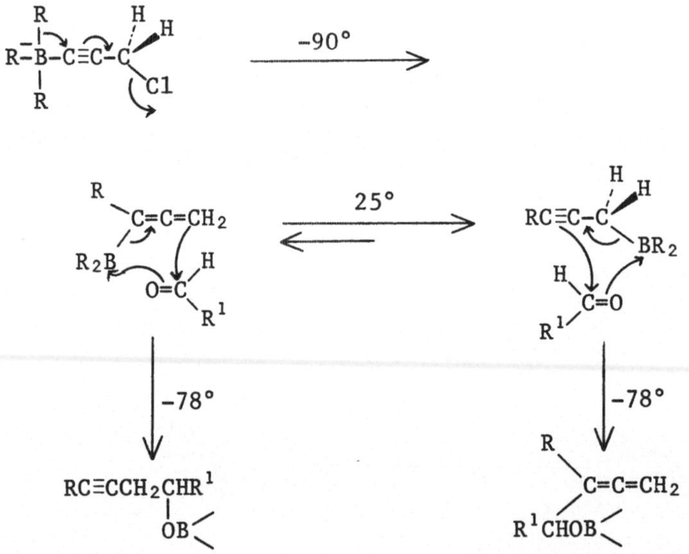

$$
\text{R–B–C}\equiv\text{C–C} \underset{\text{Cl}}{\overset{\text{H}}{\diagdown}}\text{H}
\xrightarrow{\;-90°\;}
$$

$$
\underset{R_2B}{\overset{R}{\diagdown}}C=C=CH_2
\quad\overset{25°}{\underset{\longleftarrow}{\longrightarrow}}\quad
RC\equiv C–C
\qquad (63)
$$

$$
\text{RC}\equiv\text{CCH}_2\text{CHR}^1 \qquad\qquad \underset{R^1CHOB}{\overset{R}{\diagdown}}C=C=CH_2
$$

$$
\underset{OB}{}
$$

lithium chloropropargylide at -90° initially rearranges to the allenic borane. Treatment of this with the aldehyde at -78° results in an allenic-propargylic rearrangement to give, after oxidative work-up, the homopropargylic alcohol. However, if the allenic borane is allowed to warm up to room temperature prior to addition of the aldehyde, it rearranges to the thermodynamically more stable propargylic borane. This then reacts with the aldehyde at -78° with bond transposition to produce the α-allenic alcohol.

It should be noted that utilization of ketones instead of aldehydes in the above reaction leads to mixtures of homopropargylic and α-allenic alcohols. Also, in the light of these findings, it appears that the actual precursors for the alkylallenes formed on protonolysis of the lithium chloropropargylide-trialkylborane reaction mixtures at room temperature are in fact the propargylic boranes.

The facile reaction with aldehydes appears to be a common feature of organoboranes embodying the allylic, allenic, and propargylic moieties as exemplified in equations 64-68. On the

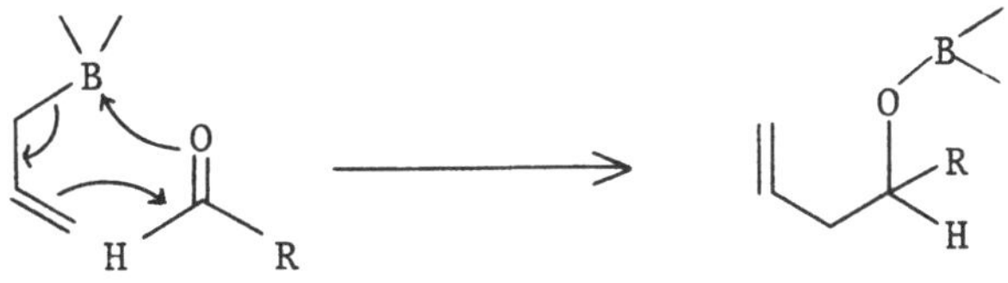

$$(64)^{31}$$

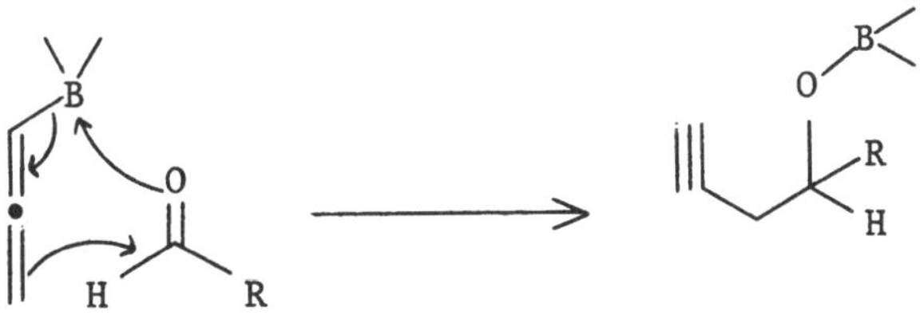

$$(65)^{30,32}$$

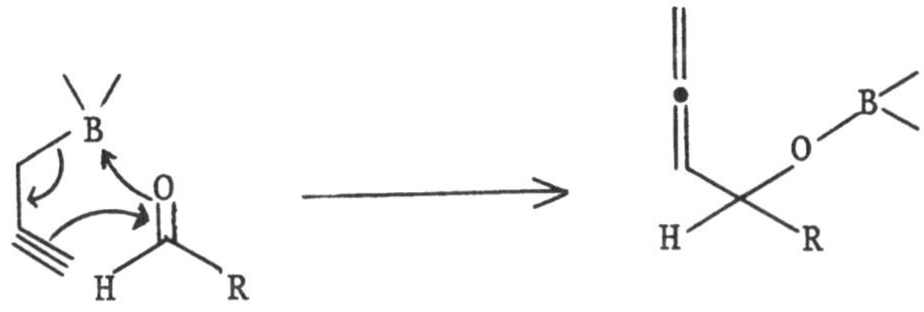

$$(66)^{30,32}$$

other hand, alkenylboranes do not react at ordinary temperature with the carbonyl groups of aldehydes and ketones. Thus, B-alkenyl-9-BBN derivatives add across the carbonyl group of simple aldehydes after refluxing for prolonged periods of time to produce, after oxidative work-up, the corresponding allylic alcohols (eq 67).[33]

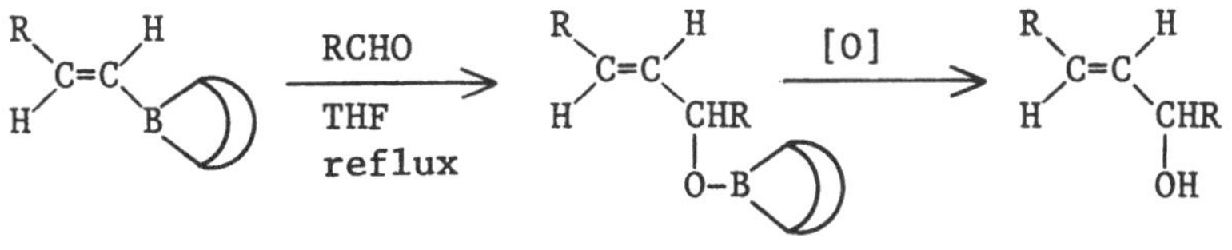

$$(67)$$

Alkenylaluminum compounds exhibit much greater reactivity toward the carbonyl groups of aldehydes and ketones than do the corresponding alkenylboron reagents. Thus, the reactions of alkenyldiisobutylalane with aldehydes or ketones produce, after work-up, the corresponding allylic alcohols[34,35], as exemplified in equations 68-70.[35] The intermolecular alkenyl group transfer proceeds in a stereospecific manner to produce the (E)-alcohols.

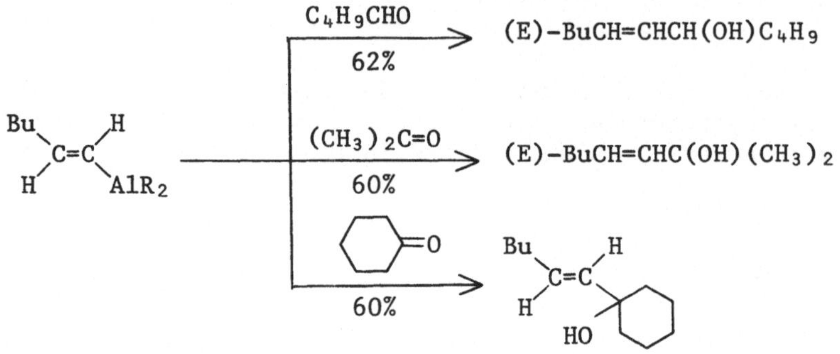

$$\xrightarrow{\underset{62\%}{C_4H_9CHO}} \quad (E)\text{-BuCH=CHCH(OH)}C_4H_9 \quad (68)$$

$$\underset{\substack{Bu \\ H}}{\overset{\substack{H \\ AlR_2}}{C=C}} \xrightarrow{\underset{60\%}{(CH_3)_2C=O}} \quad (E)\text{-BuCH=CHC(OH)}(CH_3)_2 \quad (69)$$

$$\xrightarrow[60\%]{\bigcirc=O} \quad \underset{\substack{H \\ HO}}{\overset{\substack{Bu \quad H}}{C=C}}\bigcirc \quad (70)$$

Extensions of the above reaction to produce one-carbon homologated alcohols were not successful. However, conversion of the alkenyldiisobutylalanes into the *ate* complexes with methyllithium, followed by addition of a two-fold excess of paraformaldehyde, afforded the desired allylic alcohols in good yields (eq 71).[36]

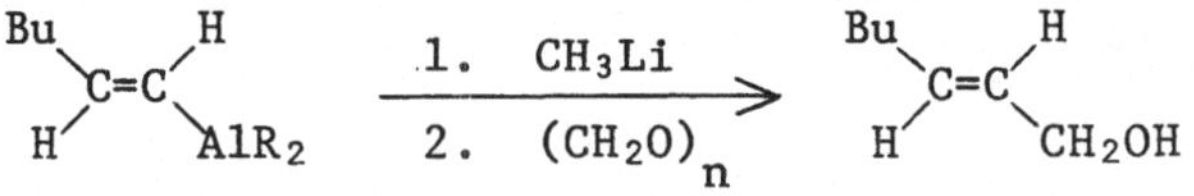

$$\underset{\substack{Bu \\ H}}{\overset{\substack{H \\ AlR_2}}{C=C}} \xrightarrow{\substack{1. \quad CH_3Li \\ 2. \quad (CH_2O)_n}} \underset{\substack{Bu \\ H}}{\overset{\substack{H \\ CH_2OH}}{C=C}} \quad (71)$$

It is evident that the transfer of a *trans*-alkenyl group from (E)-1-alken-1-yldiisobutylalanes onto carbonyl reagents is of considerable interest for organic synthesis. To delineate the scope of the alkenyl group transfer reaction, we investigated one-carbon homologation reactions of alkenylaluminum compounds using carbon dioxide, alkyl chloroformates, and chloromethyl ethyl ether as carbon electrophiles.

The treatment of *trans*-1-alken-1-yldiisobutylalanes with an excess of carbon dioxide, followed by a hydrolytic work-up, afforded the *trans*-α,β-unsaturated acids in only modest yields (30-37%).[36] However, we have observed that conversion of the vinylalanes into the *ate* complexes by addition of methyllithium enhances their reactivity toward certain of the electrophilic reagents. For example, the reaction of lithium *trans*-1-alken-1-yldiisobutylmethylalanates with carbon dioxide provides a high yield synthesis of *trans*-α,β-unsaturated acids (eq 72).[36]

$$BuC\equiv CH \quad \xrightarrow{\begin{array}{l}1. \quad R_2AlH \\ 2. \quad CH_3Li \\ 3. \quad CO_2 \\ 4. \quad H_3O^+\end{array}} \quad \underset{H}{\overset{Bu}{>}}C=C\underset{CO_2H}{\overset{H}{<}} \qquad (72)$$

Interestingly, the direct synthesis of isomerically pure *trans*-α,β-unsaturated esters from 1-alkyne derived vinylalanes and ethyl or methyl chloroformate does not require the intermediacy of the corresponding *ate* complexes (eq 73).[37] Likewise,

$$\text{cyclohexyl}-C\equiv CH \quad \xrightarrow{\begin{array}{l}1. \quad R_2AlH \\ 2. \quad ClCO_2C_2H_5 \\ 3. \quad H_3O^+\end{array}} \quad \underset{H}{\overset{\text{cyclohexyl}}{>}}C=C\underset{CO_2C_2H_5}{\overset{H}{<}} \qquad (73)$$

trans-alkenylalanes react directly with chloromethyl ethyl ether to produce *trans*-allyl ethyl ethers (eq 74).[38]

$$t\text{-}C_4H_9C\equiv CH \quad \xrightarrow{\begin{array}{l}1. \quad R_2AlH \\ 2. \quad ClCH_2OC_2H_5 \\ 3. \quad H_3O^+\end{array}} \quad \underset{H}{\overset{t\text{-}C_4H_9}{>}}C=C\underset{CH_2OC_2H_5}{\overset{H}{<}} \qquad (74)$$

The preceeding discussion has clearly supported our introductory contention that, although alkenylaluminum and alkenylboron compounds are structurally related, they exhibit very different reaction characteristics. This provides the practicing chemist with a powerful tool for the construction of complex organic molecules.

ACKNOWLEDGMENTS

I am deeply indebted to my former mentor, Professor H. C. Brown, for introducing me to the fascinating area of organoboranes, and I wish to express my gratitude to my dedicated coworkers who have made this presentation possible. Also, I would like to thank the National Science Foundation for their financial assistance in support of this work.

REFERENCES

1. G. Wilke and H. Müller, *Chem. Ber.*, __89__, 444 (1956); *Annalen*, __629__, 222 (1960).
2. H. C. Brown and G. Zweifel, *J. Am. Chem. Soc.*, __81__, 1512 (1959); *ibid.*, __83__, 3834 (1961).

3. E. Negishi, "Organoboron and Organoaluminum Compounds as Unique Nucleophiles in Organic Synthesis", ed. D. Seyferth, Elsevier, Amsterdam, 1976, pp. 93-125.

4. R. L. Miller, Ph.D. Thesis, University of California, Davis, 1971.

5. G. Zweifel, G. M. Clark, and R. A. Lynd, *Chem. Comm.*, 1593 (1971).

6. P. Binger, *Angew. Chem., Internat. Edn.*, **2**, 686 (1963).

7. G. Zweifel, G. M. Clark, and N. L. Polston, *J. Am. Chem. Soc.*, **93**, 3395 (1971).

8. G. Zweifel and R. P. Fisher, *Synthesis*, 339 (1974).

9. G. Zweifel and C. C. Whitney, *J. Am. Chem. Soc.*, **89**, 2753 (1967); G. Zweifel and H. P. On, to be published.

10. J. J. Eisch and M. W. Foxton, *J. Org. Chem.*, **36**, 3520 (1971).

11. K. Uchida, K. Utimoto, and H. Nozaki, *J. Org. Chem.*, **41**, 2215 (1976).

12. G. Zweifel and W. Lewis, *J. Org. Chem.*, **43**, 2739 (1978).

13. H. Steinmetz and R. M. Noyes, *J. Am. Chem. Soc.*, **74**, 4141 (1952).

14. Professor R. B. Miller, personal communication.

15. H. C. Brown, D. H. Bowman, S. Misumi, and M. K. Unni, *J. Am. Chem. Soc.*, **89**, 4531 (1967).

16. G. Zweifel, H. Arzoumanian, and C. C. Whitney, *J. Am. Chem. Soc.*, **89**, 3652 (1967).

17. G. Zweifel, R. P. Fisher, J. T. Snow, and C. C. Whitney, *J. Am. Chem. Soc.*, **93**, 6309 (1971).

18. G. Zweifel, J. T. Snow, and C. C. Whitney, *J. Am. Chem. Soc.*, **90**, 7139 (1968).

19. G. Zweifel, R. P. Fisher, J. T. Snow, and C. C. Whitney, *J. Am. Chem. Soc.*, **94**, 6560 (1972).

20. Reference 19, footnote 6.

21. G. Zweifel and H. Arzoumanian, *J. Am. Chem. Soc.*, **89**, 5086 (1967).

22. J. J. Eisch, H. Gopal, and S. Rhee, *J. Org. Chem.*, **40**, 2064 (1975).

23. G. Zweifel and W. Lewis, work in progress.

24. G. Zweifel and M. R. Maledy Fisher, unpublished work.

25. G. Zweifel and R. P. Fisher, unpublished work.

26. H. C. Brown and E. Negishi, *Tetrahedron*, **33**, 2331 (1977).

27. G. Zweifel and R. P. Fisher, *Synthesis*, 557 (1972).

28. G. Zweifel and S. J. Backlund, work in progress.

29. T. Leung and G. Zweifel, *J. Am. Chem. Soc.*, **96**, 5620 (1974).

30. G. Zweifel, S. J. Backlund, and T. Leung, *J. Am. Chem. Soc.*, **100**, 0000 (1978).

31. B. M. Mikhailov, *Organomet. Chem. Rev.*, [A] **8**, 1 (1972).

32. E. Favre and M. Gaudemar, *J. Organometal. Chem.*, **76**, 297 (1974).

33. P. Jacob, III and H. C. Brown, *J. Org. Chem.*, **42**, 579 (1977).

34. H. Newman, *Tetrahedron Lett.*, 4571 (1971).

35. R. A. Lynd, Ph.D. Thesis, University of California, Davis, 1973.
36. G. Zweifel and R. B. Steele, *J. Am. Chem. Soc.*, <u>89</u>, 2754 (1967).
37. G. Zweifel and R. A. Lynd, *Synthesis*, 625 (1976).
38. G. Zweifel and R. A. Lynd, *Synthesis*, 816 (1976).

NEW APPLICATIONS OF ORGANOMERCURY, -PALLADIUM AND -RHODIUM COMPOUNDS IN ORGANIC SYNTHESIS

Richard C. Larock

Department of Chemistry, Iowa State University

Ames, Iowa 50011

Organomercurials have a long history dating back over 125 years. Although initial interest in organomercurials was due primarily to their medicinal properties, they soon found widespread use in the synthesis of other organometallics. This application was relatively short-lived with the development of the more reactive organolithium and Grignard reagents at the turn of the century. Due to their low chemical reactivity towards organic substrates, organomercurials never developed the widespread synthetic utilization that these other organometallics have.

Recently, however, with increasing interest in the development of highly chemospecific organometallic reagents which will tolerate considerable functionality, organomercurials have received renewed interest as potential synthetic intermediates. In this respect, the organomercurials possess a number of very attractive features. They readily accommodate essentially all important organic functionality and exhibit remarkable chemical and thermal stability. They are readily available by a number of synthetic routes and are generally high melting crystalline solids which can be readily stored under air for long periods of time. In fact a number of organomercurials are now commercially available. The facility with which organomercurials undergo transmetallation reactions with a number of transition metal salts, particularly palladium, has greatly expanded their synthetic potential and recently provided a number of new procedures of interest to the synthetic organic chemist. It is some of these new developments which will be discussed here.

Organomercurials of all types are readily available *via* a number of different routes. These procedures have been reviewed several times in recent years.[1-4] Arylmercurials are most readily available *via* direct electrophilic mercuration of the corresponding arenes (eqs 1-3).[1,2]

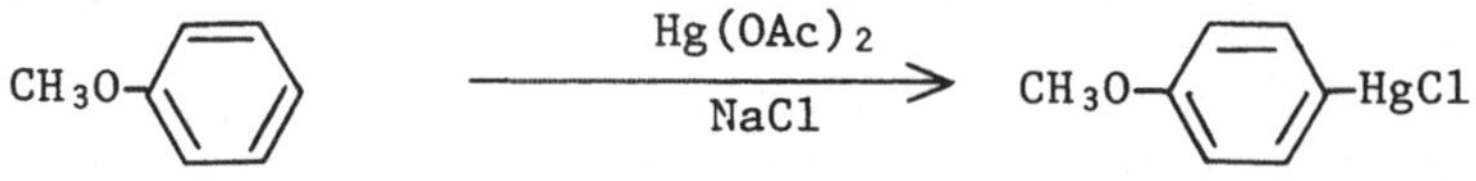

$$CH_3O-\text{⟨⟩} \xrightarrow[\text{NaCl}]{\text{Hg(OAc)}_2} CH_3O-\text{⟨⟩}-HgCl \quad (1)$$

$$\underset{NO_2}{\text{⟨⟩}} \xrightarrow[\text{NaCl}]{\text{Hg(ClO}_4)_2} \underset{NO_2}{\text{⟨⟩}}-HgCl \quad (2)$$

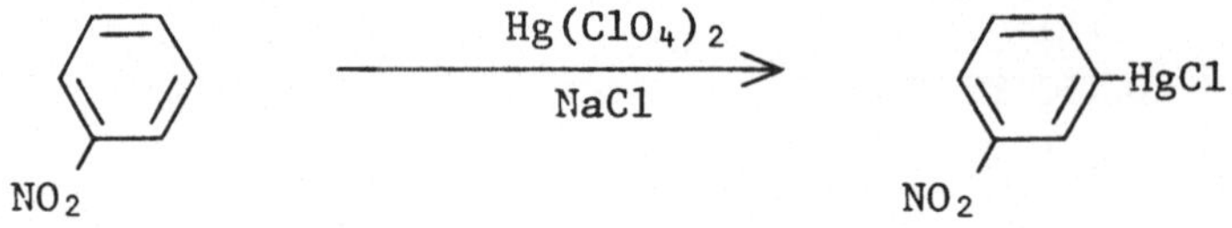

$$\underset{S}{\text{⟨⟩}} \xrightarrow{\text{HgCl}_2} \underset{S}{\text{⟨⟩}}HgCl \quad (3)$$

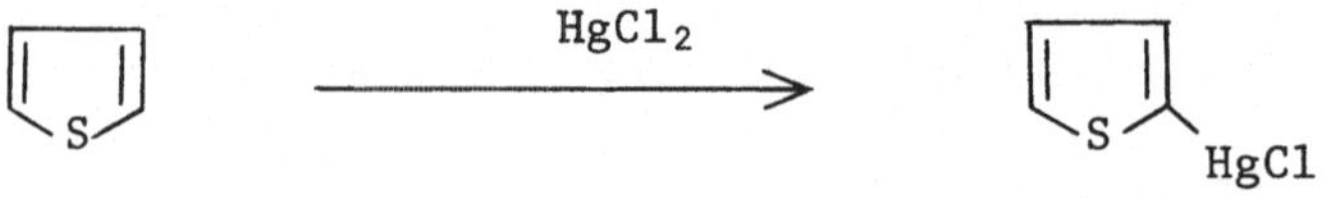

Vinylmercurials can be prepared by a variety of procedures. In our own work we have made extensive use of the hydroboration-mercuration of acetylenes using dicyclohexylborane (eqs 4,5).[5]

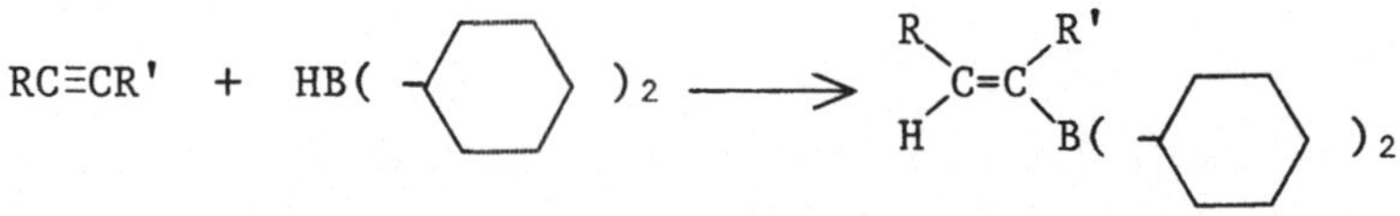

$$RC\equiv CR' + HB(\text{⟨⟩})_2 \longrightarrow \underset{H}{\overset{R}{>}}C=C\underset{B(\text{⟨⟩})_2}{\overset{R'}{<}} \quad (4)$$

$$\underset{H}{\overset{R}{>}}C=C\underset{B(\text{⟨⟩})_2}{\overset{R'}{<}} + Hg(OAc)_2 \xrightarrow[H_2O]{NaCl} \underset{H}{\overset{R}{>}}C=C\underset{HgCl}{\overset{R'}{<}} \quad (5)$$

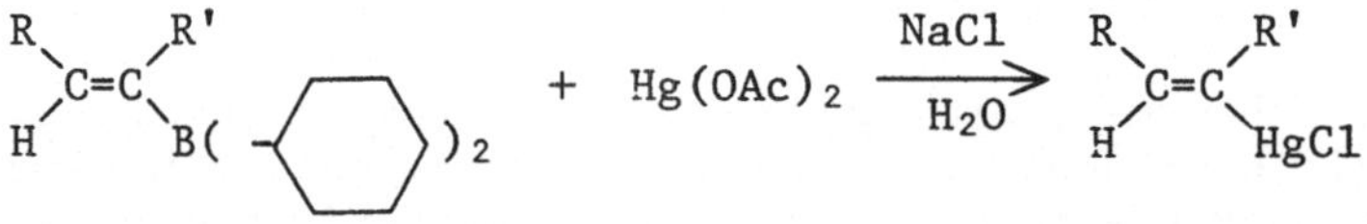

The following equations indicate the generality of this approach and provide some representative yields based on starting acetylene (eqs 6-8). This approach readily accommodates a number of

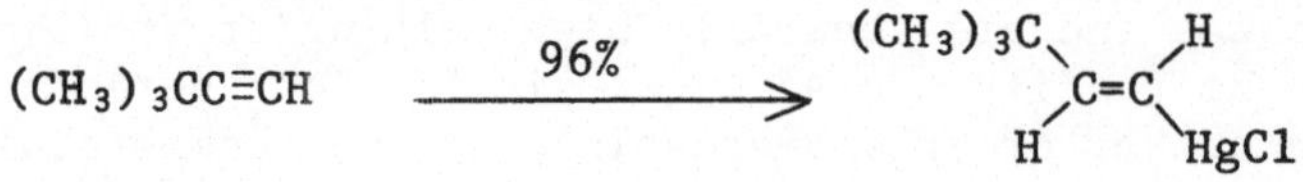

$$(CH_3)_3CC\equiv CH \xrightarrow{96\%} \underset{H}{\overset{(CH_3)_3C}{>}}C=C\underset{HgCl}{\overset{H}{<}} \quad (6)$$

$$\text{⟨⟩}-C\equiv CH \xrightarrow{98\%} \underset{H}{\overset{C_6H_5}{>}}C=C\underset{HgCl}{\overset{H}{<}} \quad (7)$$

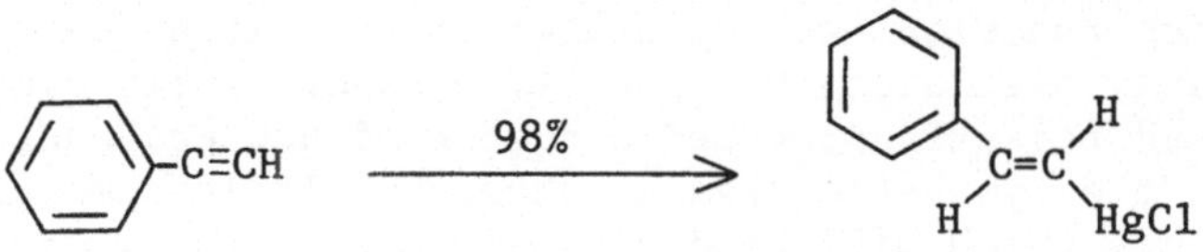

$$C_2H_5C{\equiv}CC_2H_5 \xrightarrow{59\%} \underset{H}{\overset{C_2H_5}{>}}C{=}C\underset{HgCl}{\overset{C_2H_5}{<}} \tag{8}$$

functional groups, but the yields do decrease when using internal acetylenes as indicated by the last example. It is important to note that high regioselectivity can be achieved in this reaction with internal acetylenes bearing alkyl groups of substantially different steric bulk, such as *tert*-butyl methyl acetylene. Mercuration occurs at the less sterically hindered end of the triple bond.

The yields of vinylmercurials derived from internal acetylenes can be improved by employing a hydroboration–mercuration sequence using catecholborane (eqs 9,10).[6] Some representative examples

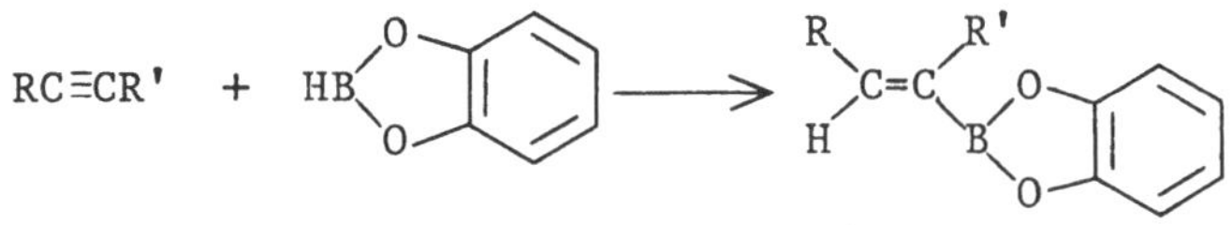

$$RC{\equiv}CR' + HB\overset{O}{\underset{O}{<}}\text{(catechol)} \longrightarrow \underset{H}{\overset{R}{>}}C{=}C\overset{R'}{<}\underset{B}{}\overset{O}{\underset{O}{<}} \tag{9}$$

$$\underset{H}{\overset{R}{>}}C{=}C\overset{R'}{<}\underset{B}{}\overset{O}{\underset{O}{<}} + Hg(OAc)_2 \xrightarrow[H_2O]{NaCl} \underset{H}{\overset{R}{>}}C{=}C\underset{HgCl}{\overset{R'}{<}} \tag{10}$$

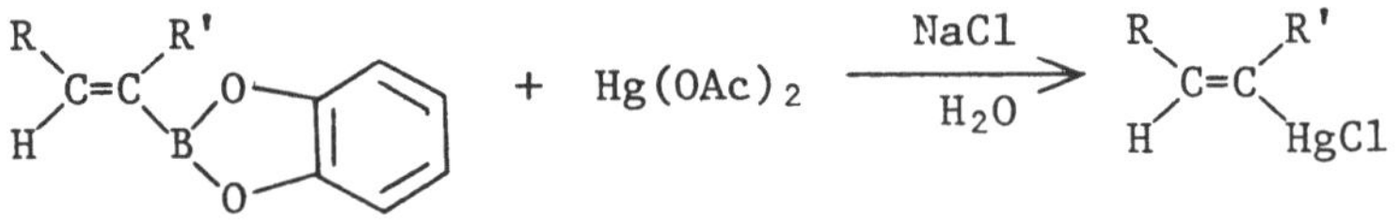

are indicated below with isolated yields of the corresponding alkenylboranes and vinylmercurials indicated (eqs 11–13). The

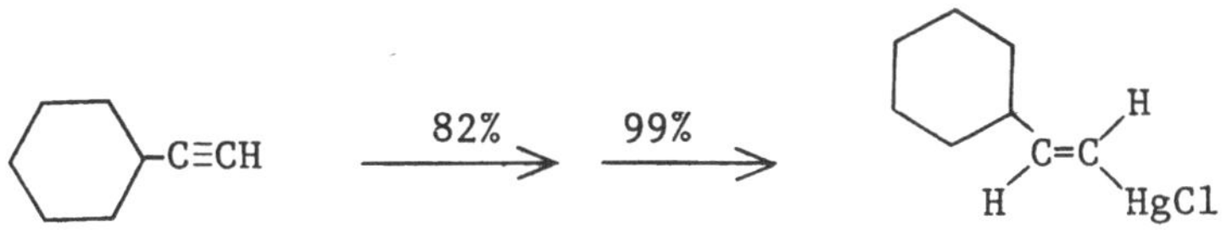

$$\text{(cyclohexyl)}{-}C{\equiv}CH \xrightarrow{82\%} \xrightarrow{99\%} \underset{H}{\overset{\text{(cyclohexyl)}}{>}}C{=}C\underset{HgCl}{\overset{H}{<}} \tag{11}$$

$$C_2H_5C{\equiv}CC_2H_5 \xrightarrow{85\%} \xrightarrow{98\%} \underset{H}{\overset{C_2H_5}{>}}C{=}C\underset{HgCl}{\overset{C_2H_5}{<}} \tag{12}$$

$$(CH_3)_3CC{\equiv}CCH_3 \xrightarrow{89\%} \xrightarrow{97\%} \underset{H}{\overset{(CH_3)_3C}{>}}C{=}C\underset{HgCl}{\overset{CH_3}{<}} \tag{13}$$

mercuration step proceeds in quantitative yields with all of these organoboranes.

Normant has also reported an approach to vinylmercurials *via* acetylene addition reactions of organocopper species (eqs 14–16).[7]

$$RMgBr \ + \ CuBr \ \xrightarrow{\hspace{3cm}} \ RCu \cdot MgBr_2 \tag{14}$$

$$RCu \cdot MgBr_2 \ + \ R'C{\equiv}CH \ \xrightarrow{\hspace{2cm}} \ \underset{R}{\overset{R'}{>}}C{=}C\underset{Cu \cdot MgBr_2}{\overset{H}{<}} \tag{15}$$

$$\underset{R}{\overset{R'}{>}}C{=}C\underset{Cu \cdot MgBr_2}{\overset{H}{<}} \ + \ HgBr_2 \ \longrightarrow \ \underset{R}{\overset{R'}{>}}C{=}C\underset{HgBr}{\overset{H}{<}} \tag{16}$$

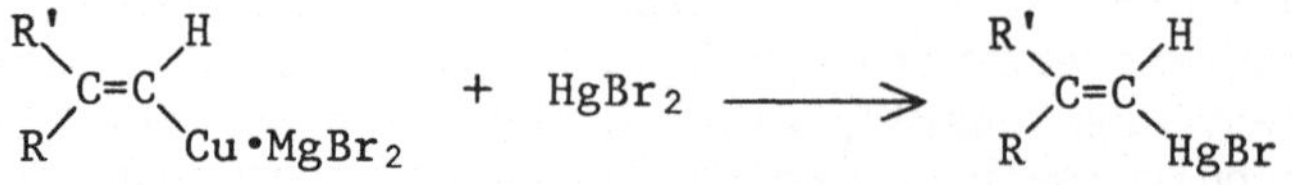

This sequence allows introduction of an alkyl group into the β-position *cis* to the mercury moiety.

β-Acetoxyvinylmercurials can also be readily obtained by simply stirring internal acetylenes with mercuric acetate in acetic acid overnight and then pouring the solution into aqueous sodium chloride (eqs 17,18).[8-10] Both *cis* and *trans* additions

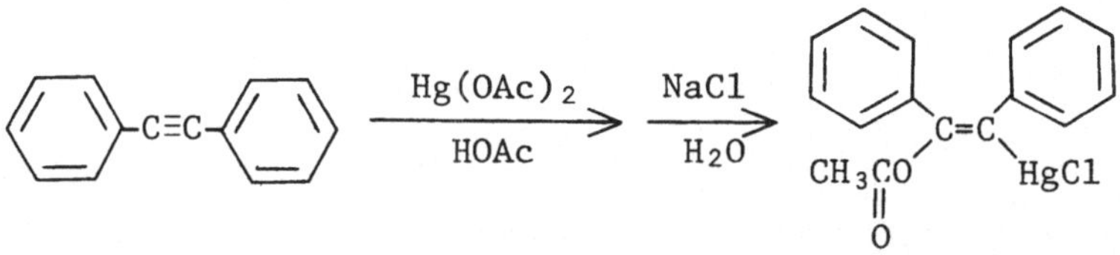

$$Ph{-}C{\equiv}C{-}Ph \ \xrightarrow[HOAc]{Hg(OAc)_2} \ \xrightarrow[H_2O]{NaCl} \ \underset{CH_3\underset{O}{\overset{\|}{C}}O}{\overset{Ph}{>}}C{=}C\underset{HgCl}{\overset{Ph}{<}} \tag{17}$$

$$CH_3C{\equiv}CCH_3 \ \longrightarrow \ \longrightarrow \ \underset{CH_3}{\overset{CH_3\overset{O}{\overset{\|}{C}}O}{>}}C{=}C\underset{HgCl}{\overset{CH_3}{<}} \tag{18}$$

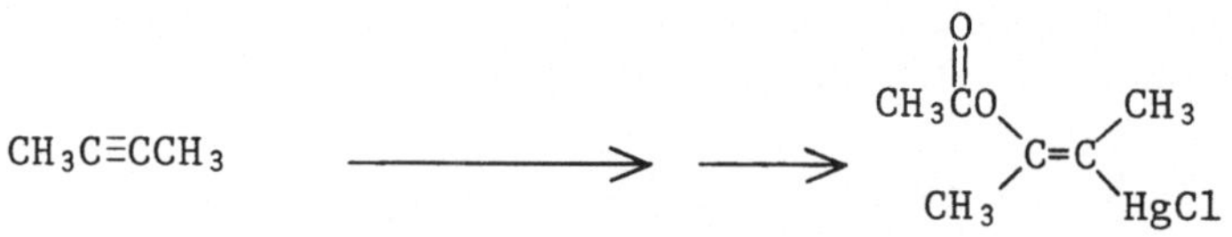

are observed. It is reported, however, that the stereochemistry of the 2-butyne addition product can be reversed by simply heating the initial addition product prior to isolation (eq 19).[10]

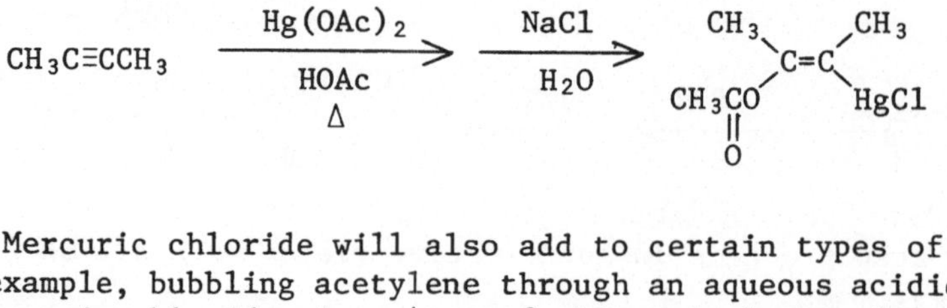

$$CH_3C{\equiv}CCH_3 \ \xrightarrow[\substack{HOAc \\ \Delta}]{Hg(OAc)_2} \ \xrightarrow[H_2O]{NaCl} \ \underset{CH_3\underset{O}{\overset{\|}{C}}O}{\overset{CH_3}{>}}C{=}C\underset{HgCl}{\overset{CH_3}{<}} \tag{19}$$

Mercuric chloride will also add to certain types of acetylenes. For example, bubbling acetylene through an aqueous acidic solution of mercuric chloride gives *trans*-β-chlorovinylmercuric chloride, easily isolated by simple filtration (eq 20). Analogous reactions have not yet been reported for other simple alkynes, but

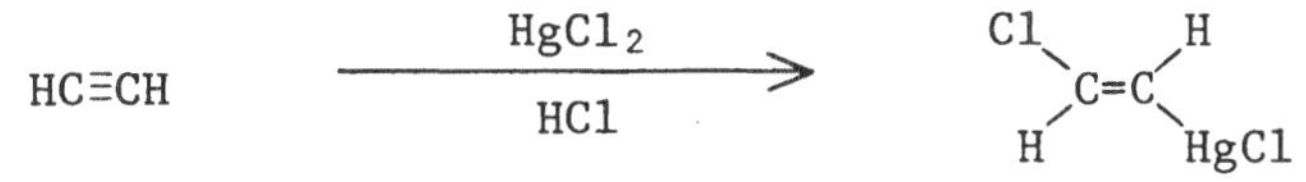

$$HC\equiv CH \xrightarrow[\text{HCl}]{\text{HgCl}_2} \begin{array}{c} Cl \\ \diagdown \\ H \end{array} C=C \begin{array}{c} H \\ \diagup \\ HgCl \end{array} \qquad (20)$$

propargylic alcohols readily add mercuric chloride when treated with saturated aqueous solutions of sodium chloride and mercuric chloride (eqs 21-23).[12] Unfortunately, the yields are often low

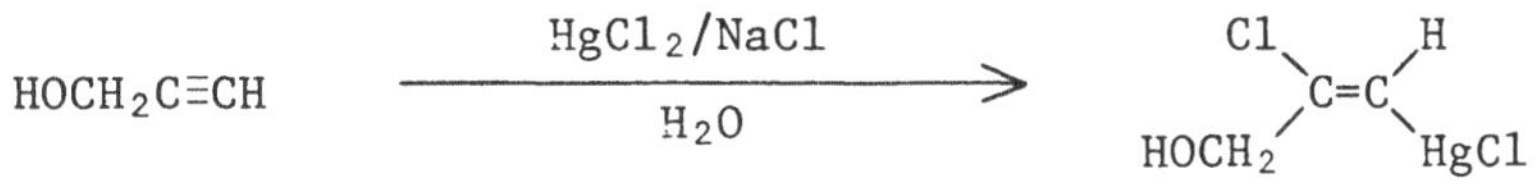

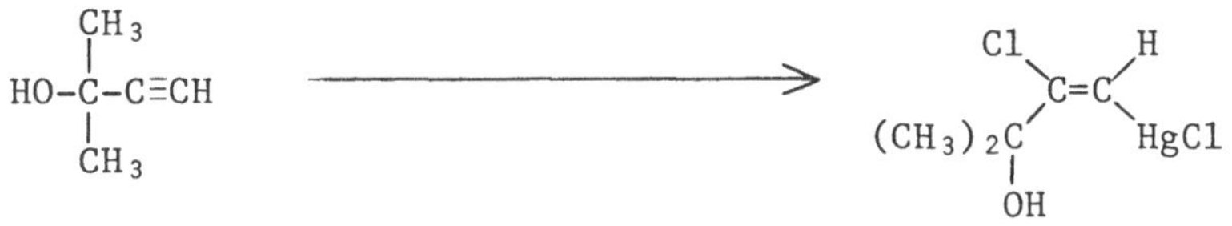

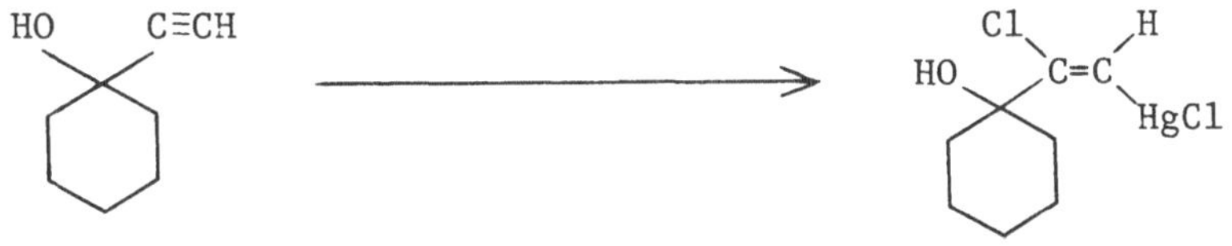

$$HOCH_2C\equiv CH \xrightarrow[\text{H}_2\text{O}]{\text{HgCl}_2/\text{NaCl}} \begin{array}{c} Cl \\ \diagdown \\ HOCH_2 \end{array} C=C \begin{array}{c} H \\ \diagup \\ HgCl \end{array} \qquad (21)$$

54%

$$\begin{array}{c} CH_3 \\ | \\ HO-C-C\equiv CH \\ | \\ CH_3 \end{array} \xrightarrow{\qquad\qquad} \begin{array}{c} Cl \\ \diagdown \\ (CH_3)_2C \\ | \\ OH \end{array} C=C \begin{array}{c} H \\ \diagup \\ HgCl \end{array} \qquad (22)$$

31%

31%

and the reaction appears to be limited to relatively low molecular weight symmetrical primary and tertiary alcohols. Nevertheless, these compounds have proven useful for the synthesis of butenolides as described later.

We have recently examined analogous mercuration reactions of propargylic amines.[13] Using conditions identical to the pro-pargylic alcohols, one obtains only highly insoluble, presumably polymeric, material. However, addition of the amines to solutions containing approximately 18% HCl and 2M $HgCl_2$ gives rise to slow formation of a powdery white precipitate. The reaction generally requires 12-24 hours to complete. Although this material general-ly possesses a very sharp melting point after simple filtration, these compounds do not appear to represent any simple stoichio-metry as judged by elemental analysis. Nevertheless, elemental analyses, NMR spectra and X-ray crystallographic data (only por-tions of the structure could be solved due to the large number of chlorine and mercury atoms present) on these compounds suggest structures of the following type (eqs 24-26). Unlike the corresponding alcohols, the amines give mercuric chloride addition

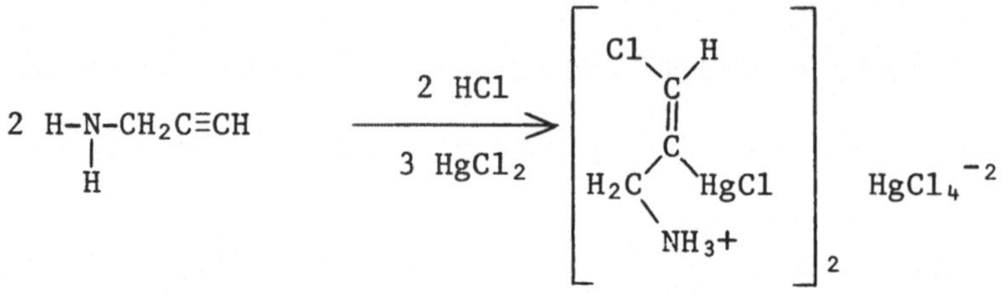

$$2 \ \underset{\underset{H}{|}}{H-N}-CH_2C\equiv CH \quad \xrightarrow[3 \ HgCl_2]{2 \ HCl} \quad \left[\begin{array}{c} \underset{Cl}{}\underset{C}{}\underset{H}{} \\ \| \\ \underset{H_2C}{}\underset{C}{}\underset{HgCl}{} \\ | \\ NH_3+ \end{array} \right]_2 HgCl_4^{-2} \tag{24}$$

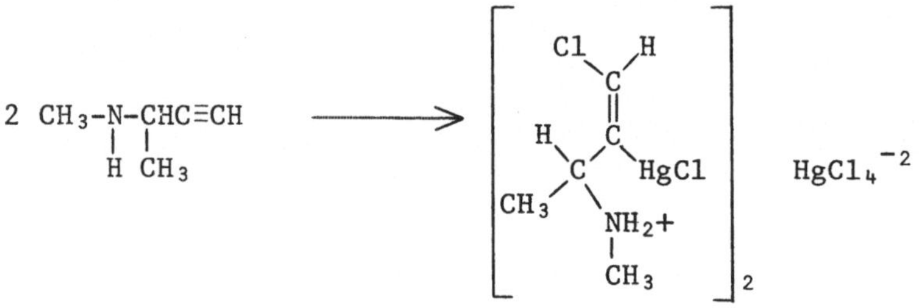

$$2 \ CH_3-\underset{\underset{H \ \ CH_3}{| \ \ |}}{N}-CHC\equiv CH \quad \longrightarrow \quad \left[\begin{array}{c} \underset{Cl}{}\underset{C}{}\underset{H}{} \\ \| \\ \underset{H}{}\underset{C}{}\underset{HgCl}{} \\ \underset{CH_3}{} \ | \\ NH_2+ \\ | \\ CH_3 \end{array} \right]_2 HgCl_4^{-2} \tag{25}$$

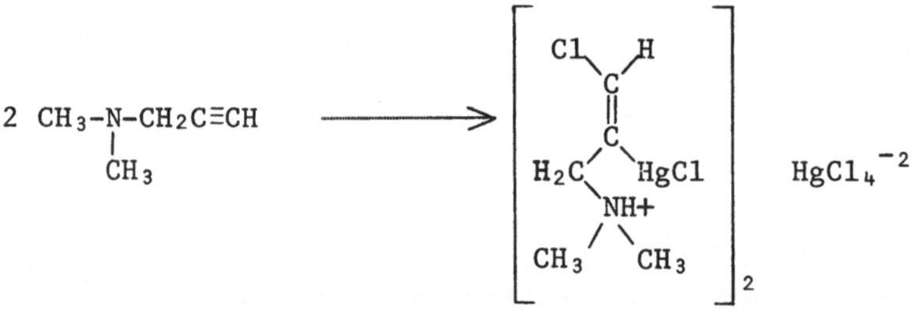

$$2 \ CH_3-\underset{\underset{CH_3}{|}}{N}-CH_2C\equiv CH \quad \longrightarrow \quad \left[\begin{array}{c} \underset{Cl}{}\underset{C}{}\underset{H}{} \\ \| \\ \underset{H_2C}{}\underset{C}{}\underset{HgCl}{} \\ | \\ NH+ \\ \underset{CH_3 \ \ CH_3}{/ \ \ \backslash} \end{array} \right]_2 HgCl_4^{-2} \tag{26}$$

products in which the mercury moiety adds to the internal carbon of the original terminal triple bond. This is presumably due to the strong electron-withdrawing effect of the ammonium ion. The impurity in these compounds would appear to be excess mercuric chloride which precipitates out with the product, especially at higher mercuric chloride concentrations. In spite of these difficulties, the reaction generally proceeds in high yield and tolerates a wide variety of structures, unlike the corresponding alcohol reactions.

With a wide variety of organomercurials readily available, we began several years ago a systematic examination of possible synthetic applications of these compounds. Our attention has focussed primarily on the aryl- and vinylmercurials because of the

diversity of structures available, and because these compounds
appear to be substantially more reactive than the alkylmercurials.
Of greatest synthetic interest have been carbon-carbon bond forming reactions, because of their great importance in organic
chemistry.

One of the first reactions we observed involves the treatment
of vinylmercurials with palladium chloride and lithium chloride
in hexamethylphosphoramide (HMPA) solvent at 0°C. This reaction
gives rise to essentially quantitative yields of the corresponding
symmetrical 1,3-dienes (eqs 27-29).[14] It works on vinylmercurials

$$
2 \quad \underset{H}{\overset{CH_3(CH_2)_2}{>}}C=C\underset{HgCl}{\overset{H}{<}} \quad \xrightarrow[\text{HMPA 0°C}]{PdCl_2/4LiCl} \quad \left(\underset{H}{\overset{CH_3(CH_2)_2}{>}}C=C\underset{}{\overset{H}{<}} \right)_2 \tag{27}
$$

98%

$$
2 \quad \underset{H}{\overset{CH_3O\overset{O}{\overset{\|}{C}}(CH_2)_8}{>}}C=C\underset{HgCl}{\overset{H}{<}} \quad \xrightarrow{\hspace{2cm}} \quad \left(\underset{H}{\overset{CH_3O\overset{O}{\overset{\|}{C}}(CH_2)_8}{>}}C=C\underset{}{\overset{H}{<}} \right)_2 \tag{28}
$$

94%

$$
2 \quad \underset{H}{\overset{}{>}}C=C\underset{HgCl}{\overset{}{<}} \quad \xrightarrow{\hspace{2cm}} \quad \left(\underset{H}{\overset{}{>}}C=C\underset{}{\overset{}{<}} \right)_2 \tag{29}
$$

100%

derived from functionally substituted acetylenes, as well as internal acetylenes. It also gives a good yield of diene when
applied to the mercuric acetate addition product of 2-butyne
(eq 30). Such dienes have previously found use in the synthesis

$$
2 \quad \underset{CH_3}{\overset{CH_3\overset{O}{\overset{\|}{C}}O}{>}}C=C\underset{HgCl}{\overset{CH_3}{<}} \quad \xrightarrow{\hspace{2cm}} \quad \left(\underset{CH_3}{\overset{CH_3\overset{O}{\overset{\|}{C}}O}{>}}C=C\underset{}{\overset{CH_3}{<}} \right)_2 \tag{30}
$$

82%

of arenes *via* Diels-Alder reactions (eq 31). The reaction is also
applicable to the synthesis of polyenes as indicated by the
following examples using vinylmercurials derived from

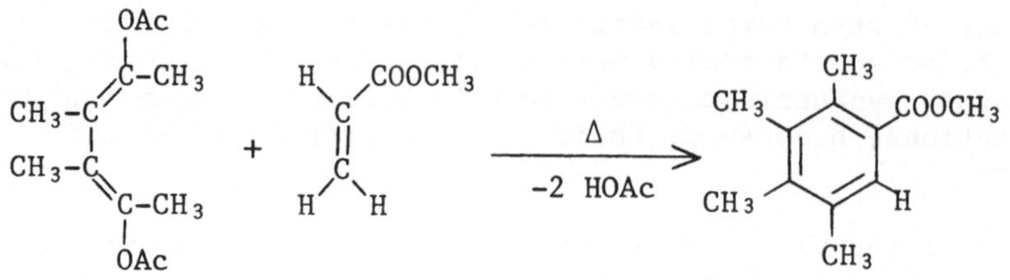

$$\text{(31)}$$

isopropenylacetylene and 2,4-hexadiyne *via* hydroboration-mercuration (eqs 32,33). These reactions are all highly dependent

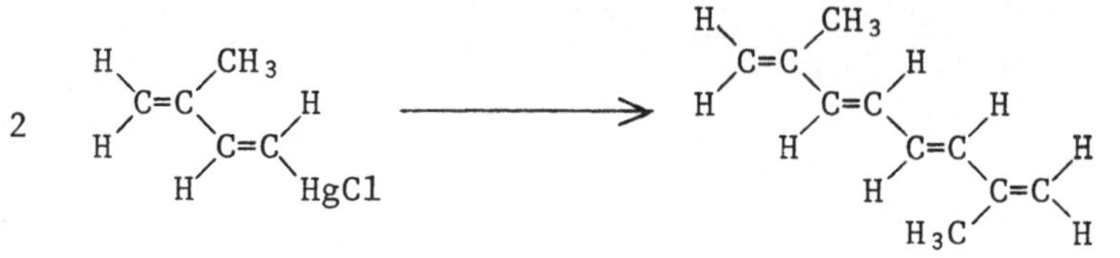

$$\text{(32)}$$

95%

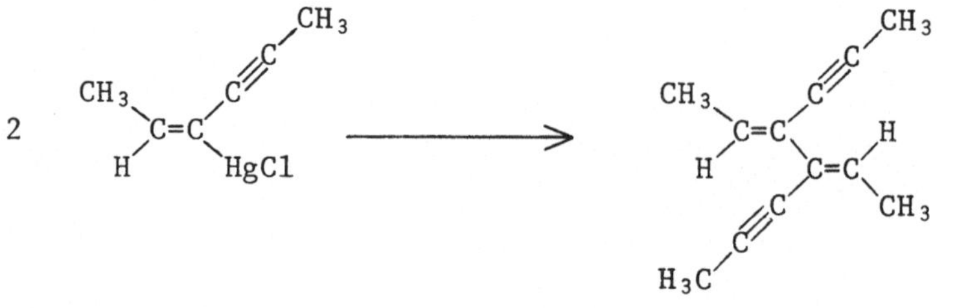

$$\text{(33)}$$

92%

on the reaction conditions. In the absence of lithium chloride or in less polar solvents than HMPA, the yield of symmetrical 1,3-diene decreases dramatically.

Although we have not pursued detailed mechanistic studies on any of our reactions due to our greater interest in extending the synthetic potential of this class of organometallics, we believe we can offer plausible mechanisms for most of the reactions we have observed. In the above-mentioned diene synthesis we suggest that the products arise by a transmetallation-reductive elimination sequence (eqs 34,35).

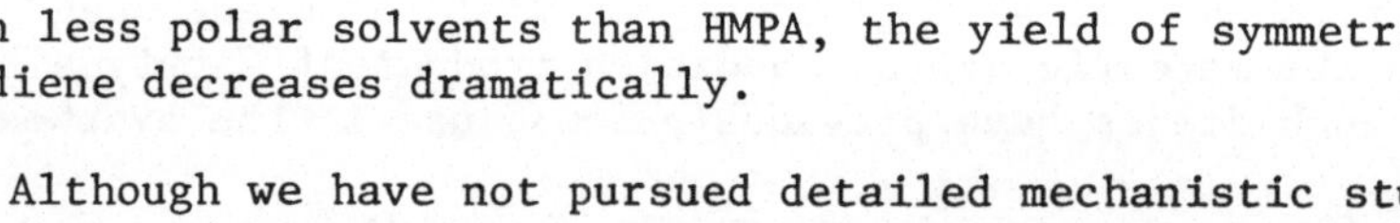

$$\text{(34)}$$

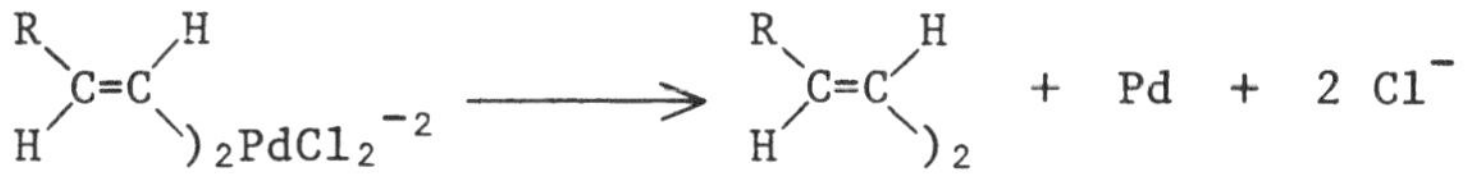

$$\underset{H}{\overset{R}{>}}C=C\underset{)_2PdCl_2^{-2}}{\overset{H}{<}} \longrightarrow \underset{H}{\overset{R}{>}}C=C\underset{)_2}{\overset{H}{<}} + Pd + 2 Cl^- \qquad (35)$$

Unfortunately, this approach to dienes has several disadvantages. It requires low temperatures, HMPA (which has since
been found to be carcinogenic) and stoichiometric amounts of expensive palladium chloride. All of these disadvantages have
recently been overcome by employing catalytic amounts of rhodium
salts.[15] As indicated by the following Table, any of a variety
of commercially available rhodium(I) or (III) salts will catalyti-
cally dimerize *trans*-1-hexenylmercuric chloride to *trans*, *trans*-
5,7-dodecadiene. Further comparison indicated that the rhodium
dicarbonyl chloride dimer in the presence of excess lithium
chloride gives the best results. It was further observed that
HMPA was no longer necessary and that a reaction time of 24 hours
gives better results. Under these conditions essentially

Table. Rhodium Catalyzed Dimerization of *trans*-1-Hexenylmercuric
        Chloride.[a]

Rhodium Complex	Diene Yield, %
$(\emptyset_3P)_3RhCl$	28
$[(CH_2=CH_2)_2RhCl]_2$	70
$(\emptyset_3P)_2Rh(CO)Cl$	74
$[(COD)RhCl]_2$	81
$[(CO)_2RhCl]_2$	87
$[(CO)_2RhCl]_2/LiCl$	95
$RhCl_3 \cdot nH_2O$	61
$RhCl_3 \cdot nH_2O/LiCl$	94

[a]All reactions were run with 10% "rhodium" per vinylmercurial for
6 hours at room temperature in HMPA.   [b]COD = 1,5-cyclooctadiene.

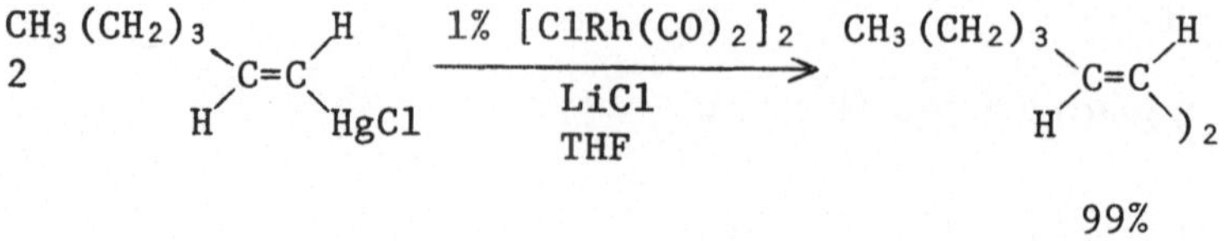

$$2 \quad \underset{H}{\overset{CH_3(CH_2)_3}{>}}C=C\underset{HgCl}{\overset{H}{<}} \quad \xrightarrow[\substack{LiCl \\ THF}]{1\% \ [ClRh(CO)_2]_2} \quad \left(\underset{H}{\overset{CH_3(CH_2)_3}{>}}C=C\underset{}{\overset{H}{<}} \right)_2 \tag{36}$$

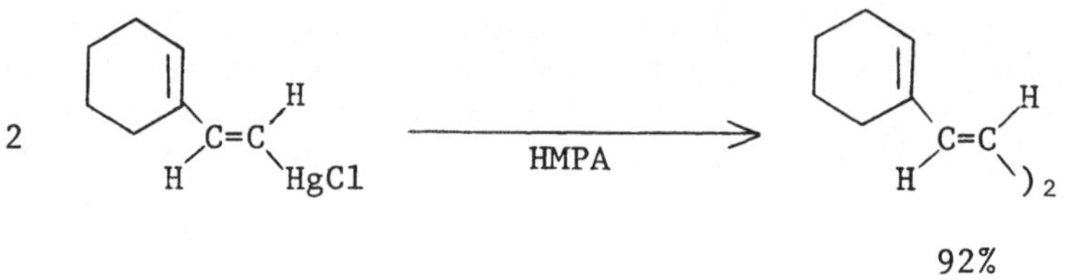

99%

$$2 \quad \underset{H}{\overset{(CH_3)_3C}{>}}C=C\underset{HgCl}{\overset{H}{<}} \quad \xrightarrow{THF} \quad \left(\underset{H}{\overset{(CH_3)_3C}{>}}C=C\underset{}{\overset{H}{<}} \right)_2 \tag{37}$$

100%

$$2 \quad \underset{H}{\overset{}{>}}C=C\underset{HgCl}{\overset{H}{<}} \quad \xrightarrow{HMPA} \quad \left(\underset{H}{\overset{}{>}}C=C\underset{}{\overset{H}{<}} \right)_2 \tag{38}$$

92%

quantitative yields can be obtained (eqs 36–38). Although this
reaction works equally well for the synthesis of polyenes,
vinylmercurials derived from internal acetylenes *via* hydroboration-
mercuration give sharply reduced yields and the palladium procedure
is superior in such cases.

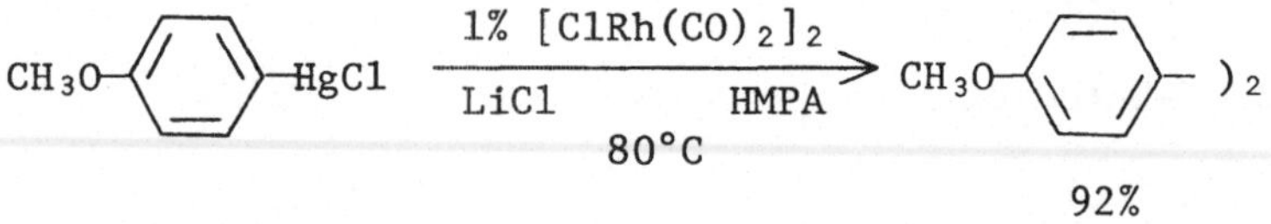

$$CH_3O\!-\!\!\langle\!\!\langle\quad\rangle\!\!\rangle\!-\!HgCl \quad \xrightarrow[\substack{LiCl \qquad HMPA \\ 80°C}]{1\% \ [ClRh(CO)_2]_2} \quad CH_3O\!-\!\!\left(\!\langle\!\!\langle\quad\rangle\!\!\rangle\!-\right)_2 \tag{39}$$

92%

$$\text{(naphthalene)}\!-\!HgCl \quad \xrightarrow{} \quad \text{(naphthalene)}\!-\!)_2 \tag{40}$$

94%

$$\text{(thiophene)}\!-\!HgCl \quad \xrightarrow{} \quad \text{(thiophene)}\!-\!)_2 \tag{41}$$

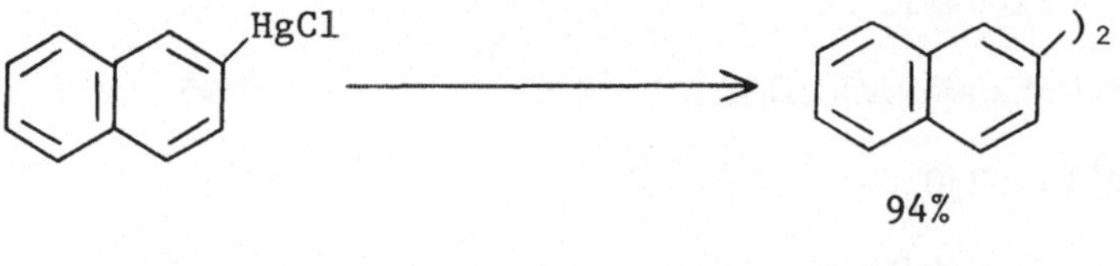

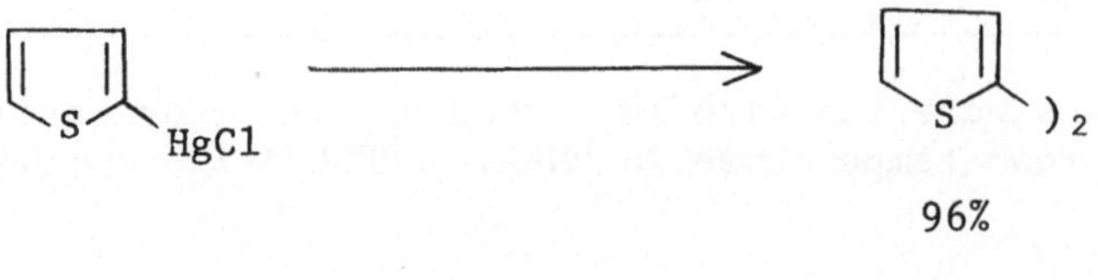

96%

The rhodium reactions work extremely well for the dimerization of arylmercurials as well (eqs 39-41). In this case, however, best results are obtained at 80°C and in the presence of HMPA. It appears that lower yields are obtained when strong electron-withdrawing groups, such as a nitro group, are attached to the aromatic ring.

In the catalytic rhodium reactions, we suggest a mechanism very similar to the palladium reactions. For rhodium(I) catalysis, we believe that a diorganorhodium(III) species is formed *via* transmetallation and oxidative addition (the order of these two reactions may well be reversed) reactions of the organomercurial and the rhodium(I) salt, and that this species then reductively eliminates the organic dimer and metallic mercury, and regenerates the rhodium(I) salt (eqs 42-44). When rhodium trichloride is employed, we believe that the diorganorhodium(III)

$$RHgCl \ + \ ClRh \ \longrightarrow \ R-Rh \ + \ HgCl_2 \qquad (42)$$

$$R-Rh \ + \ RHgCl \ \longrightarrow \ \underset{\underset{R}{|}}{R-Rh-HgCl} \qquad (43)$$

$$\underset{\underset{R}{|}}{R-Rh-HgCl} \ \longrightarrow \ R-R \ + \ ClRh \ + \ Hg \qquad (44)$$

species is formed by two transmetallation steps and that this intermediate proceeds to give the expected dimer and a rhodium(I) species which then catalyzes further dimerization as suggested in equations 42-44 above (eqs 45,46).

$$2 \ RHgCl \ + \ RhCl_3 \ \longrightarrow \ \underset{\underset{R}{|}}{R-Rh-Cl} \ + \ 2 \ HgCl_2 \qquad (45)$$

$$\underset{\underset{R}{|}}{R-Rh-Cl} \ \longrightarrow \ R-R \ + \ ClRh \qquad (46)$$

As mentioned earlier, the palladium promoted dimerization reactions require lithium chloride and a polar solvent such as HMPA. Upon reinvestigating this reaction, we have observed that a totally different mode of dimerization, namely "head-to-tail" dimerization, occurs in the absence of lithium chloride in a nonpolar solvent such as benzene (eqs 47-49).[16] Best results are obtained when two equivalents of triethylamine are added. Apparently the base reacts with the small amounts of hydrochloric

$$2 \quad \underset{H}{\overset{CH_3(CH_2)_3}{\diagdown}} C = C \underset{HgCl}{\overset{H}{\diagup}} \quad \xrightarrow[\substack{2 \ Et_3N \\ C_6H_6}]{1 \ PdCl_2} \quad \underset{H}{\overset{CH_3(CH_2)_3}{\diagdown}} C = C \underset{CH_3(CH_2)_3}{\overset{H}{\diagup}} C = C \underset{H}{\overset{H}{\diagup}} \quad (47)$$

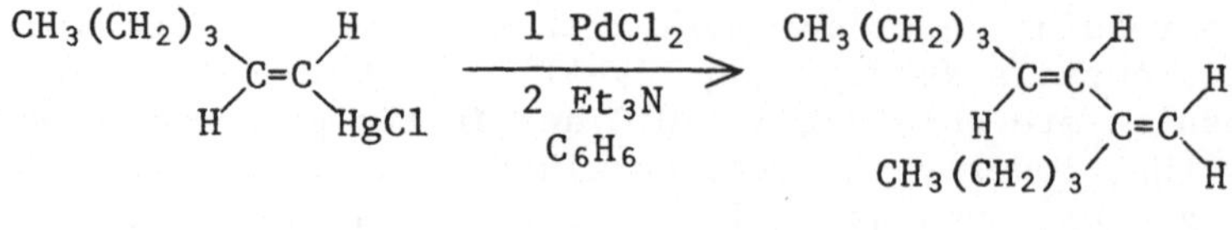

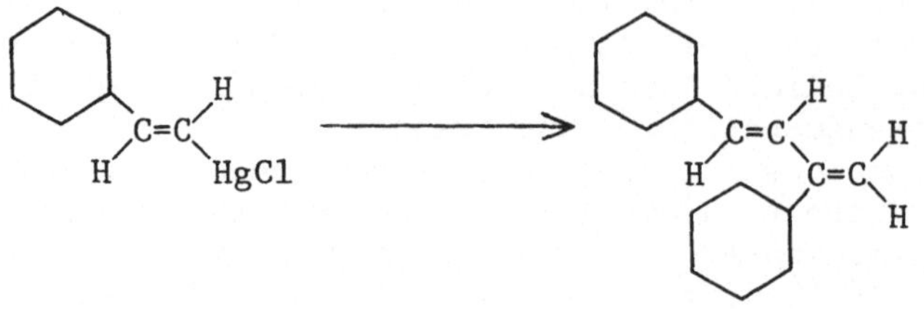

98%

$$2 \quad \text{(48)}$$

91%

$$2 \quad \underset{H}{\overset{(CH_3)_3C}{\diagdown}} C = C \underset{HgCl}{\overset{CH_3}{\diagup}} \quad \longrightarrow \quad \underset{H}{\overset{(CH_3)_3C}{\diagdown}} C = C \underset{(CH_3)_3C}{\overset{CH_3}{\diagup}} C = C \underset{CH_3}{\overset{H}{\diagup}} \quad (49)$$

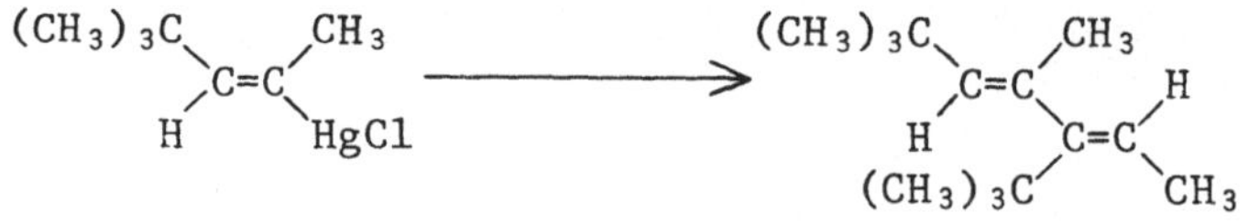

62%

acid which appear to be generated during the course of this reaction. This remarkable reaction gives very high yields of dienes with generally less than 3% of the corresponding symmetrical 1,3-dienes. This is even true with the vinylmercurial derived from *tert*-butyl methyl acetylene (eq 49), where there appears to be no obvious driving force for formation of this more highlv hindered diene.

Although the reaction appears catalytic in palladium as described above, the catalyst turnover is low. With the addition of anhydrous cupric chloride, however, the reaction can be carried out in good yield with only 10% palladium chloride (eqs 50,51).

$$2 \quad \xrightarrow[\substack{2 \ Et_3N \\ C_6H_6}]{\substack{0.1 \ PdCl_2 \\ 2 \ CuCl_2}} \quad \text{(50)}$$

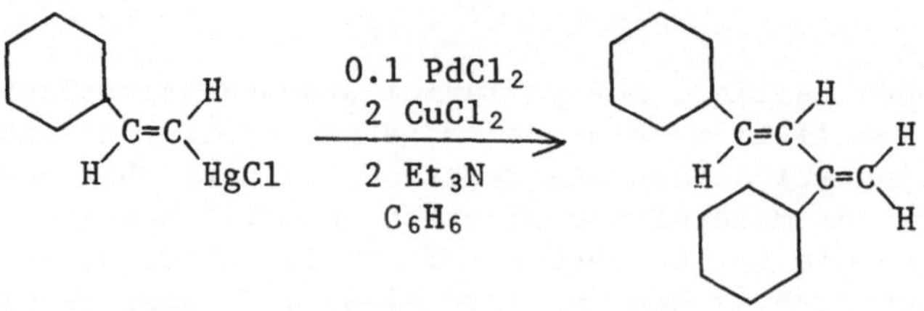

98%

$$2 \quad \underset{H}{\overset{CH_3(CH_2)_3}{>}}C=C\underset{HgCl}{\overset{H}{<}} \longrightarrow \underset{H}{\overset{CH_3(CH_2)_3}{>}}C=C\underset{CH_3(CH_2)_3}{\overset{H}{<}}C=C\underset{H}{\overset{H}{<}} \qquad (51)$$

$$89\%$$

At present no simple mechanism appears to adequately describe this reaction. We feel, however, that the rearranged vinyl group probably arises by some sort of palladium hydride rearrangement of an intermediate vinylpalladium species (eq 52). The hydrochloric acid persumably arises from the same intermediate by

$$\underset{H}{\overset{R}{>}}C=C\underset{PdCl}{\overset{H}{<}} \longrightarrow \left[\begin{array}{c} RC\equiv CH \\ \vdots \\ HPdCl \end{array}\right] \longrightarrow \underset{ClPd}{\overset{R}{>}}C=C\underset{H}{\overset{H}{<}} \qquad (52)$$

elimination of a palladium hydride which subsequently decomposes to palladium(0) and hydrochloric acid (eq 53).

$$HPdCl \longrightarrow HCl + Pd \qquad (53)$$

The two palladium dimerization reactions nicely complement previous literature procedures for the formation of *cis,cis*-[17] and *cis,trans*-[18-20] 1,4-disubstituted-1,3-dienes via vinylborane reactions (eqs 54,55).

$$2 \ RC\equiv CH \longrightarrow \underset{H}{\overset{R}{>}}C=C\underset{H}{\overset{H}{<}}\underset{R}{\overset{H}{>}}C=C\underset{R}{\overset{H}{<}} \qquad (54)$$

$$2 \ RC\equiv CH \longrightarrow \underset{H}{\overset{R}{>}}C=C\underset{H}{\overset{H}{<}}\underset{H}{\overset{H}{>}}C=C\underset{H}{\overset{R}{<}} \qquad (55)$$

During the course of these studies, we became curious as to what other palladium salts might do in this reaction. To our surprise, palladium acetate gives a totally different reaction, providing a new stereospecific route to enol acetates (eqs 56-58).[21] This reaction is highly stereospecific and can be effected using as little as 1% Pd(OAc)$_2$ if one equivalent of mercuric acetate is added. The reaction can also be used to prepare *trans*-enediacetates as indicated in equation 58. Using

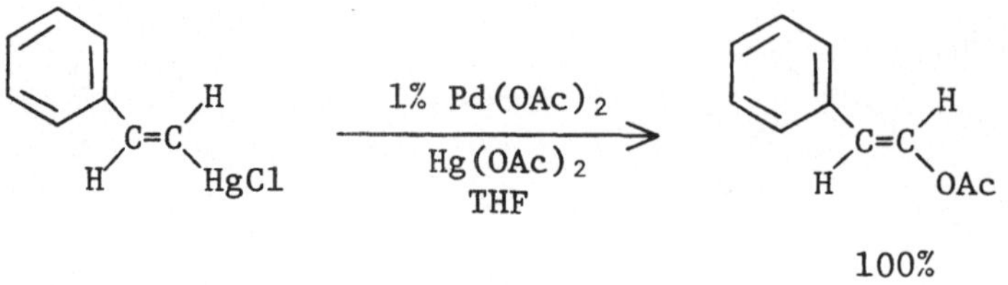

$$\text{C}_6\text{H}_5\text{CH=CH-HgCl} \xrightarrow[\substack{\text{Hg(OAc)}_2 \\ \text{THF}}]{1\% \text{ Pd(OAc)}_2} \text{C}_6\text{H}_5\text{CH=CH-OAc} \tag{56}$$

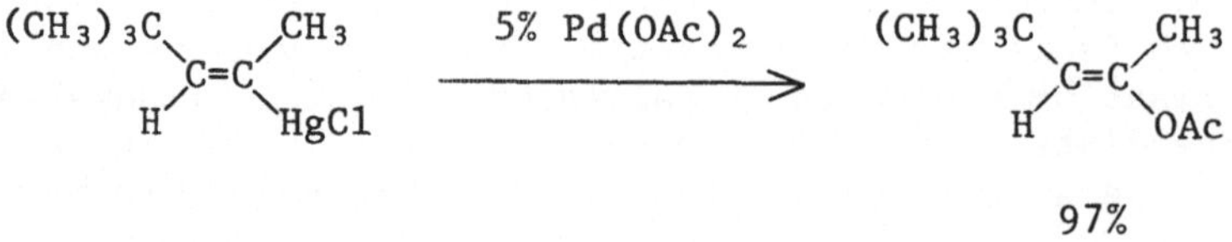

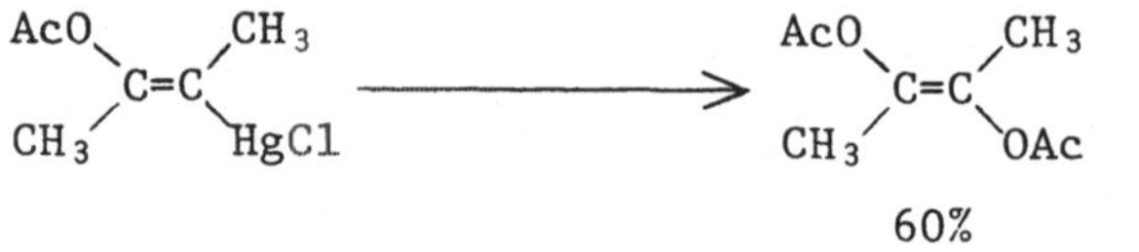

100%

$$(\text{CH}_3)_3\text{C}(\text{CH}_3)\text{C=CH-HgCl} \xrightarrow{5\% \text{ Pd(OAc)}_2} (\text{CH}_3)_3\text{C}(\text{CH}_3)\text{C=CH-OAc} \tag{57}$$

97%

$$\text{AcO}(\text{CH}_3)\text{C=C}(\text{CH}_3)\text{HgCl} \longrightarrow \text{AcO}(\text{CH}_3)\text{C=C}(\text{CH}_3)\text{OAc} \tag{58}$$

60%

other mercuric carboxylates one can prepare the corresponding enol esters as well.

Mechanistically this reaction probably involves (1) formation of a vinylpalladium acetate, (2) demetallation and (3) oxidation of the resulting palladium metal (eqs 59-62).

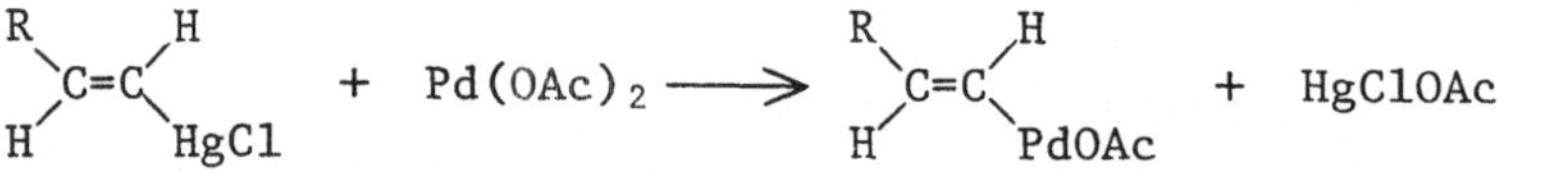

$$\text{RCH=CH-HgCl} + \text{Pd(OAc)}_2 \longrightarrow \text{RCH=CH-PdOAc} + \text{HgClOAc} \tag{59}$$

$$\text{RCH=CH-PdOAc} \longrightarrow \text{RCH=CH-OAc} + \text{Pd} \tag{60}$$

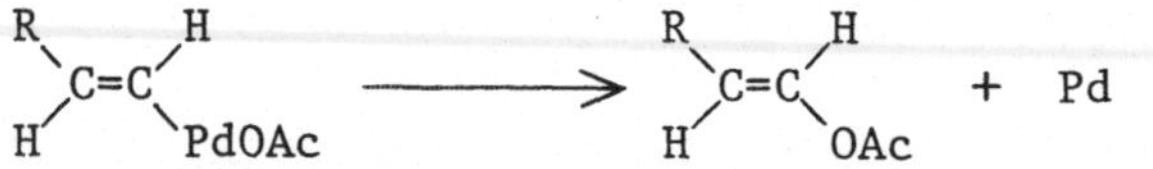

$$\text{Pd} + \text{Hg(OAc)}_2 \longrightarrow \text{Pd(OAc)}_2 + \text{Hg} \tag{61}$$

$$\text{Hg} + \text{HgClOAc} \longrightarrow \text{Hg}_2\text{ClOAc} \tag{62}$$

Wishing to extend our earlier diene syntheses to completely unsymmetrical 1,3-dienes, we examined the following reaction (eq 63). Ample literature precedent from the work of Heck suggested that this reaction should proceed as indicated. To our great surprise, we obtained instead a π-allylpalladium compound

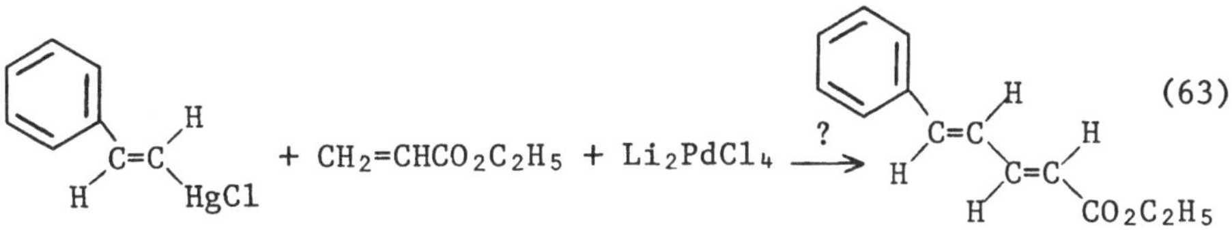

(63)

(eq 64).[22] This reaction works equally well for a variety of
other vinylmercurials and alkenes, although giving best results
with terminal alkenes (eqs 64-68).

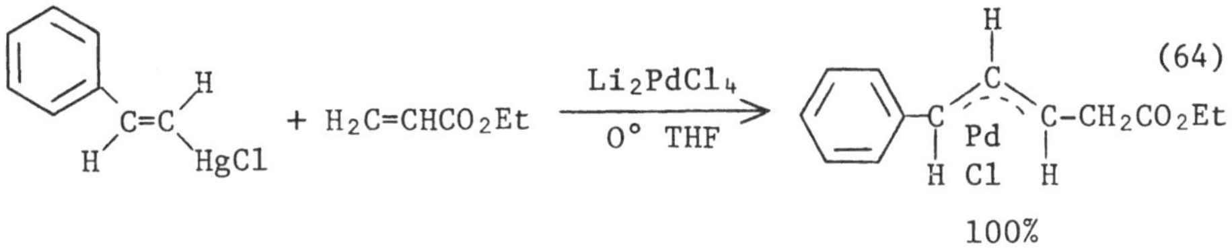

(64)

100%

(65)

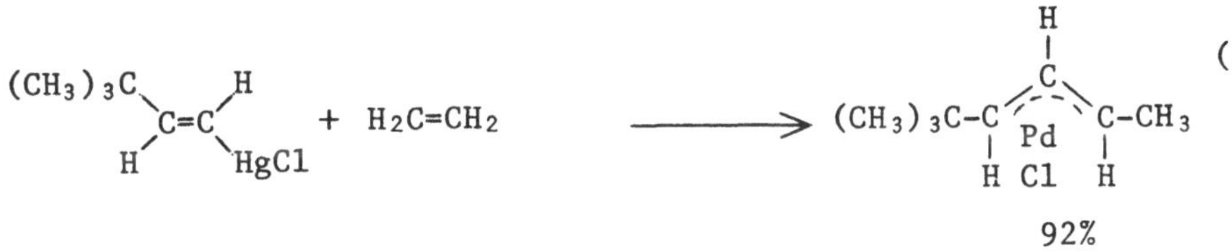

, 97%

(66)

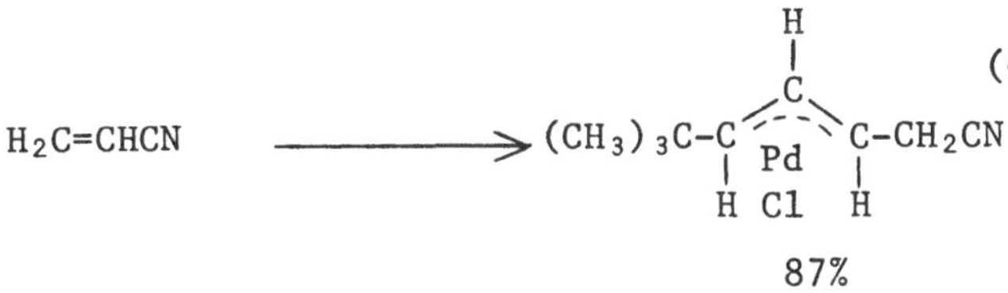

92%

(67)

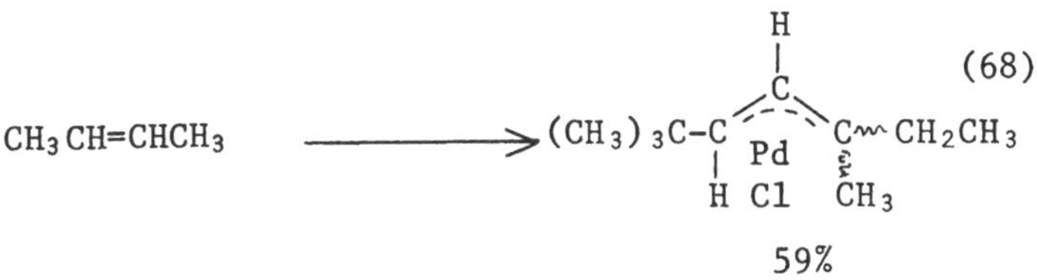

87%

(68)

59%

When examining this reaction with the bicyclic olefin norbornene, we obtained totally different organopalladium compounds (eqs 69-71).[23] We are currently examining applications of this

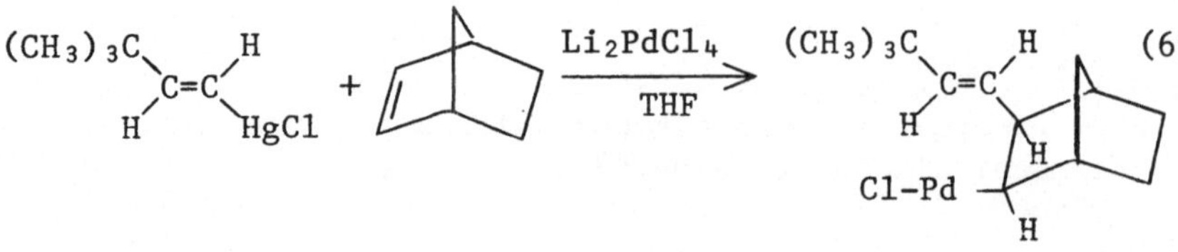

(69)

89%

(70)

88%

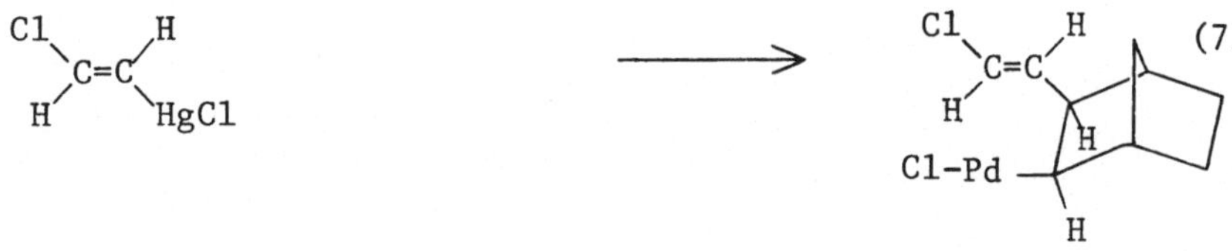

(71)

88%

reaction to the synthesis of bicyclic prostaglandin analogs (eq 72). We are also investigating analogous π-allylpalladium addition

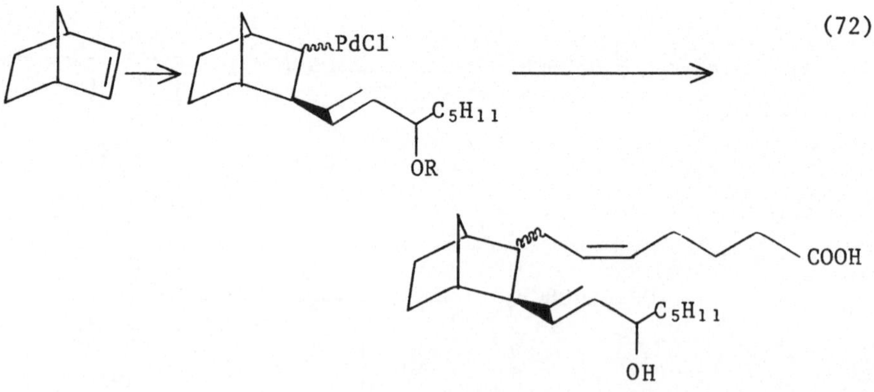

(72)

reactions to norbornene and related bicyclic alkenes as a possible route to prostaglandin analogs (eq 73).

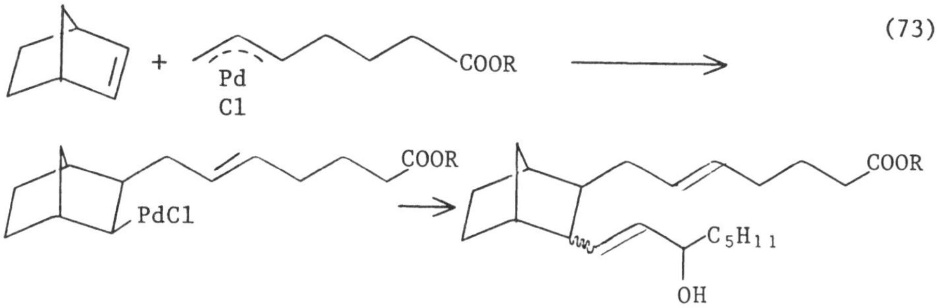

(73)

All of these reactions are consistent with the following mechanistic scheme (eqs 74-78). With norbornene the initial olefin

(74)

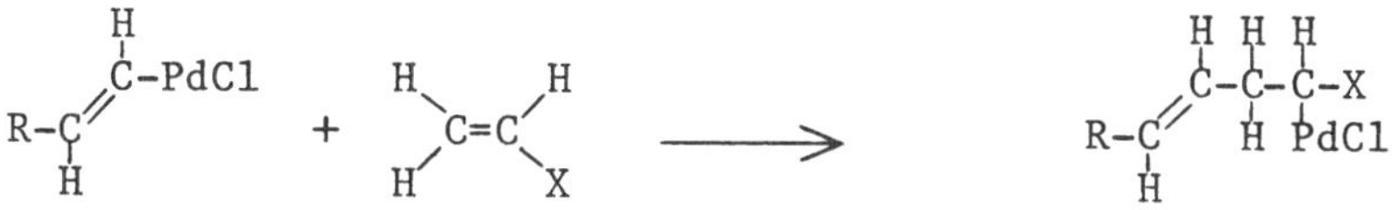

(75)

(76)

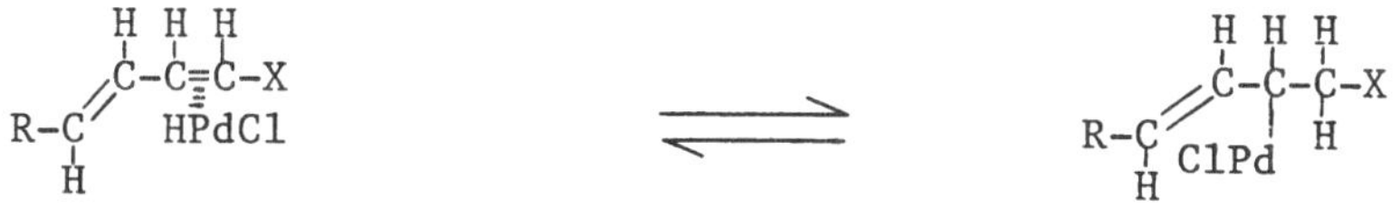

(77)

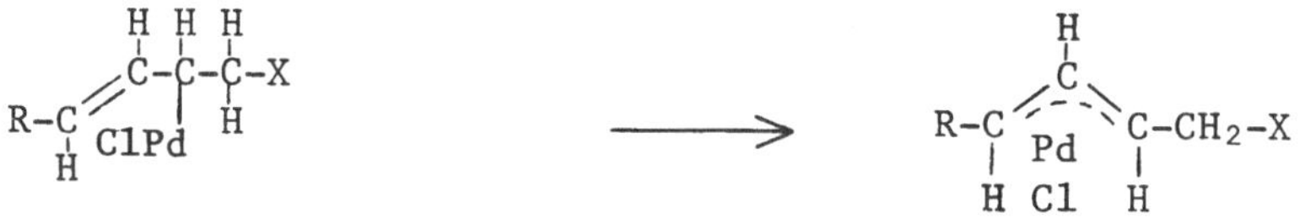

(78)

addition product (eq 75) is quite stable and can be isolated. Its stability presumably arises because of π-coordination with the adjacent carbon-carbon double bond, and the absence of a β-hydrogen *cis* to the palladium which can readily undergo β-hydride elimination (elimination would also lead to a highly strained bridgehead double bond). With acyclic olefins β-hydride elimination is facile and presumably reversible. Eventually the palladium moiety ends up in the allylic position and collapses to the highly stable π-allylpalladium product.

   In examining analogous reactions with allylic halides, 1,4-dienes and not π-allylpalladium compounds are obtained.[24] The reaction requires lithium chloride and an excess of the allylic halide for best results, and can in certain cases be accomplished using only catalytic amounts of palladium chloride (eqs 79,80).

$$
\underset{H}{\overset{(CH_3)_3C}{>}}C=C\underset{HgCl}{\overset{H}{<}} \;+\; 5\;\; CH_2=CHCH_2Cl \;\xrightarrow{10\%\;Li_2PdCl_4}\; \underset{H}{\overset{(CH_3)_3C}{>}}C=C\underset{CH_2CH=CH_2}{\overset{H}{<}} \quad (79)
$$

                                                                                96%

$$ \quad\xrightarrow{\hspace{2cm}}\quad $$ (80)

                                                                                100%

In most cases, however, stoichiometric amounts of palladium chloride are required and even then increasing substitution about the double bond of the allylic halide results in sharply reduced yields (eqs 81–83).

$$
\underset{H}{\overset{(CH_3)_3C}{>}}C=C\underset{HgCl}{\overset{H}{<}} \;+\; 10\;\; CH_2=CHCH_2Cl \;\xrightarrow{Li_2PdCl_4}\; \underset{H}{\overset{(CH_3)_3C}{>}}C=C\underset{CH_2CH=CH_2}{\overset{H}{<}} \quad (81)
$$

                                                                                100%

$$
\overset{Cl}{\underset{}{CH_2=CHCHCH_3}} \;\xrightarrow{\hspace{2cm}}\; \underset{H}{\overset{(CH_3)_3C}{>}}C=C\underset{CH_2CH=CHCH_3}{\overset{H}{<}} \quad (82)
$$

                                                                                49%

$$
CH_3CH=CHCH_2Cl \;\xrightarrow{\hspace{2cm}}\; \underset{H}{\overset{(CH_3)_3C}{>}}C=C\underset{\underset{CH_3}{\overset{|}{CHCH=CH_2}}}{\overset{H}{<}} \quad (83)
$$

                                                                                32%

Mechanistically, these reactions are undoubtedly very closely related to the other olefin reactions (eqs 84–86). The initial

$$R\!\!\diagdown\!\!{}_H\!\!C\!=\!C\!\!\diagup\!\!{}^H_{HgCl} + PdCl_4{}^{-2} \longrightarrow R\!\!\diagdown\!\!{}_H\!\!C\!=\!C\!\!\diagup\!\!{}^H_{PdCl_3}{}^{-2} + HgCl_2 \qquad (84)$$

$$R\!\!\diagdown\!\!{}_H\!\!C\!=\!C\!\!\diagup\!\!{}^H_{PdCl_3}{}^{-2} + CH_2\!=\!CH\!-\!CH\!-\!CH_3{}_{|\ Cl} \longrightarrow \left[R\!\!\diagdown\!\!{}_H\!\!C\!=\!C\!\!\diagup\!\!{}^H_{CH_2\!-\!CH\!-\!CH\!-\!CH_3}{}_{|\quad|\atop Cl_3Pd\ Cl} \right]^{-2} \qquad (85)$$

$$\left[R\!\!\diagdown\!\!{}_H\!\!C\!=\!C\!\!\diagup\!\!{}^H_{CH_2\!-\!CH\!-\!CH\!-\!CH_3}{}_{|\quad|\atop Cl_3Pd\ Cl} \right]^{-2} \longrightarrow R\!\!\diagdown\!\!{}_H\!\!C\!=\!C\!\!\diagup\!\!{}^H_{CH_2\!-\!CH\!=\!CH\!-\!CH_3} + PdCl_4{}^{-2} \qquad (86)$$

olefin addition product apparently more readily eliminates a β-chloride than the β-hydride. This results in S_N2' displacement of the chloride and transposition of the double bond. It also regenerates palladium(II) salts which can then reenter the cycle to promote catalytic coupling.

Recently we have become quite interested in developing more general coupling procedures.[25] For example, we would like to be able to find suitable catalysts to promote methyl iodide-vinylmercurial coupling (eq 87). We reasoned that if we could generate mixed methyl vinyl transition metal species, we might

$$R\!\!\diagdown\!\!{}_H\!\!C\!=\!C\!\!\diagup\!\!{}^H_{HgCl} + ICH_3 \xrightarrow{\text{catalyst}} R\!\!\diagdown\!\!{}_H\!\!C\!=\!C\!\!\diagup\!\!{}^H_{CH_3} + ClHgI \qquad (87)$$

observe reductive elimination to the corresponding methyl olefin. We began our studies using the thermally stable, green crystalline methylrhodium(III) species derived from oxidative addition of methyl iodide to Wilkinson's catalyst (eq 88). We felt that

$$2\ CH_3I + RhCl(P\varnothing_3)_3 \longrightarrow {}^{CH_3}\!\!\diagdown\!\!{}_I\!\!\diagup RhI(P\varnothing_3)_2\!\cdot\!CH_3Cl + P\varnothing_3 \qquad (88)$$

transmetallation with a vinylmercurial and reductive elimination should provide the corresponding olefin (eqs 89,90). Indeed,

$$\begin{matrix}CH_3\\ \quad\diagdown\\ \quad RhI(P\phi_3)_2\cdot CH_3Cl\\ \quad\diagup\\ I\end{matrix} + RCH{=}CHHgCl \longrightarrow \begin{matrix}CH_3\\ \quad\diagdown\\ \quad RhI(P\phi_3)_2\cdot CH_3Cl\\ \quad\diagup\\ RCH{=}CH\end{matrix} + ClHgI \qquad (89)$$

$$\begin{matrix}CH_3\\ \quad\diagdown\\ \quad RhI(P\phi_3)_2\cdot CH_3Cl\\ \quad\diagup\\ RCH{=}CH\end{matrix} + P\phi_3 \longrightarrow RCH{=}CHCH_3 + RhI(P\phi_3)_3 + CH_3Cl \qquad (90)$$

this turned out to be true (eqs 91-93). Again these reactions require excess lithium chloride and HMPA. The reactions also work

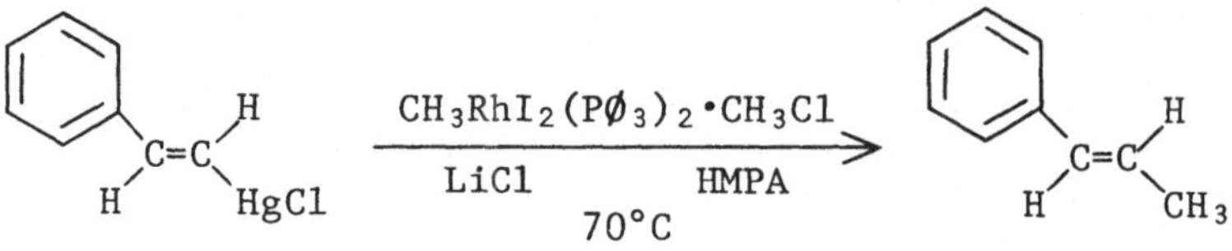

95% (91)

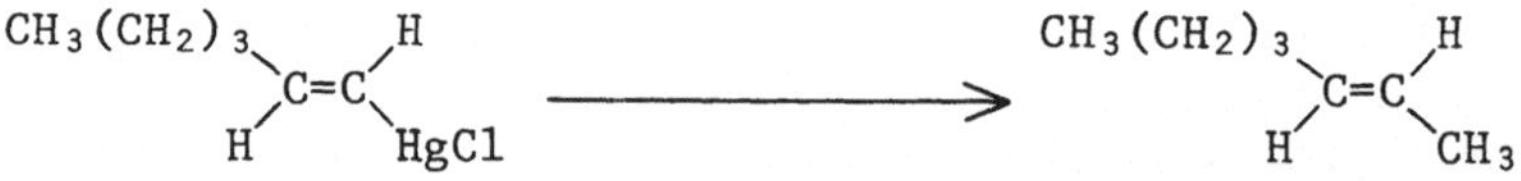

94% (92)

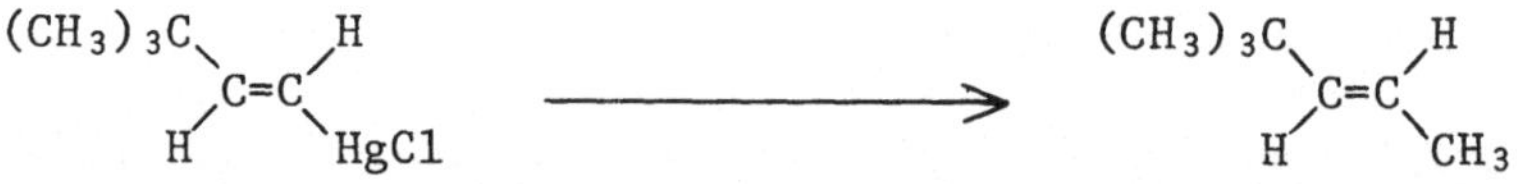

93% (93)

well with aryl- and alkynylmercurials (eqs 94-96). We have recently observed that these reactions can be effected using only

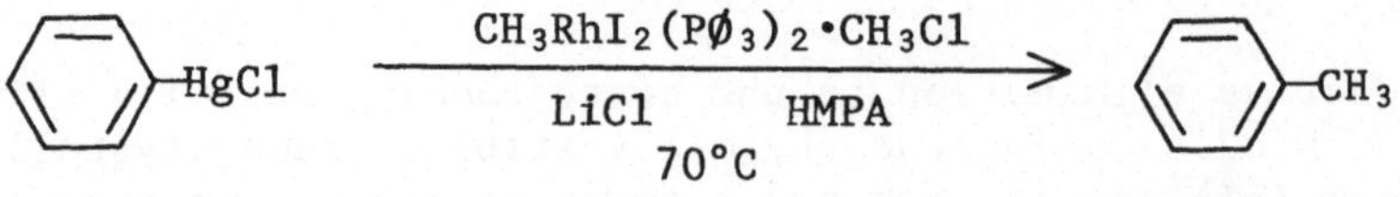

99% (94)

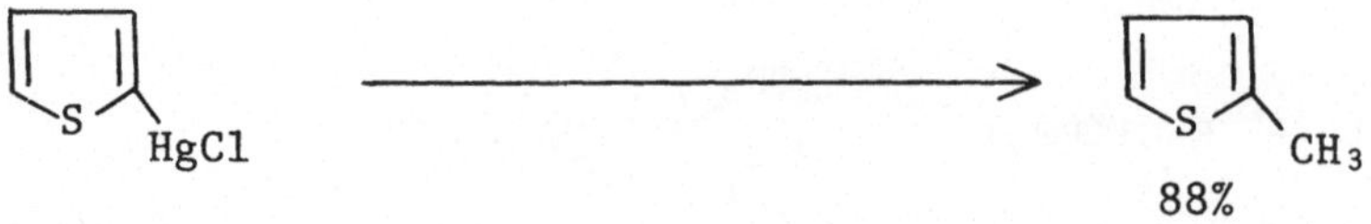

88% (95)

$$[(CH_3)_3CC{\equiv}C]_2Hg \longrightarrow 2\ (CH_3)_3CC{\equiv}CCH_3 \qquad (96)$$

100%

catalytic amounts of the methylrhodium complex, if methyl iodide is added to the reaction mixture.

We have also briefly examined possible coupling reactions with organorhodium(III) species obtained by rhodium hydride addition to alkenes and alkynes (eq 97)[26], and by decarbonylation of aryl acid chlorides (eq 98).[27] Both vinyl- and

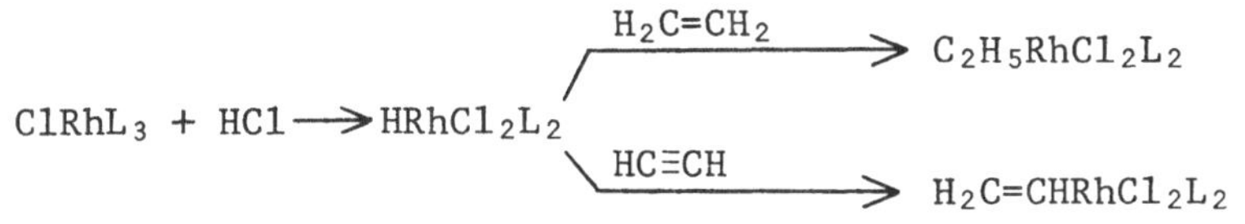

$$ClRhL_3 + HCl \longrightarrow HRhCl_2L_2 \begin{array}{c} \xrightarrow{H_2C=CH_2} C_2H_5RhCl_2L_2 \\ \\ \xrightarrow{HC\equiv CH} H_2C=CHRhCl_2L_2 \end{array} \qquad (97)$$

$$ClRhL_3 + Cl\overset{O}{\overset{\|}{C}}Ar \longrightarrow [Ar\overset{O}{\overset{\|}{C}}RhCl_2L_2] \longrightarrow ArRhCl_2(CO)L_2 \qquad (98)$$

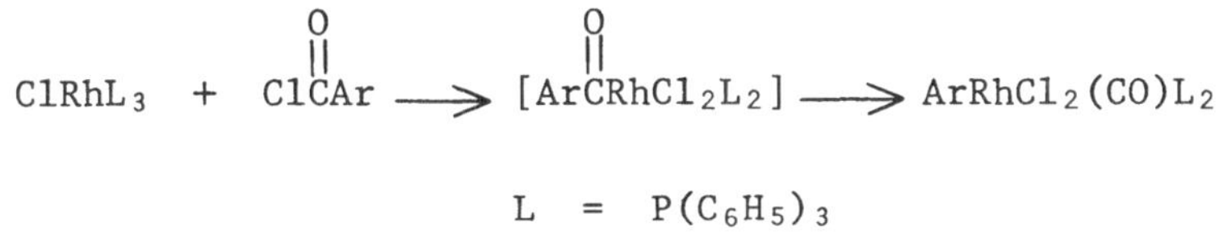

$$L = P(C_6H_5)_3$$

phenylrhodium(III) species couple with organomercurials, but the yields are generally low. We have also observed that this type of reaction can be extended to give direct alkyne addition reactions (eqs 99,100).

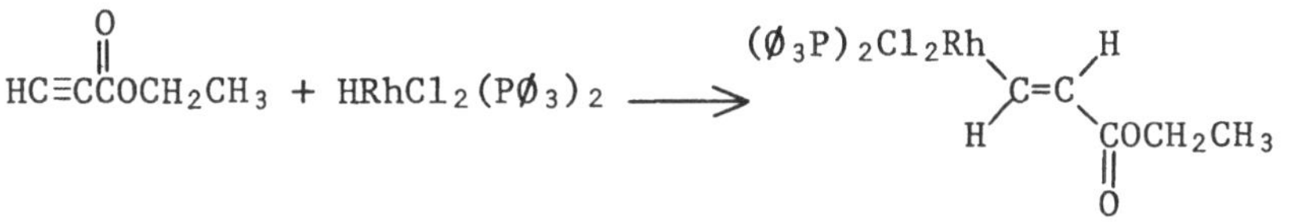

$$HC\equiv C\overset{O}{\overset{\|}{C}}OCH_2CH_3 + HRhCl_2(P\emptyset_3)_2 \longrightarrow \begin{array}{c} (\emptyset_3P)_2Cl_2Rh \\ \diagdown \\ H \end{array} C=C \begin{array}{c} H \\ \diagup \\ COCH_2CH_3 \\ \overset{\|}{O} \end{array} \qquad (99)$$

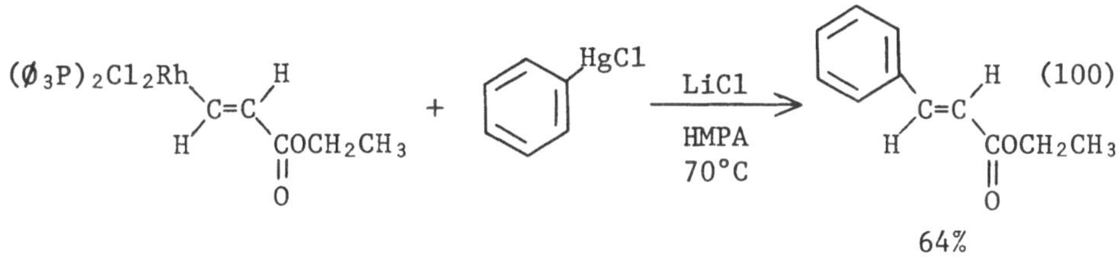

$$\begin{array}{c} (\emptyset_3P)_2Cl_2Rh \\ \diagdown \\ H \end{array} C=C \begin{array}{c} H \\ \diagup \\ COCH_2CH_3 \\ \overset{\|}{O} \end{array} + \bigcirc\!\!\!\!-HgCl \xrightarrow[\substack{HMPA \\ 70°C}]{LiCl} \bigcirc\!\!\!\!- C=C \begin{array}{c} H \\ \diagup \\ COCH_2CH_3 \\ \overset{\|}{O} \end{array} \qquad (100)$$

$$64\%$$

The acylation of vinylmercurials was among the first reactions we ever examined. By simply stirring vinylmercurials and acid chlorides with aluminum chloride for 5 minutes in methylene chloride, we are able to obtain excellent yields of the corresponding α,β-unsaturated ketones (eqs 101-103).[28] The reaction

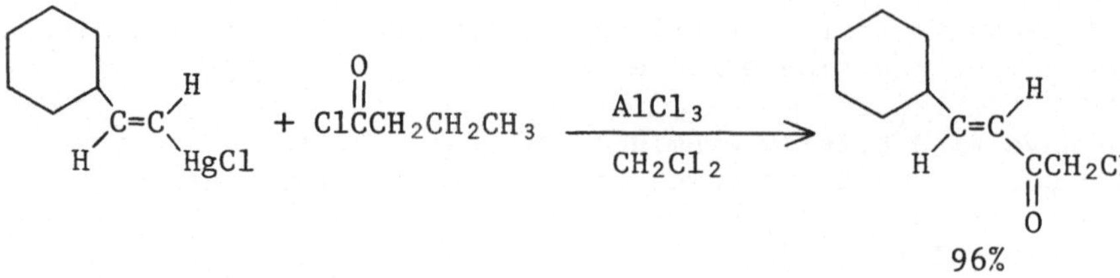

$$96\%$$ (101)

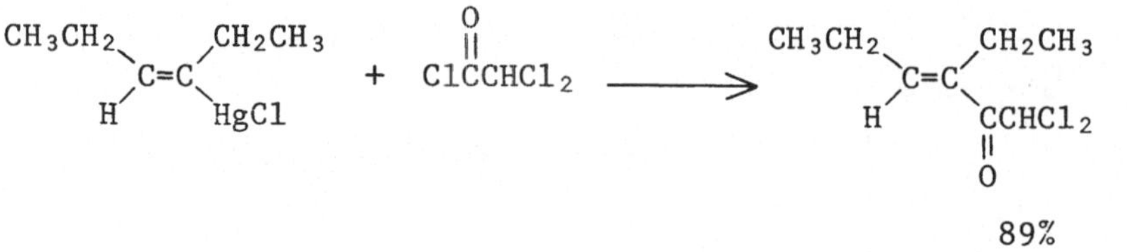

$$89\%$$ (102)

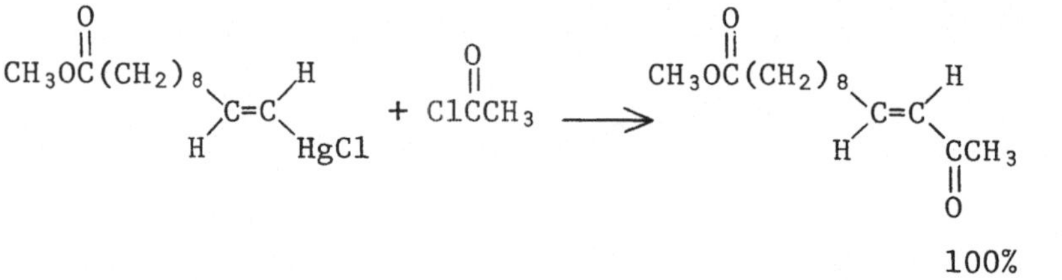

$$100\%$$ (103)

proceeds with high stereospecificity and can also be applied to
the synthesis of unsymmetrical dienones (eq 104). These reactions
do not work well with aromatic acid chlorides however. These same

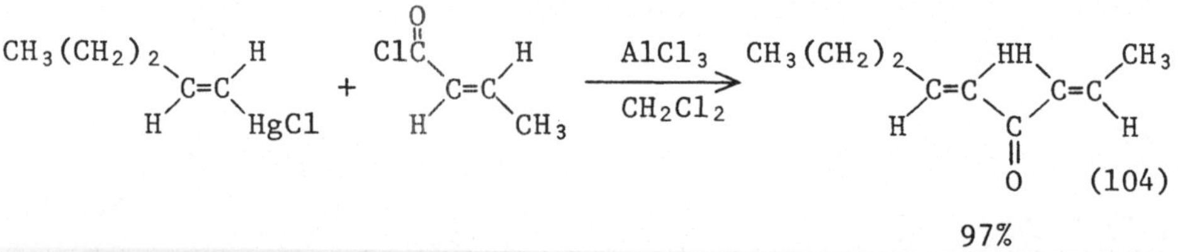

$$97\%$$ (104)

reactions can also be effected by using aluminum powder or foil
(eq 105) or catalytic amounts of tetrakis(triphenylphosphine)-
palladium(0) (eq 106), but the yields are much lower. When one

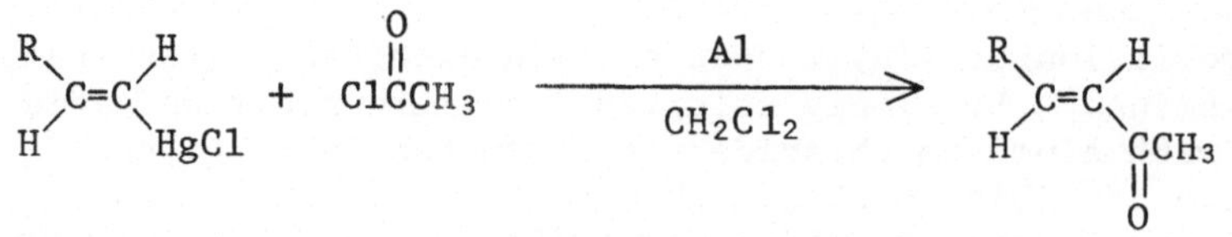

(105)

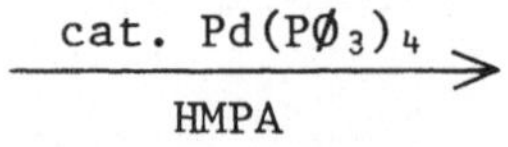

(106)

employs titanium tetrachloride to promote acylation, a substantial amount of the stereochemically inverted enone is observed (eq 107). Unfortunately, this reaction has not been very reproducible.

$$CH_3(CH_2)_3 \underset{H}{\overset{H}{>}}C=C\underset{HgCl}{\overset{H}{<}} + \underset{}{\overset{O}{\underset{}{Cl\overset{\|}{C}CH_3}}} \xrightarrow[-78°C]{\underset{CH_2Cl_2}{TiCl_4}} CH_3(CH_2)_3\underset{H}{\overset{\overset{O}{\overset{\|}{CCH_3}}}{>}}C=C\underset{H}{<} \qquad (107)$$

Several mechanisms seem plausible for the aluminum chloride promoted reactions. The three most probable would appear to be (1) transmetallation with aluminum chloride to give a more reactive vinylaluminum species which then undergoes acylation (eq 108); (2) direct carbon-mercury bond cleavage (eq 109); or (3) addition of the Lewis acid complexed acid chloride across the carbon-carbon double bond of the vinylmercurial, followed by mercuric chloride elimination (eq 110). We presently have no evidence that

$$\underset{H}{\overset{R}{>}}C=C\underset{HgCl}{\overset{H}{<}} \xrightarrow{AlCl_3} \underset{H}{\overset{R}{>}}C=C\underset{AlCl_2}{\overset{H}{<}} \longrightarrow Products \qquad (108)$$

$$\underset{H}{\overset{R}{>}}C=C\underset{HgCl}{\overset{H}{<}} \xrightarrow[substitution]{direct} \underset{H}{\overset{R}{>}}C=C\underset{\underset{O}{\overset{|}{\overset{\|}{CR'}}}}{\overset{H}{<}} \qquad (109)$$

$$\underset{H}{\overset{R}{>}}C=C\underset{HgCl}{\overset{H}{<}} \longrightarrow \underset{\underset{H}{|}\underset{HgCl}{|}}{\overset{\overset{R}{|}\overset{H}{|}\overset{O}{\|}}{Cl-C-C-CR'}} \xrightarrow{-HgCl_2} \underset{H}{\overset{R}{>}}C=C\underset{\underset{O}{\overset{|}{\overset{\|}{CR'}}}}{\overset{H}{<}} \qquad (110)$$

vinylalanes are formed by transmetallation of vinylmercurials by aluminum chloride. Hydrolysis studies of such reactions produce only minor amounts of the alkene which would be anticipated upon hydrolysis of the vinylalanes. This does not of course rule out the possibility that only minor amounts of the vinylalanes are formed at any one time in these reactions. In the aluminum metal promoted reactions, however, we do feel that we are in fact getting a transmetallation between the mercurials and aluminum metal to get the more reactive vinylalanes, which subsequently undergo acylation. Vinylmercurials react quite readily with aluminum foil.

It is much more difficult to distinguish between the other two mechanistic possibilities. Side reactions involving molecular rearrangement have been observed during the course of these studies which suggest that a positive charge is built up on the β-carbon atom of the vinylmercurial. This suggests that perhaps the reactions proceed as in equation 110. We cannot be sure however that this is not simply a side reaction unrelated to the usual reaction leading to α,β-unsaturated ketones. The unusual inversion of stereochemistry observed in the titanium tetrachloride reactions also suggests an addition–elimination mechanism. *Trans* addition and *trans* elimination (or *cis* and *cis*) would result in overall inversion of configuration. Once again, however, we cannot tell if this is related to the aluminum chloride reaction or a totally different reaction. Present evidence does not allow us to distinguish between the latter two mechanistic possibilities.

We have recently spent considerable time examining carbonylation reactions of vinylmercurials. We have observed that these mercurials are readily converted to α,β–unsaturated carboxylic acids upon treatment with 1 atmosphere of carbon monoxide, 1–5% aqueous tetrahydrofuran (THF) and dilithium tetrachloropalladate (eq 111).[29] The reagents are simply mixed at -78°C and allowed

$$
\underset{H}{\overset{R}{}}C=C\underset{HgCl}{\overset{H}{}} \quad \xrightarrow[\text{1-5\% H}_2\text{O-THF}]{\overset{\text{CO}}{\text{Li}_2\text{PdCl}_4}} \quad \underset{H}{\overset{R}{}}C=C\underset{\underset{O}{\overset{\|}{\text{COH}}}}{\overset{H}{}} \qquad (111)
$$

to warm to room temperature overnight. Although the reaction appears to be fairly sensitive to the amount of water present in the reaction mixture, it readily accommodates functional groups and generally gives very good yields (eqs 112–114).

$$
\xrightarrow{\text{2\% H}_2\text{O}} \qquad (112)
$$

72%

$$
\underset{H}{\overset{NC(CH_2)_3}{}}C=C\underset{HgCl}{\overset{H}{}} \quad \xrightarrow{\text{5\% H}_2\text{O}} \quad \underset{H}{\overset{NC(CH_2)_3}{}}C=C\underset{\underset{O}{\overset{\|}{\text{COH}}}}{\overset{H}{}} \qquad (113)
$$

72%

$$
\begin{array}{c}
C_2H_5 \\ \\ H
\end{array}\!\!>\!C=C\!<\!\!\begin{array}{c}C_2H_5 \\ \\ HgCl\end{array}
\quad\xrightarrow{\ 5\%\ H_2O\ }\quad
\begin{array}{c}C_2H_5 \\ \\ H\end{array}\!\!>\!C=C\!<\!\!\begin{array}{c}C_2H_5 \\ \\ \underset{\underset{O}{\|}}{C}OH\end{array}
\tag{114}
$$

$$85\%$$

This reaction can also be effected using only catalytic amounts of either palladium chloride or palladium on carbon, if two equivalents of cupric chloride are added as a reoxidant for palladium (eq 115). This approach gives equally high yields of

$$
\begin{array}{c}R \\ \\ H\end{array}\!\!>\!C=C\!<\!\!\begin{array}{c}H \\ \\ HgCl\end{array}
+ CO + H_2O + 2\ CuCl_2 + 2\ LiCl \longrightarrow
\begin{array}{c}R \\ \\ H\end{array}\!\!>\!C=C\!<\!\!\begin{array}{c}H \\ \\ \underset{\underset{O}{\|}}{C}OH\end{array}
\tag{115}
$$

$$\text{catalyst:}\quad PdCl_2 \text{ or } Pd/C$$

α,β-unsaturated carboxylic acids (eq 116).

$$
\begin{array}{c}CH_3(CH_2)_3 \\ \\ H\end{array}\!\!>\!C=C\!<\!\!\begin{array}{c}H \\ \\ HgCl\end{array}
\qquad
\begin{array}{c}
\overset{\displaystyle 5\%\ H_2O}{\overset{\displaystyle 10\%\ PdCl_2}{\nearrow}}\\[2em]
\underset{\displaystyle 10\%\ Pd/C}{\underset{\displaystyle 5\%\ H_2O}{\searrow}}
\end{array}
\qquad
\begin{array}{c}CH_3(CH_2)_3 \\ \\ H\end{array}\!\!>\!C=C\!<\!\!\begin{array}{c}H \\ \\ \underset{\underset{O}{\|}}{C}OH\end{array}
\tag{116}
$$

$$98\%$$

This approach is readily applicable to the synthesis of α,β-unsaturated esters as well (eq 117).[29] A procedure identical

$$
\begin{array}{c}R \\ \\ H\end{array}\!\!>\!C=C\!<\!\!\begin{array}{c}H \\ \\ HgCl\end{array}
\quad\xrightarrow[\displaystyle ROH]{\displaystyle \overset{CO}{Li_2PdCl_4}}\quad
\begin{array}{c}R \\ \\ H\end{array}\!\!>\!C=C\!<\!\!\begin{array}{c}H \\ \\ \underset{\underset{O}{\|}}{C}OR\end{array}
\tag{117}
$$

to that described above, only using alcohols as the solvent, works well, and a variety of functional groups are readily accommodated (eqs 118-120). Once again the reaction can be effected using only

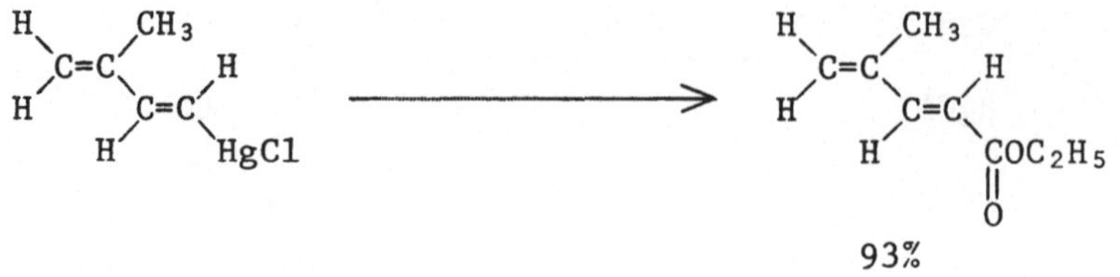

$$
\begin{array}{c}
\text{(structures)}
\end{array}
\qquad (118)
$$

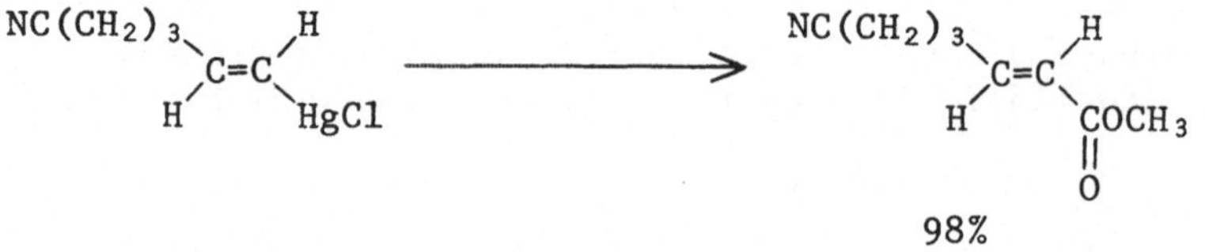

93%

$$
(119)
$$

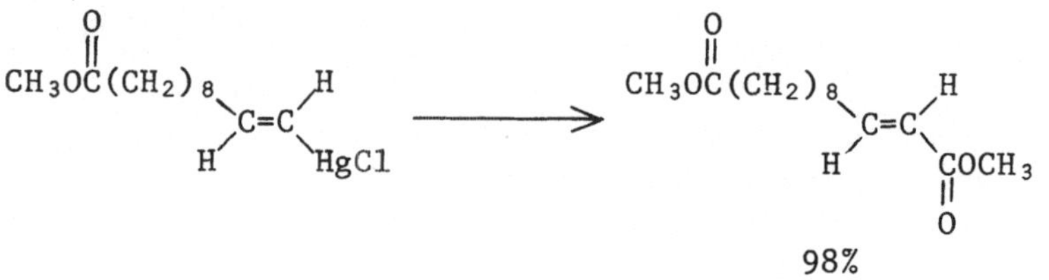

98%

$$
(120)
$$

98%

catalytic amounts of palladium chloride or palladium on carbon
(eqs 121,122). The last example illustrates an extremely simple
method for the preparation of methyl *trans*-β-chloroacrylate from
acetylene via carbonylation of the corresponding vinylmercurial.

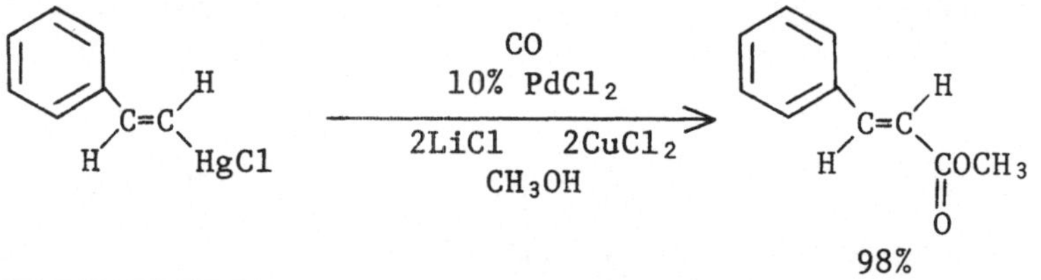

$$
\xrightarrow[\begin{array}{c} 2\text{LiCl} \quad 2\text{CuCl}_2 \\ \text{CH}_3\text{OH} \end{array}]{\begin{array}{c} \text{CO} \\ 10\% \ \text{PdCl}_2 \end{array}} \qquad (121)
$$

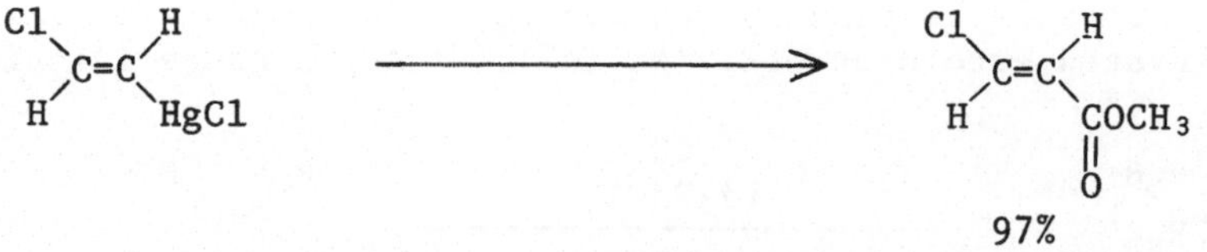

98%

$$
\longrightarrow \qquad (122)
$$

97%

The palladium-promoted carbonylation of vinylmercurials
undoubtedly proceeds by an initial mercury-palladium exchange
reaction (eq 123), carbon monoxide insertion into the resultant
vinylpalladium compound (eq 124), and subsequent solvolysis to
give the α,β-unsaturated acid or ester and palladium metal (eq 125).
In the catalytic reactions the palladium metal is reoxidized to
palladium(II) by cupric chloride (eq 126). Support for this

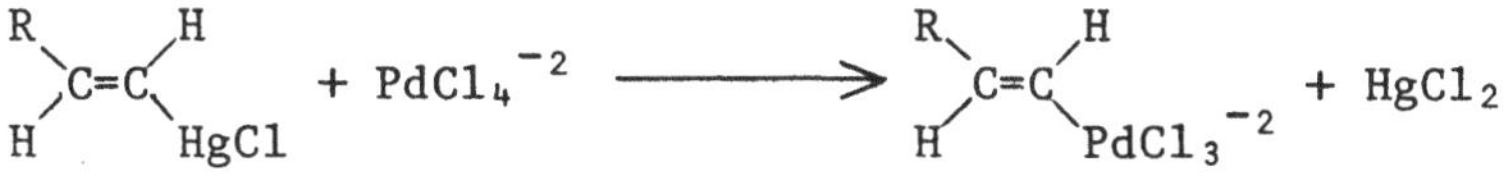

$$\begin{array}{c} R \\ H \end{array}\!\!\!C\!=\!C\!\!\!\begin{array}{c} H \\ HgCl \end{array} + PdCl_4^{-2} \longrightarrow \begin{array}{c} R \\ H \end{array}\!\!\!C\!=\!C\!\!\!\begin{array}{c} H \\ PdCl_3 \end{array}^{-2} + HgCl_2 \quad (123)$$

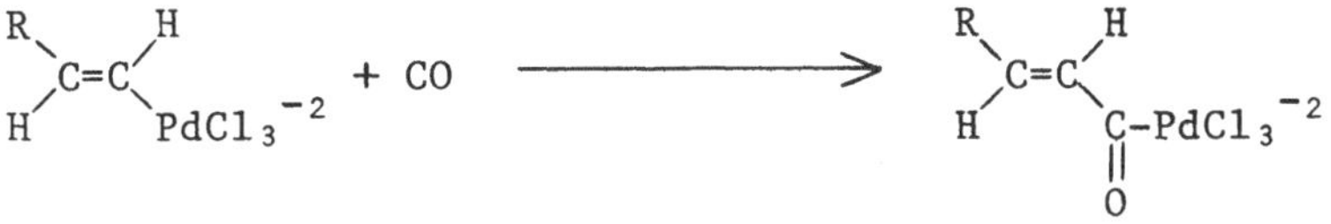

$$\begin{array}{c} R \\ H \end{array}\!\!\!C\!=\!C\!\!\!\begin{array}{c} H \\ PdCl_3 \end{array}^{-2} + CO \longrightarrow \begin{array}{c} R \\ H \end{array}\!\!\!C\!=\!C\!\!\!\begin{array}{c} H \\ C-PdCl_3 \\ \| \\ O \end{array}^{-2} \quad (124)$$

$$\begin{array}{c} R \\ H \end{array}\!\!\!C\!=\!C\!\!\!\begin{array}{c} H \\ C-PdCl_3 \\ \| \\ O \end{array}^{-2} + ROH \longrightarrow \begin{array}{c} R \\ H \end{array}\!\!\!C\!=\!C\!\!\!\begin{array}{c} H \\ COR \\ \| \\ O \end{array} + HCl + Pd + 2Cl^- \quad (125)$$

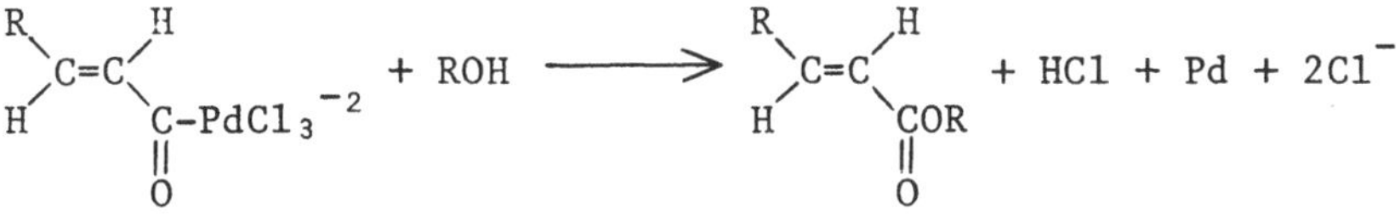

$$Pd + 2 CuCl_2 \longrightarrow PdCl_2 + 2 CuCl \quad (126)$$

mechanism is found in the many analogous reactions reported previously in the literature.

We have recently extended these carbonylation reactions to the synthesis of several biologically active ring systems. The butenolide ring system appears in a number of natural products and possesses a wide range of biological activity. We have observed that the vinylmercurials derived from propargylic alcohols can be readily carbonylated to give β-chloro-Δα,β-butenolides in essentially quantitative yields (eqs 127-129).[30] In these reactions

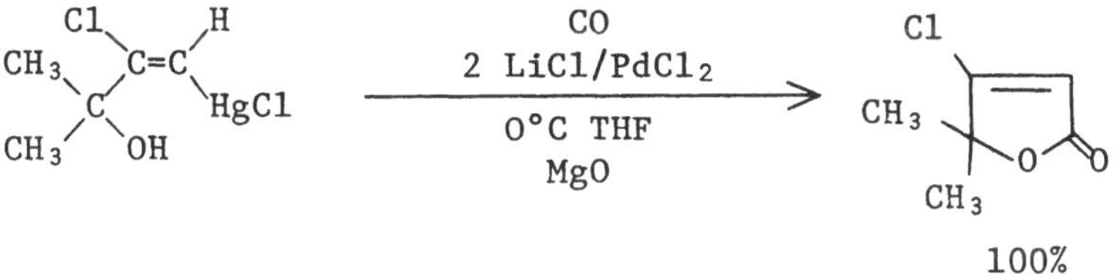

$$\qquad\qquad\qquad\qquad (127)$$

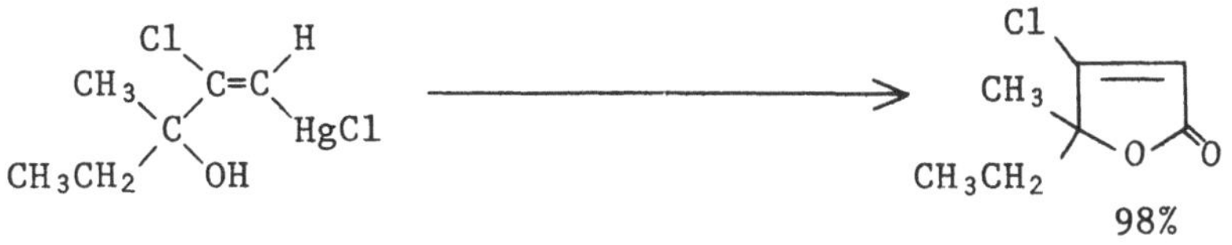

100%

$$\qquad\qquad\qquad\qquad (128)$$

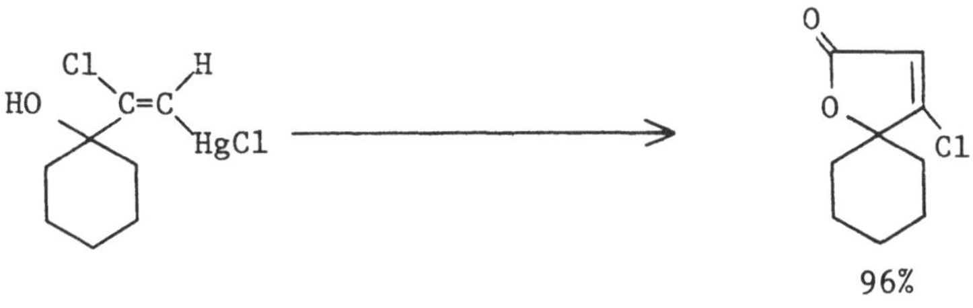

98%

$$\qquad\qquad\qquad\qquad (129)$$

96%

we have found it advantageous to run the reaction in THF at 0°C
for 16-24 hours with one equivalent of magnesium oxide added.
The inorganic base apparently traps the HCl generated upon lactone
formation and prevents it from destroying the acid-sensitive
vinylmercurial. These reactions become catalytic when anhydrous
cupric chloride is added, but the solvent must be changed to
benzene for best results (eqs 130-132). These reactions can now
be run at room temperature as well.

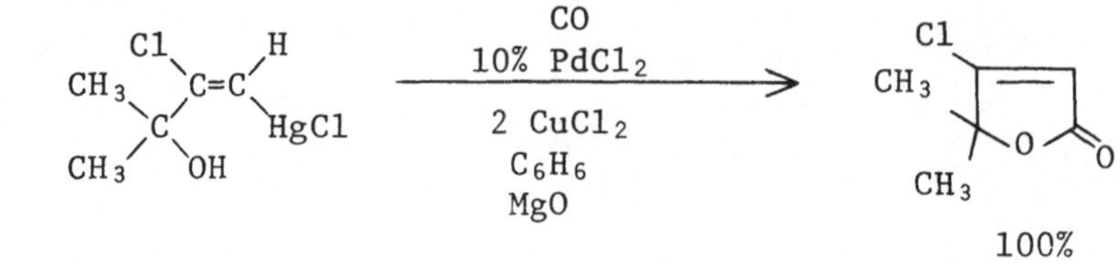

$$
\underset{\substack{\text{CH}_3 \\ \text{CH}_3}}{\overset{\text{Cl}}{\underset{\text{OH}}{\text{C}}}}\overset{\text{H}}{\underset{\text{HgCl}}{\text{C=C}}} \xrightarrow[\substack{2\ \text{CuCl}_2 \\ \text{C}_6\text{H}_6 \\ \text{MgO}}]{\substack{\text{CO} \\ 10\%\ \text{PdCl}_2}} \qquad \qquad 100\% \tag{130}
$$

$$
\underset{\substack{\text{CH}_3 \\ \text{CH}_3\text{CH}_2}}{\overset{\text{Cl}}{\underset{\text{OH}}{\text{C}}}}\overset{\text{H}}{\underset{\text{HgCl}}{\text{C=C}}} \xrightarrow{\hspace{2cm}} \qquad \qquad 100\% \tag{131}
$$

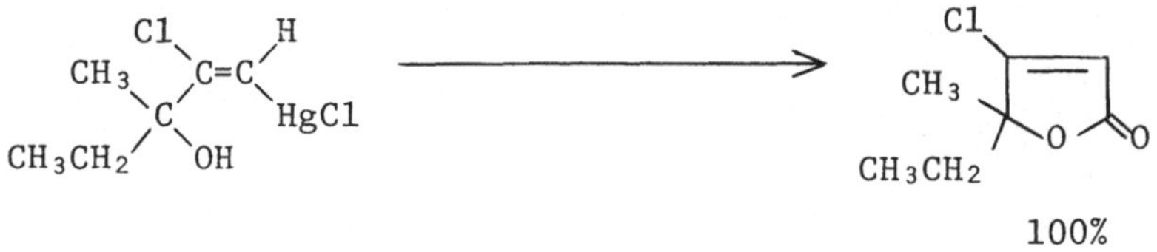

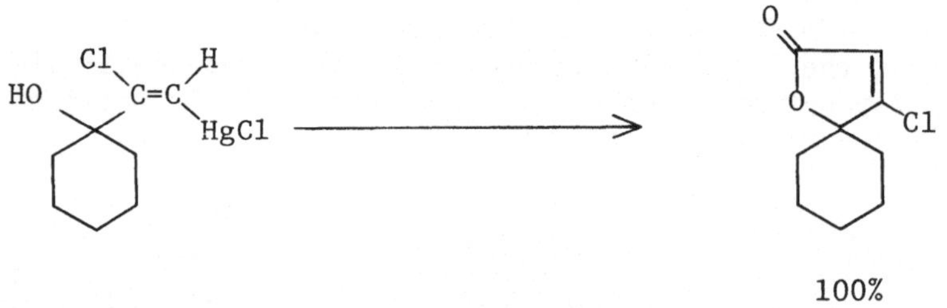

$$
\tag{132} \qquad 100\%
$$

At present we are examining a more direct approach to the
butenolides, since the earlier approach is limited by the in-
accessibility of the requisite mercurials. We have found that
these propargylic alcohols can be directly carbonylated by simply
refluxing with all the reagents used in our earlier two step
approach (eq 133). After refluxing 5 hours one obtains the

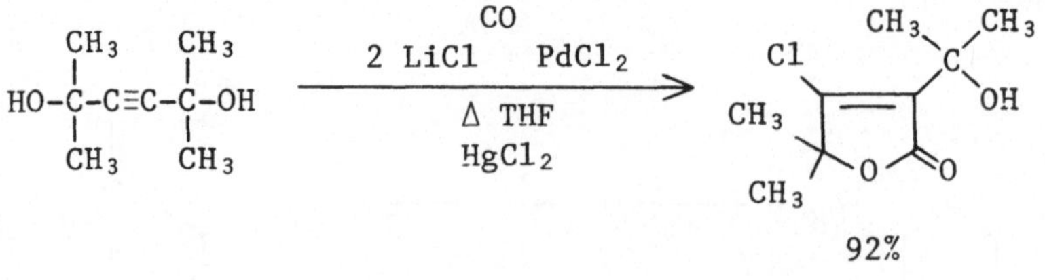

$$
\underset{\substack{\text{CH}_3 \\ \text{HO-C-C}\equiv\text{C-C-OH} \\ \text{CH}_3}}{\overset{\text{CH}_3}{}}\ \xrightarrow[\substack{\Delta\ \text{THF} \\ \text{HgCl}_2}]{\substack{\text{CO} \\ 2\ \text{LiCl}\quad\text{PdCl}_2}} \qquad \qquad 92\% \tag{133}
$$

butenolide in 92% yield. If the mercuric chloride is omitted, one can still obtain a 70% yield after 20 hours of reflux. Apparently, the palladium salt itself is able to add directly to the triple bond.

Although very few natural products possess a β-chloride, we have observed that this group can be substituted by using organocopper reagents (eq 134). Unfortunately, this reaction

$$
\text{Cl} \diagup\!\!\diagup\!\!\diagdown_{O}\diagdown_{O} \;+\; R_2\text{CuLi}\cdot\text{PBu}_3 \;\xrightarrow[\text{THF}]{-78^\circ\text{C}}\; \text{R}\diagup\!\!\diagup\!\!\diagdown_{O}\diagdown_{O}
\qquad (134)
$$

is not presently as general as we would like and further work is required.

Carbonylation of the vinylmercurials derived from propargylic amines offers promise of a novel route to α-methylene-β-lactams (eq 135).[13] In this reaction we appear to get the expected lactam

$$
\left[
\begin{array}{c}
\text{Cl} \quad \text{H} \\
\text{C}=\text{C} \\
\text{H} \quad \text{C} \quad \text{HgCl} \\
\text{CH}_3 \quad \text{NH}_2+ \\
\text{CH}_3
\end{array}
\right]_2 \text{HgCl}_4{}^{-2}
\xrightarrow[\substack{\text{NaHCO}_3 \\ \text{CH}_3\text{OH}}]{\substack{\text{CO} \\ \text{PdCl}_2}}
\xrightarrow{\text{NaBH}_4}
\quad
\begin{array}{c}
\text{O} \\
\parallel \\
\text{CH}_3\text{OCO} \quad \text{H} \\
\text{C} \\
\parallel \\
\text{H} \quad \text{C} \\
\text{C} \quad \text{C}=\text{O} \\
\text{CH}_3 \quad \text{N} \\
\text{CH}_3
\end{array}
\qquad (135)
$$

with the β-chloride substituted by a methyl carbonate group. Sodium borohydride is used to remove the mercury from the product and appears in some way to possibly promote cyclization.

We have recently extended the organopalladium carbonylation reactions to the synthesis of aryl lactones, anhydrides and imides (eqs 136-139).[31] By taking advantage of McKillop and Taylor's[32,33] heteroatom directed ortho thallation reactions and the facile transmetallation of arylthallium compounds by palladium salts, we have been able to obtain arylpalladium compounds which readily carbonylate and cyclize to the corresponding cyclic derivatives. These carbonylation reactions are catalytic in palladium chloride. Apparently the thallium(III) salt generated upon transmetallation is a sufficiently strong oxidant to re-oxidize palladium(0) back to palladium(II) so that no additional oxidant needs to be added.

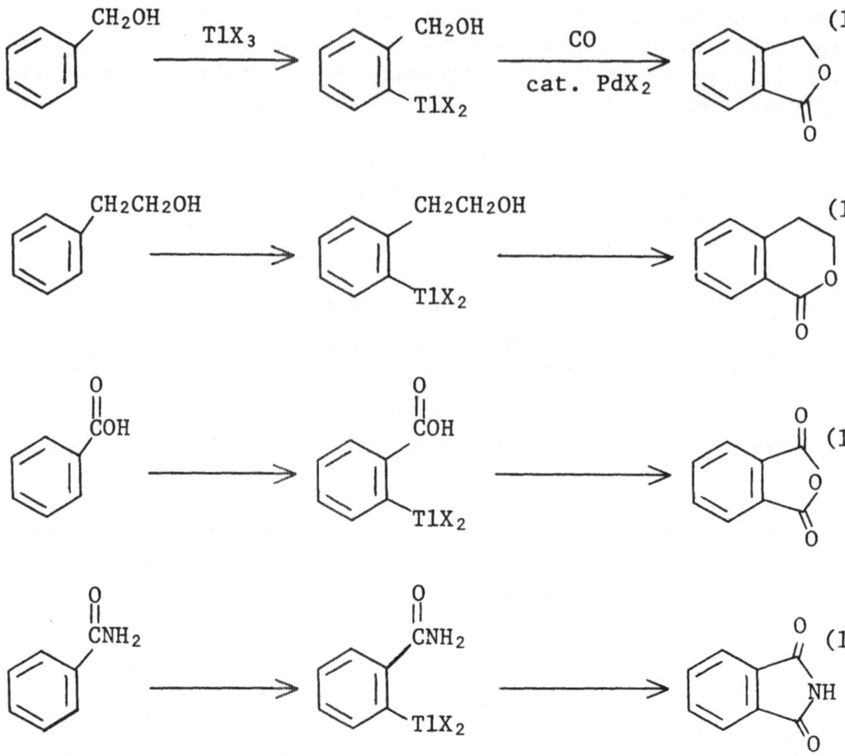

(136)

(137)

(138)

(139)

One further cyclization reaction promoted by transition metal catalysts that we have been examining is the intramolecular hydroacylation of 4-enals to cyclopentanones (eqs 140-142).[34] This reaction is best effected by rhodium(I) phosphine complexes in methylene chloride solvent under an ethylene atmosphere. At present the tri-p-anisylphosphine ligand appears to give the best results. When attempting to extend this reaction to the synthesis

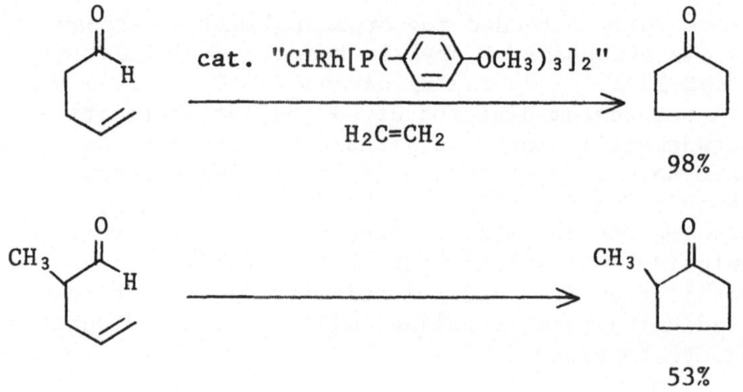

(140)

98%

(141)

53%

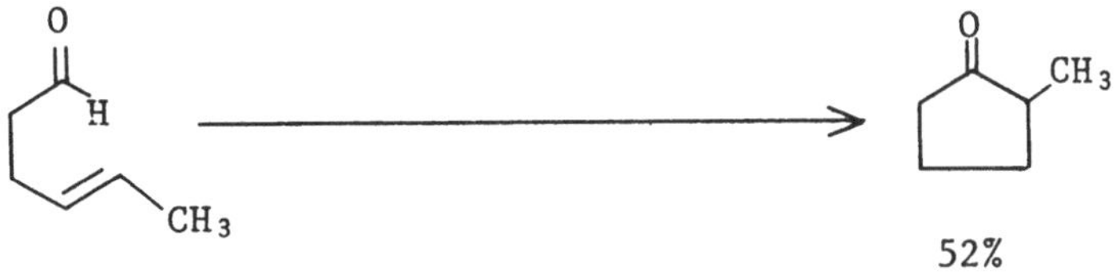

$$(142)$$

$$52\%$$

of spiro ketones, we observed exclusive addition of ethylene to the aldehyde (eq 143). We are currently examining further applications of this reaction to organic synthesis.

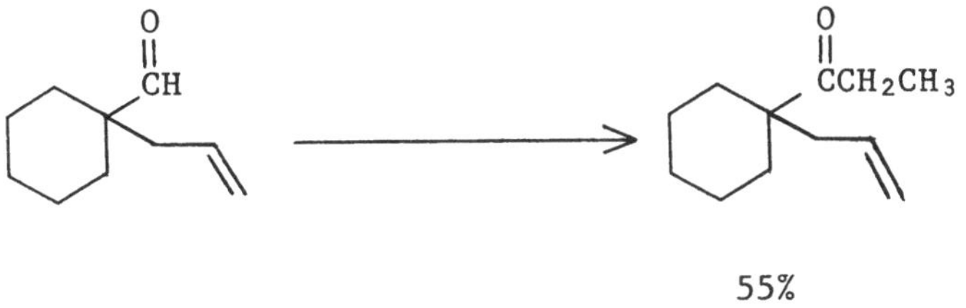

$$(143)$$

$$55\%$$

Mechanistically, these reactions appear to proceed by (1) oxidative addition of the aldehyde to the rhodium(I) complex (eq 144), (2) rhodium hydride addition to the carbon-carbon double bond (eq 145), and (3) reductive elimination to regenerate the rhodium(I) catalyst and the cyclopentanone (eq 146).

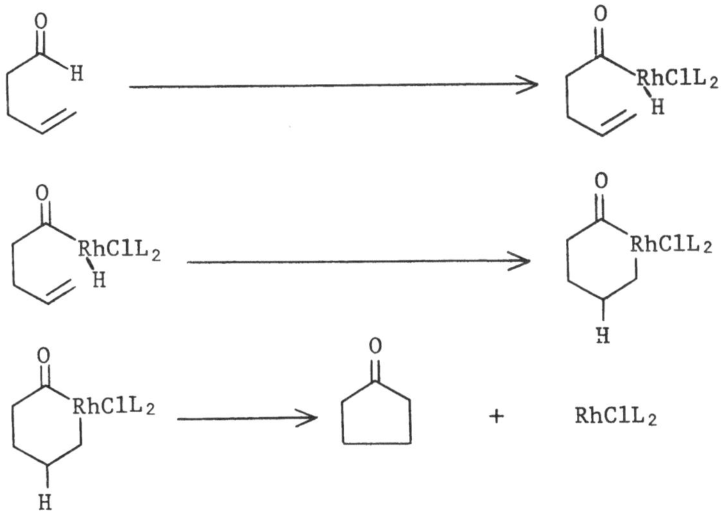

$$(144)$$

$$(145)$$

$$+ \quad RhClL_2 \qquad (146)$$

It is hoped that the work outlined here has indicated some of the promise evident in the use of both organomercurials and organotransition metal compounds in organic synthesis. Much has been accomplished in recent years and much more can be expected.

ACKNOWLEDGMENTS

I wish to acknowledge Professor George Zweifel for having inspired my initial interest in organometallic chemistry, Professor H. C. Brown for having so carefully cultivated and nurtured that interest, and finally my own graduate students whose names are mentioned in the references for having provided the hard work necessary to bring it all to fruition. The excellent technical assistance of Ms. Helen Tranter and the generous financial support of the Petroleum Research Fund, Iowa State Research Foundation, Research Corporation, National Science Foundation, DuPont, National Institutes of Health, Matthey Bishop and Engelhard Industries are also deeply appreciated.

REFERENCES

1. L. G. Makarova and A. N. Nesmeyanov, "The Organic Compounds of Mercury", North Holland, Amsterdam, 1967.
2. H. Staub, K. P. Zeller and H. Leditsche in Houben-Weyl: Methoden der Organischen Chemie, 4th ed., Vol. 13/2b, Thieme, Stuttgart, 1974.
3. Richard C. Larock, "New Applications of Organometallic Compounds in Organic Synthesis", ed. D. Seyferth, Elsevier, Amsterdam, 1976, pp. 257-303.
4. Richard C. Larock, *Angew. Chem.*, 90, 28 (1978); *Angew Chem. Int. Ed. Eng.*, 17, 27 (1978).
5. R. C. Larock and H. C. Brown, *J. Organometal. Chem.*, 36, 1 (1972).
6. R. C. Larock, S. K. Gupta and H. C. Brown, *J. Am. Chem. Soc.*, 94, 4371 (1972).
7. J. F. Normant, C. Chuit, G. Cahiez and J. Villieras, *Synthesis*, 803 (1974).
8. G. Drefahl, G. Heublein and A. Wintzer, *Angew. Chem.*, 70, 166 (1958).
9. A. N. Nesmeyanov, A. E. Borisov, I. S. Savel'eva and M. A. Osipova, *Izv. Akad. Nauk SSSR, Otdel. Khim. Nauk*, 1249 (1961).
10. A. E. Borisov, V. D. Vil'chevskaya and A. N. Nesmeyanov, *Izv. Akad. Nauk SSSR, Otdel. Khim. Nauk*, 1008 (1954).
11. A. N. Nesmeyanov and R. Kh. Friedlina, *Dokl. Akad. Nauk SSSR*, 24, 59 (1940).
12. R. C. Larock, B. Riefling and C. A. Fellows, *J. Org. Chem.*, 43, 131 (1978).
13. R. C. Larock, L. D. Burns and M.-C. Cheng, work in progress.
14. Richard C. Larock, *J. Org. Chem.*, 41, 2241 (1976).
15. R. C. Larock and J. C. Bernhardt, *J. Org. Chem.*, 42, 1680 (1972).
16. R. C. Larock and B. Riefling, *J. Org. Chem.*, 43, 1468 (1978).
17. G. Zweifel and N. L. Polston, *J. Am. Chem. Soc.*, 92, 4068 (1970).

18. G. Zweifel, N. L. Polston and C. C. Whitney, *J. Am. Chem. Soc.*, 90, 6243 (1968).
19. H. C. Brown and N. Ravindran, *J. Org. Chem.*, 38, 1617 (1973).
20. E. Negishi, G. Lew and T. Yoshida, *J. Chem. Soc.*, *Chem. Commun.*, 874 (1973).
21. R. C. Larock and K. M. Beatty, work in progress.
22. R. C. Larock and M. A. Mitchell, *J. Am. Chem. Soc.*, 100, 180 (1978).
23. R. C. Larock, M. A. Mitchell and S. A. Smith, work in progress.
24. R. C. Larock, J. C. Bernhardt and R. J. Driggs, *J. Organometal. Chem.*, in press.
25. R. C. Larock, S. A. Smith and K. M. Beatty, work in progress.
26. M. C. Baird, J. T. Mague, J. A. Osborn and G. Wilkinson, *J. Chem. Soc.*, A, 1347 (1967).
27. J. K. Stille and M. T. Regan, *J. Am. Chem. Soc.*, 96, 1508 (1974).
28. R. C. Larock and J. C. Bernhardt, *J. Org. Chem.*, 43, 710 (1978).
29. Richard C. Larock, *J. Org. Chem.*, 40, 3237 (1975).
30. R. C. Larock, B. Riefling and C. A. Fellows, *J. Org. Chem.*, 43, 131 (1978).
31. R. C. Larock and C. A. Fellows, work in progress.
32. A. McKillop, J. D. Hunt, M. J. Zelesko, J. S. Fowler, E. C. Taylor, G. McGillivray and F. Kienzle, *J. Am. Chem. Soc.*, 93, 4841 (1971).
33. E. C. Taylor, F. Kienzle, R. L. Robey, A. McKillop and J. D. Hunt, *J. Am. Chem. Soc.*, 93, 4845 (1971).
34. R. C. Larock and G. F. Potter, work in progress.

SELECTIVE CARBON-CARBON BOND FORMATION *via* TRANSITION METAL

CATALYSIS: IS NICKEL OR PALLADIUM BETTER THAN COPPER?

Ei-ichi Negishi

Department of Chemistry, Syracuse University

Syracuse, New York 13210

[I wish to dedicate this paper to my former mentor, Professor H. C. Brown of Purdue University, on the occasion of his 66th birthday.]

INTRODUCTION

The cross-coupling reaction involving the interaction of organometallic species, such as those containing lithium and magnesium, with organic halides and related electrophilic derivatives represents one of the most straightforward methods of carbon-carbon bond formation (eq 1).

$$R^1M \; + \; R^2X \; \longrightarrow \; R^1\text{--}R^2 \; + \; MX \qquad (1)$$

Despite its inherent simplicity, however, its synthetic utility had been rather limited until the mid-1960s, due to various complications, such as competitive elimination and halogen-metal exchange reactions and the general lack of chemoselectivity.

Advances in organocopper chemistry made mainly within the last decade or so have revolutionized the art of cross coupling. Thus, the reaction of organocoppers with organic halides and sulfonates has provided a number of procedures permitting many different types of cross coupling which had been difficult to achieve previously with organolithiums and Grignard reagents.[1] Even so, not all problems of cross coupling have been solved by the introduction of the organocopper reactions. For example, substitution at tertiary carbon centers still remains a difficult

process. The reaction of unsaturated organocopper reagents with
unsaturated organic halides is also plagued with some serious
difficulties. More specifically, selective alkenyl-alkenyl or aryl-
aryl cross coupling is still generally difficult to achieve with
organocoppers, and the cross-coupling reactions of alkynylcoppers
are generally quite sluggish, requiring drastic reaction conditions.

The success and difficulties observed with organocoppers have
triggered investigations of other transition metal compounds as
reagents for selective carbon-carbon bond formation. In 1967,
Corey introduced the reaction of π-allylnickel derivatives with
organic halides (Corey-Semmelhack-Hegedus reaction) as a selective
procedure for cross coupling involving allylic groups[2] (eq 2).

$$\text{(structure)} \quad + \quad RX \quad \longrightarrow \quad \text{(structure)} R \qquad (2)$$

The reaction is stoichiometric with respect to nickel, and its
scope is limited to allylation of organic halides. Note that the
π-allylnickel derivative in equation 2 is used as a Grignard-like
reagent, i.e., as a nucleophile, at least in a formal sense. Subsequent attempts to utilize aryl- and alkenylnickel compounds as
Grignard-like reagents, mainly by Semmelhack and his associates[3],
led to the development of certain procedures suitable for homo
coupling of aryl or alkenyl groups. However, these procedures are
not satisfactory for cross coupling.

In 1965, Tsuji[4] reported an allylation reaction with π-allyl-
palladium derivatives, the scope of which has been greatly expanded by subsequent studies by Trost and others[5] (eq 3).

$$\text{(structure)} \quad + \quad MHC \overset{COY}{\underset{Z}{\diagup}} \quad \longrightarrow \quad \text{(structure)} CH \overset{COY}{\underset{Z}{\diagup}} \qquad (3)$$

M = alkali metal. Y = OR, organic group. Z = COY, SO$_2$R.

In the great majority of the reported examples the reaction is
stoichiometric with respect to palladium, although Trost[6] has
recently reported that the allylation reaction can be achieved
with catalytic amounts of palladium complexes. An interesting
point to be noted here is that, unlike π-allylnickel compounds,
π-allylpalladium derivatives are used, at least in a formal sense,
as electrophilic reagents.

Another organopalladium reaction worth mentioning here is the
addition-elimination reaction of organopalladium compounds with

olefins and acetylenes developed by Heck[7] (Heck reaction) (eq 4).

$$RPdL_n \; + \; \overset{H}{\underset{}{}}C=C \longrightarrow R-\overset{|}{\underset{H}{C}}-\overset{|}{\underset{PdL_n}{C}}- \longrightarrow \overset{}{\underset{R}{}}C=C \; + \; HPdL_n \quad (4)$$

The organopalladium reagent is usually generated *in situ* either from another organometallic compound, such as an organomercurial, *via* transmetallation or from an organic halide *via* oxidative addition. In the former case the reaction is stoichiometric with respect to palladium, whereas the latter is catalytic. Although the reaction is of considerable synthetic utility, it tends not to be regio- and/or stereoselective.

In 1972, Kumada[8] and Corriu[9] independently reported that the cross-coupling reaction of Grignard reagents and organolithiums with alkenyl and aryl halides could be markedly catalyzed by certain Ni-phosphine complexes (eq 5).

$$R^1M \; + \; R^2X \; \xrightarrow{\text{cat. } NiL_n} \; R^1-R^2 \; + \; MX \quad (5)$$

M = MgX or Li. R^2 = aryl, alkenyl.

Although this reaction falls into the category of the so-called Kharasch reaction,[10] i.e., the transition-metal-catalyzed reaction of Grignard reagents with organic halides, the use of metal-phosphine complexes as catalysts had not been explored in any detail until 1972. While the precise mechanism of the reaction remains to be further clarified, the following scheme, consisting of an oxidative addition-transmetallation reductive elimination sequence proposed by Kumada[8] and scrutinized by Kochi[11] and Parshall[12], appears plausible in many cases.

The products of the Ni-catalyzed cross-coupling reaction reported in the papers by Kumada and Corriu were those which could

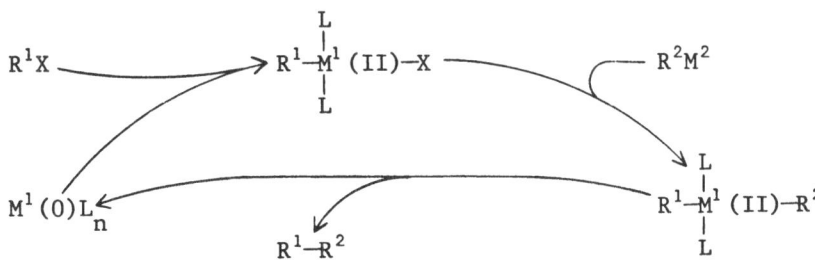

Scheme I

also be prepared readily by the corresponding organocopper reactions. Thus, the unique advantages of this new cross-coupling reaction were, until recently, essentially unknown. Furthermore, it was evident that the use of organolithiums and Grignard reagents would severely limit the range of functionalities that could be tolerated in this reaction. Nonetheless, we were strongly attracted by the possibility that Scheme I might represent a general and useful approach to selective carbon-carbon bond formation *via* cross coupling, the scope of which might possibly be far wider than indicated by the known results. Some of the specific questions to which we addressed ourselves were the following:

1. What might be the full scope of Scheme I with respect to M^1?
   What about Pd, Pt, Rh, etc.? If other transition metals should participate in similar cross-coupling reactions, do any of these offer any synthetic advantages over Ni? Although Murahashi,[13] Cassar,[14] Sonogashira[15] and others have reported some Pd-catalyzed cross-coupling reactions since 1975, at the outset of our investigation this ability of Pd was virtually unknown.

2. What might be the full scope of Scheme I with respect to M^2?
   Specifically, we were attracted by the possibility of utilizing B, Al, Si, Sn and Zr. Organometallics containing these metals are known to be far more compatible with various electrophilic functionalities, such as esters, amides, cyano, and nitro groups, than those containing Li and Mg. More exciting to us was the possibility of generating stereo- and regio-defined alkenylmetals containing these metals *via* the corresponding hydrometallation reactions and transferring the alkenyl groups from these metals to carbon (eq 6).

$$R^1C \equiv CH \xrightarrow{\text{HML}_n} \underset{H}{\overset{R^1}{>}}C=C\underset{ML_n}{\overset{H}{<}} \xrightarrow[M'L_n]{R^2X} \underset{H}{\overset{R^1}{>}}C=C\underset{R^2}{\overset{H}{<}} \qquad (6)$$

3. What might be the full scope of Scheme I with respect to the R^1 and R^2 groups? Specifically, could we provide some solutions to many pending problems, such as development of selective alkenyl-alkenyl and aryl-aryl coupling procedures?

THE Ni-CATALYZED REACTION OF (E)-1-ALKENYLALANES WITH ARYL HALIDES

We have found that alkenylalanes can be activated towards aryl bromides and iodides by certain Ni-phosphine complexes, such as $Ni(PPh_3)_4$, so as to form arylated *trans*-alkenes in high yields[16] (Table 1).

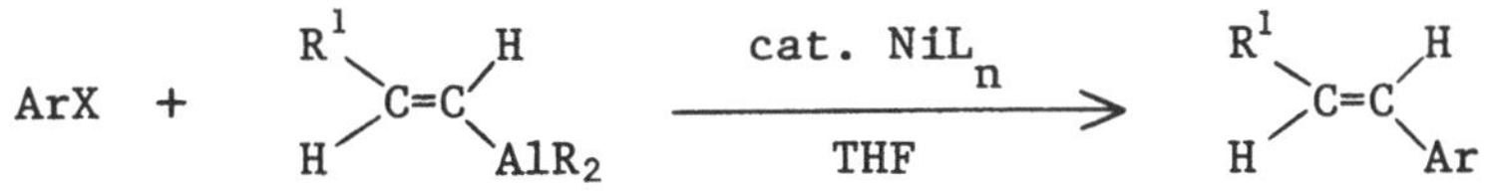

$$\text{ArX} + \underset{H}{\overset{R^1}{>}}C=C\underset{AlR_2}{\overset{H}{<}} \xrightarrow[\text{THF}]{\text{cat. } NiL_n} \underset{H}{\overset{R^1}{>}}C=C\underset{Ar}{\overset{H}{<}} \quad (7)$$

This appears to represent the first paper describing the use of organoalanes in the Ni-catalyzed cross-coupling reaction. Some of the salient features of the reaction are as follows. (1) The reaction is catalytic with respect to the nickel reagent. (2) 1-Naphthylbromobis(triphenylphosphine)nickel (1), obtainable as an isolable species *via* oxidative addition of 1-bromonaphthalene to Ni(PPh$_3$)$_4$, reacts with an equimolar amount of *trans*-1-hexenylalane (2) to form 3 in 65% yield within 1 hr at 25°. Under otherwise comparable conditions, the corresponding catalytic reaction using 5 mole % of Ni(PPh$_3$)$_4$ requires 12–24 hr for completion. It is therefore likely that 1 is an intermediate in this reaction.

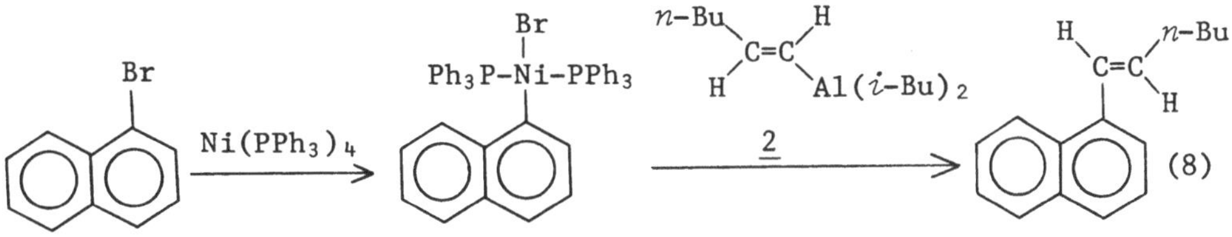

$$(8)$$

1   3

(3) The stereospecificity of the reaction is >99%. (4) Use of Ni-phosphine complexes appears essential, since other types of Ni-complexes, such as bis(cyclooctadiene)nickel, are not effective catalysts. (5) The reaction of alkenylaluminate complexes produces the desired cross-coupled products only in low yields, while alkenylboranes and their "ate" complexes have failed to undergo the desired cross coupling.

Table 1

The Ni-Catalyzed Alkenyl-Aryl Coupling Reaction

ArX	R^1 of (E)-R^1CH=CHAl(i-Bu)$_2$	Yield of (E)-R^1CH=CHAr (%)
PhI	*n*-Bu	89
PhBr	*n*-Bu	85
p-NCPhBr	*n*-Bu	64
1-Naphthyl bromide	*n*-Bu	93
p-MePhBr	Cyclohexyl	75

THE Pd-CATALYZED ALKENYL-ALKENYL CROSS-COUPLING REACTION

Our finding that alkenylalanes can participate in the Ni-catalyzed cross coupling prompted us to develop a selective procedure for the synthesis of conjugated dienes, which had been difficult to prepare in a selective manner by the reaction of an alkenylmetal derivative with an alkenyl halide. As expected, the Ni(PPh$_3$)$_4$-catalyzed reaction of (E)-1-heptenylalane with (E)-1-hexenyl iodide produced 5,7-tridecadiene in 70% yield, along with 5,7-dodecadiene and 6,8-tetradecadiene, formed in 2 and 15% yields, respectively.[17] A careful stereochemical examination has revealed that the cross-coupled product is 95% E,E, contaminated with one or more stereoisomers which appear to be largely the (E,Z)-isomers. In the corresponding reaction of (Z)-1-hexenyl iodide, the stereo-specificity was only 90%.

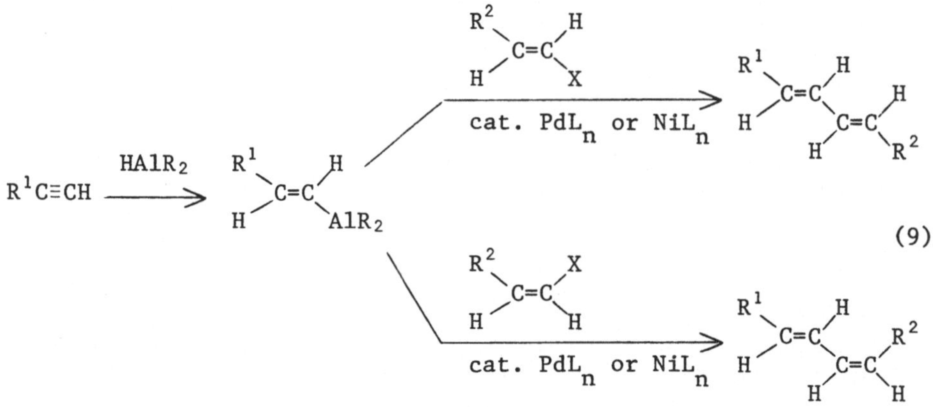

$$\text{(9)}$$

The above results reveal a serious limitation associated with the Ni-catalyzed cross coupling, although it is not clear at what stage the stereochemical scrambling takes place. Fortunately, we soon found that a Pd-phosphine catalyst generated *in situ* by treating Cl$_2$Pd(PPh$_3$)$_2$[18] with 2 molar equivalents of diisobutylaluminum hydride (DIBAH) was superior to Ni(PPh$_3$)$_4$.[17] Using this Pd-catalyst both (E,E)- and (E,Z)-conjugated dienes were obtained as stereochemically pure products. An added bonus was that no more than traces of homo-coupled products were formed in all of the Pd-catalyzed alkenyl-alkenyl cross-coupling reactions examined to date. Some typical examples of the Pd-catalyzed alkenyl-alkenyl cross-coupling reaction are shown in eqs 10-12.

$$\text{(10)}$$

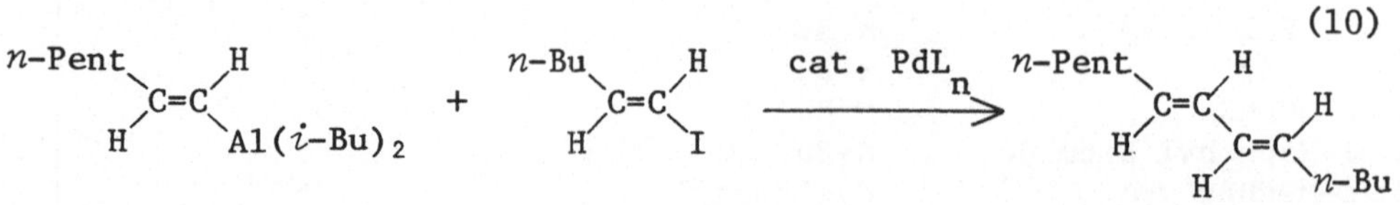

75% (>99% E,E)

(11)

$$n\text{-Pent}\diagdown \diagup H$$
$$\underset{H}{\overset{}{C=C}}\diagdown Al(i\text{-Bu})_2$$
+
$$\underset{H}{\overset{n\text{-Bu}}{C=C}}\diagup I$$
$$\xrightarrow{\text{cat. PdL}_n}$$
$$n\text{-Pent}\diagdown \diagup H$$

55% (>99% E,Z)

(12)

$$n\text{-Bu}\diagdown \diagup H$$
$$\underset{H}{\overset{}{C=C}}\diagdown Al(i\text{-Bu})_2$$
+
$$\underset{H}{\overset{Br}{C=C}}\diagup Me$$
$$\xrightarrow{\text{cat. PdL}_n}$$
$$n\text{-Bu}\diagdown \diagup H$$

75% (97% E,E)

Pd-CATALYZED ALKYNYL-ARYL CROSS COUPLING — COMPARISON OF VARIOUS METALS (M^2) IN THE ORGANOMETALLIC REAGENTS (R^2M^2)

As clearly demonstrated in the case of alkenyl-alkenyl coupling, the Pd-catalyzed cross coupling offers unique advantages over the organocopper-organic halide reactions. We therefore felt it desirable to examine the scope of the Pd-catalyzed cross-coupling with respect to the metal (M^2) of organometallic reagents (R^2M^2). For this purpose, we chose the reaction of alkynylmetals with aryl halides, since a wide variety of alkynylmetals and of aryl halides were readily available. Accordingly, we treated 1-heptyne with n-butyllithium, converted the 1-heptynyllithium thus formed into 1-heptynylmetal derivatives containing various metals, and reacted them with o-tolyl iodide in the presence of either Pd(PPh₃)₄ or the product of the Cl₂Pd(PPh₃)₂-DIBAH reaction. The results, summarized in Table 2[19], reveal the following noteworthy points.

1. Quite unexpectedly, 1-heptynylmetal derivatives containing Li and Na proved to be ineffective in providing the expected cross-coupled product 6.
2. On the other hand, those derivatives containing Zn, B (borate) and Sn produced 6 in excellent yields. The Mg- and Al-containing derivatives also give 6 in moderate yields.
3. The Si- and Zr-containing derivatives appear inert.
4. On the other hand, 1-heptynylmercuric chloride appears to undergo an undesirable redox reaction with the Pd(0)-phosphine complex, although this point remains to be clarified.

Our subsequent investigations have indicated that the alkynyl-aryl coupling reaction is not an ideal model for examining the effects of various metals. For example, whereas alkynylborates can readily participate in the Pd-catalyzed cross coupling, all

Table 2 - Reaction of 1-Heptynyllithium with o-Tolyl Iodide (Pd-Phosphine Catalyst Present)

$$n\text{-PentC}{\equiv}\text{CM} + o\text{-Tol-I} \xrightarrow[\text{THF}]{\text{PdL}_n} o\text{-Tol-C}{\equiv}\text{C-}n\text{-Pent} + (n\text{-PentC}{\equiv}\text{C-})_2 + (o\text{-Tol-})_2$$
$$\underline{4}\qquad\quad \underline{5}\qquad\qquad\qquad\qquad \underline{6}\qquad\qquad\qquad \underline{7}\qquad\qquad \underline{8}$$

M of 4	Catalyst[a]	Time[b] (hr)	5 (mmoles)	Product (mmoles) 6	7	8
Li	A	1	8.8	trace	0	0
Na[c]	B	24[d]	8.0	0.3	0	0
MgBr[e]	A	24[d]	4.1	5.8	0	0
	A	1	5.5	2.9	0	0
	A	24	3.3	4.9	0	0
ZnCl	A	1	0.8	9.1	0	trace
	A	3	0.2	8.8	0	trace
HgCl	A	1	9.2	trace	0	0
⁻B(n-Bu)₃	A	6[d]	8.8	trace	0	7
	A	3	7.6	1.0	trace	trace
⁻Al(i-Bu)₂(n-Bu)	A	1[d]	0.5	9.2	trace	trace
	A	3	8.0	0.4	0	0
Al(i-Bu)₂	A	1[d]	1.0	3.8[f]	trace	trace
	A	3	4.6	4.9	0	0
SiMe₃	A	1[d]	9.4	trace	0	0
Sn(n-Bu)₃	A	1	1.4	7.5	trace	3
	A	6	0.6	8.3	trace	3
ZrCp₂Cl	A	1	9.1	0	0	0
	A	3[d]	8.0	0	0	0

The initial amount of 4 or 5 = 10 mmoles. [a]5 mole % of A or B. A=Cl₂Pd(PPh₃)₂ + 2 i-Bu₂AlH. B=Pd(PPh₃)₄. [b]The reaction was run at 20-22° unless otherwise mentioned. [c]Prepared by treating 1-heptyne with NaOMe. This reaction was carried out under the conditions used by Cassar [L. Cassar, J. Organometal. Chem., 93, 253 (1975)]. [d]Reflux. [e]Prepared by treating 1-heptyne with i-PrMgBr. [f]A minor amount (<8%) of o-butyltoluene and an appreciable amount of toluene were also present.

other types of organoborates examined to date, which include the alkyl, alkenyl and aryl derivatives, are unreactive. On the other hand, in marked contrast with the unreactive alkynylzirconium derivative, the corresponding alkenylzirconium derivatives have exhibited a high reactivity in the Pd-catalyzed cross coupling, as discussed later in detail. At present, there is no satisfactory explanation for the reversal of the relative reactivities observed with alkenyl and alkynyl derivatives containing B and Zr.

The results presented in Table 2 provide a few other interesting puzzles. The inability of 1-heptynyllithium to produce the expected product is noteworthy and puzzling. As it is inconceivable that 1-heptynyllithium remains unreactive under the reaction conditions, it is likely that the reagent participates in some undesirable side reactions which inactivate the catalyst, e.g., "ate" complexation. The unexpectedly high reactivity of the tin and borate derivatives is also somewhat puzzling in view of the well-known low nucleophilicity of the C—Sn and C—B bonds. Although there is no evidence, it is not inconceivable that these reactions proceed via some mechanisms not involving transmetallation. One likely candidate is the addition-elimination mechanism similar to that proposed for the Heck reaction. It should be noted that such an addition-elimination mechanism would be greatly facilitated by the presence of the SnR_3 or $^-BR_3$ group through σ-π conjugation.[20]

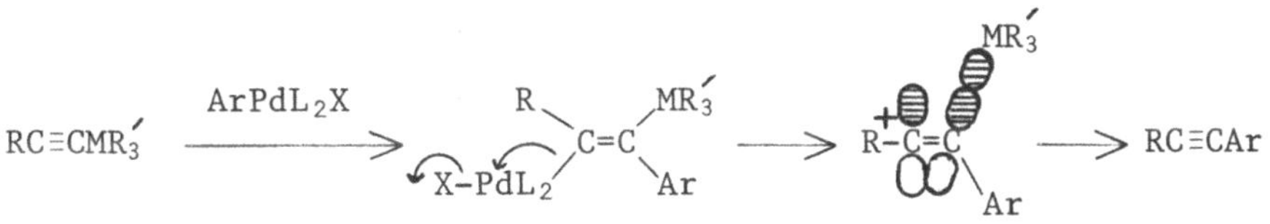

$$RC\equiv CMR_3' \xrightarrow{ArPdL_2X} \ \xrightarrow{} \ \xrightarrow{} RC\equiv CAr$$

M = Sn or $^-$B

Scheme II

From the synthetic viewpoint, the Pd-catalyzed reaction of alkynylzinc halides with aryl iodides or bromides (eq 13) provides probably the most general and selective procedure known today for the preparation of arylalkynes via cross coupling.

$$RC\equiv CZnCl \ + \ ArX \ \xrightarrow{\text{cat. } PdL_n} \ RC\equiv CAr \qquad (13)$$

It offers some distinct advantages over the alkynylcopper reaction of Castro[21] (eq 14) and a few Pd-catalyzed alkynyl-aryl coupling reactions reported in 1975 by Cassar[14] (eq 15), Heck[22] (eq 16) and Sonogashira[15] (eq 17).

$$RC{\equiv}CCu \quad + \quad ArX \quad \xrightarrow[\underline{>}100°]{C_5H_5N} \quad\quad\quad RC{\equiv}CAr \quad (14)$$

$$RC{\equiv}CNa \quad + \quad ArX \quad \xrightarrow{cat.\ PdL_n,\ 50°} \quad RC{\equiv}CAr \quad (15)$$

$$RC{\equiv}CH \quad + \quad ArX \quad \xrightarrow[Et_3N,\ 100°]{cat.\ Pd(OAc)_2(PPh_3)_2} \quad RC{\equiv}CAr \quad (16)$$

$$RC{\equiv}CH \quad + \quad ArX \quad \xrightarrow[Et_2NH]{cat.\ CuI-Cl_2Pd(PPh_3)_2} \quad RC{\equiv}CAr \quad (17)$$

With the possible exception of the Sonogashira procedure (eq 17), these reactions are generally far more sluggish than the Zn reaction (eq 13). Moreover, none of the procedures shown in eqs 14–17 appears suited for the direct preparation of terminal arylalkynes.[23] On the other hand, the mild conditions required in the reaction shown in equation 13 and the relatively high thermal stability of ethynylzinc chloride, readily obtainable by treating ethynyllithium with $ZnCl_2$, permits a direct preparation of terminal arylalkynes.[18] Some representative examples of arylalkyne syntheses are summarized in Table 3.

Table 3

Preparation of Arylalkynes by the Pd-Catalyzed Reaction
of Alkynylzinc Reagents with Aryl Halides[a]

R of $RC{\equiv}CZnCl$	Aryl halide Ar	X	Catalyst[b] (mole %)	Time hr	Yield[c] of $ArC{\equiv}CR$ %
H	o-Tol	I	A(5)	3	71
H	p-Anisyl	I	A(5)	1	66(56)
	Ph	I	A(1)	1	67
Methyl	2-Thienyl	I	A(1)	1	92(82)
n-Pentyl	2-Thienyl	I	B(1)	2	85(70)
	2-Thienyl	Br	A(5)	48	75
	p-NCPh	Br	A(1)	4	93
	p-O$_2$NPh	I	B(1)	3	94(64)
	o-Tol	I	B(5)	1	88
Phenyl	m-Tol	I	B(5)	1.5	87(80)

[a]All reactions were run at room temperature in THF. The $RC{\equiv}CZnCl/ArX$ ratio is 2 for the cases where R=H and 1 for R≠H. [b]A=Pd(PPh$_3$)$_4$ B=Cl$_2$Pd(PPh$_3$)$_2$ + 2 i-Bu$_1$AlH. [c]By GLC. The numbers in parentheses are isolated yields.

SCOPE OF THE Pd-CATALYZED CROSS COUPLING INVOLVING ALKYNYL REAGENTS

The alkynyl-aryl cross-coupling procedure developed in the preceding section was then applied to the alkynyl-alkenyl and alkynyl-alkynyl cases.

Alkynyl-Alkenyl Coupling. The conjugate enyne unit is characteristic of a number of natural products. It is also readily convertible into the conjugated diene by known procedures.[24] Despite various procedures developed recently for enyne synthesis,[25] certain types of enynes, e.g., terminal *cis*-enynes, are not yet readily accessible in an efficient and selective manner.

We were pleased to find that there was no difficulty in applying the Pd-catalyzed alkynyl-aryl coupling procedure developed above to the reaction of alkynylzinc chlorides with alkenyl halides[26] (eq 18).

$$\underset{R^2}{\overset{R^1}{\diagdown}}C=C\underset{X}{\overset{R^3}{\diagup}} \quad + \quad ClZnC\equiv CR^4 \quad \xrightarrow[0-25°]{\text{cat. PdL}_n} \quad \underset{R^2}{\overset{R^1}{\diagdown}}C=C\underset{C\equiv CR^4}{\overset{R^3}{\diagup}} \qquad (18)$$

Some representative results are summarized in Table 4. The following observations and facts are worth noting. As in the alkynyl-aryl coupling, the alkali metal derivatives of alkynes do not give enynes in any significant yields under comparable conditions. Terminal enynes are readily obtainable using ethynylzinc chloride. The enyne products are formed without the concomitant production of other undesirable by-products, such as dienes and diynes, in any significant yields (<5%). As expected, the reaction is highly stereospecific (≥97%).

The Cross-Coupling Reactions Involving Alkynyl Halides. The success observed in the alkynyl-aryl and alkynyl-alkenyl coupling experiments prompted us to examine the corresponding alkynyl-alkynyl cross-coupling. Although the Cadiot-Chodkiewicz reaction[27] has been successfully used in the preparation of a number of conjugated diynes, it has also been plagued with various difficulties. For example, it cannot readily be adapted to the preparation of terminal diynes. Although some indirect routes[28] have been developed, they are generally rather cumbersome. An organoborate procedure[29] developed recently is promising, but its synthetic scope has not yet been well delineated. In view of the fact that a large number of natural products contain conjugated diyne and poly-yne units,[30] it is desirable to develop additional procedures which would alleviate the existing difficulties.

Quite disappointingly, the Pd-catalyzed reaction of 1-heptynylzinc chloride with 1-hexynyl iodide or bromide produces a nearly

Table 4

The Pd-Catalyzed Reaction of Alkynylzinc
Chlorides with Alkenyl Halides

R^1	R^2	R^3	R^4	X	Catalyst[a]	Yield[b] (%)
H	n-Bu	H	H	I	A	83(65)
n-Bu	H	H	H	I	A	76(71)
n-Bu	Et	H	H	I	A	70
H	n-Bu	H	n-Pent	I	A	82
n-Bu	H	H	n-Pent	I	A	87(76)
n-Bu	H	H	n-Pent	I	B	82
n-Bu	Et	H	n-Pent	I	B	80
COOMe	Me	H	n-Bu	Br	A	87(65)

[a] $A=Pd(PPh_3)_4$; $B=Cl_2Pd(PPh_3)_2 + 2$ i-Bu$_2$AlH.
[b] By GLC. The numbers in parentheses are isolated yields.

statistical (1:2:1) mixture of C_{12}, C_{13} and C_{14} diynes, albeit in
nearly quantitative overall yield (eq 19).

$$\text{(19)}$$
$$R^1C{\equiv}CZnCl + XC{\equiv}CR^2 \xrightarrow{\text{cat. PdL}_n} R^1C{\equiv}C{-}C{\equiv}CR^2 + R^1C{\equiv}C{-})_2 + R^2C{\equiv}C{-})_2$$

Various attempts to improve the results by changing reaction parameters, such as metal, solvent and temperature, have so far
failed to circumvent the difficulty. These results are quite un-
expected and puzzling in view of the fact that all the Csp^2-Csp^2
and Csp^2-Csp cross-coupling reactions discussed earlier display
high cross-/homo- coupling ratios.

Assuming that the reaction proceeds via oxidative addition-
transmetallation-reductive elimination, any of these micro-steps
may involve scrambling of the two alkynyl groups. Alternatively,
the two alkynyl reagents can undergo halogen-metal exchange under
the reaction conditions. In an attempt to identify the source of
the difficulty, we examined the Pd-catalyzed reaction of 1-hexynyl
iodide with o-tolylzinc chloride and ($\underline{E}$)-1-heptenylzinc chloride.
Interestingly, the expected cross-coupled products were obtained
in excellent yields[19] (eqs 20 and 21).

$$\text{90\%}\qquad\text{(20)}$$

$$n\text{-Pent}\diagdown_{C=C}\diagup^{H} \quad + \quad IC\equiv C\text{-}n\text{-Bu} \quad \xrightarrow{\text{cat. PdL}_n} \quad n\text{-Pent}\diagdown_{C=C}\diagup^{H} \qquad (21)$$

$$85\%$$

These results suggest that the oxidative addition reaction of 1-hexynyl iodide with the Pd catalyst, if it were involved, proceeds cleanly, at least in the absence of an alkynylmetal derivative. From the synthetic viewpoint, the reactions shown in equations 20 and 21 promise to provide highly satisfactory procedures for the preparation of arylalkynes and enynes. Their synthetic scope and their advantages and disadvantages over the corresponding organocopper reactions[26b],[31] remain to be delineated.

The extensive scrambling observed in the Pd-catalyzed alkynyl-alkynyl cross coupling prompted us to seek an indirect method for the synthesis of diynes. Since a variety of conjugated enynes is now readily obtainable, all that is required is to synthesize those enynes which could readily be converted into diynes. After several trials and errors, we have developed the following two-step procedure for the synthesis of hitherto difficulty obtainable terminal diynes involving the use of (E)-1,2-chloroiodoethylene as a key reagent[32] (eq 22).

$$RC\equiv CZnCl \quad + \quad {}^{I}\diagdown_{C=C}\diagup^{H} \quad \xrightarrow{\text{cat. PdL}_n} \quad {}^{RC\equiv C}\diagdown_{C=C}\diagup^{H}$$

$$(22)$$

$$\xrightarrow[\text{2. } H_3O^+]{\text{1. } NaNH_2,\ NH_3} \quad RC\equiv CC\equiv CH$$

When the R group is n-pentyl, the yields for the coupling and elimination steps are 75 and 80%, respectively. 1,2-Chloro-iodoethylene can be prepared in high yield by treating acetylene with iodine chloride.[33] As terminal diynes can be converted into various unsymmetrically disubstituted diynes, this procedure provides a convenient route to both terminal and internal conjugated diynes.

Finally, all our attempts to promote cross-coupling reactions involving alkynylmetals and/or alkynyl halides with catalytic amounts of Ni-phosphine complexes have resulted in low yields (<50%) of the products. At least in a few cases, there were indications that the acetylenic reagents and/or products are competitively oligomerized under the reaction conditions. It thus appears that certain Pd-phosphine complexes, e.g., Pd(PPh₃)₄, are

distinctly superior to the corresponding Ni-phosphine complexes
in cases where one or both of the reactants are alkyne derivatives.

UNSYMMETRICAL BIARYLS

One of the difficulties associated with the organocopper
approach to cross coupling is its inability to provide a general
and selective aryl-aryl cross-coupling procedure.[1] Although
various other unsymmetrical biaryl syntheses[34] have been developed,
there are few which are general and selective except for some intramolecular biaryl syntheses. It was therefore of interest to
develop a highly satisfactory procedure for aryl-aryl cross
coupling involving Ni- or Pd-catalysis.

In order to find a class of arylmetals that is highly suited
for the desired aryl-aryl cross-coupling, phenyllithium, phenyl-
magnesium bromide, phenylzinc chloride and phenyldiisobutylalane
were treated with p-iodoanisole in the presence of 5 mole % of
the Pd catalyst prepared by the reaction of $Cl_2Pd(PPh_3)_2$ and
DIBAH.[35] The results indicate that phenyllithium is unsatis-
factory, since it undergoes an extensive halogen-metal exchange
reaction leading to the formation of all three possible biaryls.
Although all the other phenylmetals gave high yields of 4-methoxy-
biphenyl [71% (Mg), 72% (Al) and 87% (Zn)], the amount of biphenyl
formed as a byproduct during the reaction was significantly less
with the Zn reagent (3%) than with the Mg and Al reagents (16-25%).
These results, together with the ability of arylzinc derivatives
to tolerate various electrophilic functional groups, such as
nitrile, ester and nitro groups, make the organozinc derivatives
the reagents of choice in this case as well. Some representative
results are summarized in Table 5.

Both Ni-phosphine and Pd-phosphine complexes are satisfactory
in most cases. However, the Pd complexes provide a unique advantage in that they are compatible with the nitro group, which
destroys the catalytic ability of the Ni complexes. During the
course of our development of the aryl-aryl cross-coupling procedure, a few related Ni- or Pd-catalyzed aryl-aryl coupling
reactions came to our attention. The Ni-catalyzed reaction of
arylmagnesium halides with aryl halides was investigated as a
route to *ortho*-substituted unsymmetrical biaryls[36] including 1,8-
diarylnaphthalenes.[37] In an analogous manner, chloro-, fluoro-
and *o*-methyl-substituted unsymmetrical biaryls were prepared by
the Pd-catalyzed reaction of arylmagnesium halides with aryl
halides.[38] None of these papers explicitly discusses the synthetic scope and limitations. Judging from the results presented in
these papers, however, these procedures, as such, seem to be of
limited applicability as selective routes to unsymmetrical biaryls.

Table 5

Preparation of Biaryl by the Ni- or Pd-Catalyzed
Reaction of Arylzinc Chloride with Aryl Halides

$$Ar^1ZnCl \quad + \quad Ar^2X \quad \xrightarrow[\text{25°, 1-2 hr}]{\text{cat. NiL}_n \text{ or PdL}_n} \quad Ar^1-Ar^2$$

Ar^1	Ar^2	X	Catalyst[a]	Yield[b] (%)
Ph	p-MeOPh	I	A	85
Ph	p-MeOPh	I	B	87
Ph	p-NCPh	Br	A	90
Ph	p-MeO$_2$CPh	Br	A	70
Ph	p-O$_2$NPh	I	B	90(74)
m-Tol	m-MePh	I	A	95

[a]A=Ni(PPh$_3$)$_4$, B=Cl$_2$Pd(PPh$_3$)$_2$ + 2 i-Bu$_2$AlH.
[b]By GLC. The number in parentheses is an isolated yield.

THE Ni- AND Pd-CATALYZED CROSS-COUPLING REACTION
OF (E)-1-ALKENYLZIRCONIUM DERIVATIVES

Although the alkenyl-aryl[16] and alkenyl-alkenyl[17] coupling
reactions of (E)-1-alkenylalanes have provided convenient and
selective entries into the arylalkene and conjugated diene series,
their synthetic applicability is severely limited by the fact that
the hydroalumination reaction is often incompatible with various
hetero-functional groups. Even ether groups tend to seriously
interfere with this reaction.[39] To circumvent this difficulty,
we have turned our attention to the hydrozirconation reaction
developed recently by Schwartz.[40] At the outset of this investigation, the application of organozirconium compounds to carbon-
carbon bond formation was limited to only two reactions, i.e.,
carbonylation[40] and the reaction with acyl halides.[41] It was
therefore of particular interest to develop a new type of carbon-
carbon bond–forming reaction of organozirconium compounds. It
should be recalled that all our attempts to induce the Ni- or
Pd-catalyzed reaction of alkenylboron compounds have so far failed.

We have, indeed, found that (E)-1-alkenylzirconium compounds
derived from terminal alkynes and Cl(H)ZrCp$_2$ react smoothly with
aryl[42] and alkenyl[43] halides in the presence of a catalytic
amount of a Ni or Pd complex (eq 23).

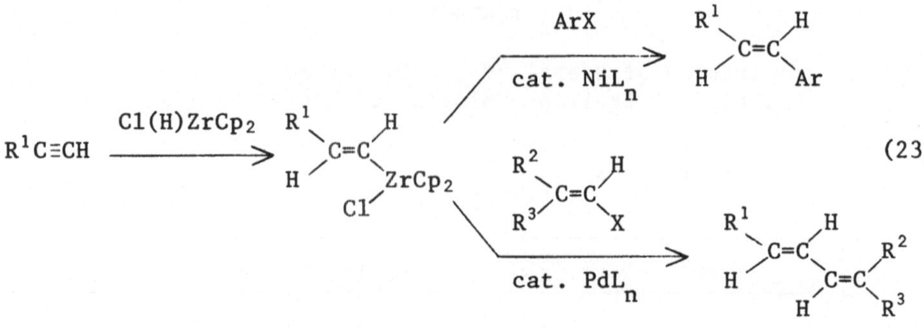

$$R^1C{\equiv}CH \xrightarrow{Cl(H)ZrCp_2} \underset{H}{\overset{R^1}{>}}C{=}C\underset{Cl}{\overset{H}{<}}ZrCp_2$$

$$\xrightarrow[\text{cat. NiL}_n]{ArX} \underset{H}{\overset{R^1}{>}}C{=}C\underset{Ar}{\overset{H}{<}}$$

$$\xrightarrow[\text{cat. PdL}_n]{\underset{R^3}{\overset{R^2}{>}}C{=}C\underset{X}{\overset{H}{<}}} \underset{H}{\overset{R^1}{>}}C{=}C\underset{H}{\overset{H}{<}}C{=}C\underset{R^3}{\overset{R^2}{<}}$$

(23)

Some representative results are summarized in Tables 6 and 7.

The following findings are worth noting. (1) In contrast to hydroalumination, hydrozirconation can readily tolerate certain common ether functionalities, such as OEt and OTHP groups. (2) (E)-1-Alkenylzirconium compounds display a reactivity comparable to that of the corresponding alkenylalanes. (3) The stereospecificities of the alkenyl-aryl and alkenyl-alkenyl coupling reactions are ≥98 and ≥97%, respectively. In light of the results obtained with alkenylalanes, Pd complexes were used in the alkenyl-alkenyl coupling.

Table 6

The Ni-Catalyzed Reaction of Alkenylzirconium
Derivatives with Aryl Halides[a]

R^1	Ar	X	Yield[b] (%)
n-Pent	Ph	I	96
n-Pent	1-Naph	Br	70
EtO	Ph	I	99(76)
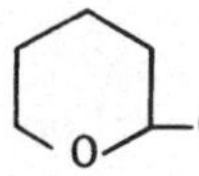 [tetrahydropyran]—O-CH$_2$CH$_2$	Ph	I	84
n-Bu	p-ClPh	I	95
n-Bu	p-MeOPh	I	80
n-Bu	p-NCPh	Br	92
n-Bu	p-MeO$_2$CPh	Br	92

[a] Ni(PPh$_3$)$_4$ was used as a catalyst (10 mole %).
[b] By GLC. The number in parentheses is an isolated yield.

Table 7

The Pd-Catalyzed Reaction of Alkenylzirconium
Derivatives with Alkenyl Halides

R^1	R^2	R^3	X	Catalyst[a]	Yield[b] (%)
n-Pent	n-Bu	H	I	A	91
n-Bu	COOMe	Me	Br	B	75
(tetrahydropyranyl)—OCH$_2$	COOMe	Me	Br	B	~(70)
(tetrahydropyranyl)—OCH$_2$	H	H	Br	A	~(77)

[a] A=Pd(PPh$_3$)$_4$. B=Cl$_2$Pd(PPh$_3$)$_2$ + 2 i-Bu$_2$AlH. The amount of a catalyst in each case was 5 mole %.
[b] By GLC. The numbers in parentheses are isolated yields.

     Closely related to our alkenyl-alkenyl cross-coupling reac-
tions of alkenylalanes[16,17] and alkenylzirconium derivatives[42,43]
is a recent study of the corresponding reaction of alkenylmagnesium halides with alkenyl iodides in the presence of a catalytic amount of Pd(PPh$_3$)$_4$.[44] The results are comparable to those
observed with the Al and Zr derivatives. Unfortunately, however,
there is no direct and stereoselective route to alkenylmagnesium
halides at present. The chemoselectivity of the Mg procedure is
obviously lower than that of the Al or Zr procedure. An interesting piece of information described in this paper is the formation
of an isomeric product in the synthesis of a (Z,Z)-diene. It is
not clear at present if this is a general problem or not. It is
also reported in this paper that the reaction of (E)- or (Z)-1-
iodo-1-octene with propenylcopper reagents in THF is accompanied
by extensive stereochemical scrambling (20-30%).

     DOUBLE-METAL-CATALYZED CROSS COUPLING —— A STEREOSELECTIVE
SYNTHESIS OF TRISUBSTITUTED OLEFINS via Zr-CATALYZED CARBOALUMINATION

     Our efforts to expand the applicability of the olefin syn-
thesis via Ni- or Pd-catalyzed cross coupling encountered another
serious difficulty. Thus, the reaction of alkenylalanes and
alkenylzirconium compounds, which proceeds readily in cases where
these reagents are β-monoorgano-substituted alkenylmetal derivatives, has either failed to produce the desired product or pro-
ceeded very sluggishly in cases where these alkenylmetal reagents

are $\alpha,\beta-$ and $\underline{\beta},\underline{\beta}-$diorgano-substituted. Typically, the reaction of (E)-3-hexenyldiisobutylalane with m-iodotoluene in the presence of 5 mole % of $Pd(PPh_3)_4$ did not produce any more than a trace of the desired coupled product even after one week at room temperature, whereas the corresponding reaction of (E)-1-hexenyldiisobutyl-alane gave the expected product (75%) in 12 hr at room temperature. Particularly frustrating was the generally low reactivity of the $\underline{\beta},\underline{\beta}-$diorgano-substituted alkenylaluminum reagents, which had just become available via Zr-catalyzed carboalumination of terminal acetylenes (eq 24),[45] in the Pd-catalyzed cross coupling.

$$R^1C{\equiv}CH \;+\; R^2_3Al \xrightarrow[20-25°]{Cl_2ZrCp_2} \underset{R^2}{\overset{R^1}{\diagdown}}C{=}C\underset{AlR^2_2}{\overset{H}{\diagup}} \qquad (24)$$

Unlike the carbocupration reaction of terminal acetylenes developed by Normant,[46] which proceeds sluggishly with methyl-copper reagents, the Zr-catalyzed carboalumination proceeds readily with Me_3Al, thereby providing a facile and essentially 100% stereoselective entry into methyl-substituted (E)-trisub-stituted olefins $\underline{4}$.

$$\underset{Me}{\overset{R^1}{\diagdown}}C{=}C\underset{AlMe_2}{\overset{H}{\diagup}} \qquad \underset{Me}{\overset{R^1}{\diagdown}}C{=}C\underset{\diagdown C{=}C\diagup}{\overset{H}{\diagup}} \qquad \underset{Me}{\overset{R^1}{\diagdown}}C{=}C\underset{C{\equiv}CR}{\overset{H}{\diagup}}$$

$$\underline{4} \qquad\qquad\qquad \underline{5} \qquad\qquad\qquad \underline{6}$$

It is then evident that, if $\underline{4}$ could be readily converted into $\underline{5}$ or $\underline{6}$, we would have a facile and stereoselective entry into carote-noids and other polyenic natural products.

Although a number of procedures are available for the syn-thesis of vitamin A ($\underline{7}$) via β-ionone ($\underline{8}$), none appears to be highly stereoselective.[47]

$$\underline{7} \qquad\qquad\qquad\qquad\qquad\qquad \underline{8}$$

The main difficulty lies in the lack of a stereoselective pro-cedure for the olefination of ketones. It then occurred to us that, at least in the case of methyl ketone olefination (which is undoubtedly by far the most important case from the viewpoint of natural products synthesis), the problem could be circumvented by first converting methyl ketones into the corresponding terminal acetylenes and then into $\underline{5}$ and/or $\underline{6}$ *via* $\underline{4}$ (Scheme III). It therefore became highly desirable for us to be able to successfully utilize $\underline{4}$ in the alkenyl-alkenyl or alkenyl-alkynyl coupling.

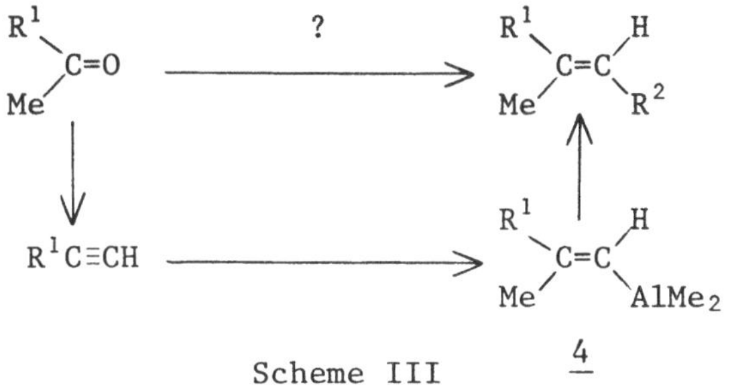

$$\begin{array}{ccc}
R^1\!\!\diagdown \\
C=O \\
Me\!\!\diagup
\end{array}
\xrightarrow{\;\;?\;\;}
\begin{array}{c}
R^1\!\!\diagdown\diagup H \\
C=C \\
Me\!\!\diagup\diagdown R^2
\end{array}$$

Scheme III $\qquad$ $\underline{4}$

To alleviate the above-mentioned difficulty, we turned our attention to a simple but little-tested possibility of doubly or multiply catalyzing the Pd- or Ni-catalyzed cross coupling which presumably proceeds *via* oxidative addition-transmetallation-reductive elimination (Scheme I). Since the difficulty appeared largely steric in origin, and since steric acceleration rather than steric hindrance was expected in the reductive elimination step, the difficulty appeared to lie in the transmetallation step. Note that the oxidative addition step does not involve alkenylmetal reagents. We further reasoned that, in a catalytic reaction of this nature, the difficulty must be kinetic rather than thermodynamic, and that such a kinetically unfavorable transmetallation process could, in principle, be facilitated by the use of one or more coordinatively unsaturated compounds of low steric requirements containing metals whose electronegativities are either comparable to or between those of the two metals in question. On these bases, we tested various metal salts containing Cu, Zn, Cd, In, Sn and Ti.

We were quite pleased to find that many of the Pd- or Ni-catalyzed reactions of the alkenylaluminum or alkenylzirconium compounds having two alkyl substituents ($\underline{\alpha},\underline{\beta}$ or $\underline{\beta},\underline{\beta}$) with alkenyl, aryl or alkynyl halides (which either fail to give the cross-coupled products, or produce them only in low yields) were significantly promoted to give the desired products, in high yields, by the addition of a Zn or Cd salt, such as $ZnCl_2$.[48]

Typically, the $Pd(PPh_3)_4$ catalyzed reaction of $(\underline{E})$-3-hexenyl-diisobutylalane with *m*-iodotoluene mentioned above, which failed to give the desired product even after one week, was complete within 1 hr at room temperature in the simultaneous presence of a catalytic amount of $Pd(PPh_3)_4$ and one equivalent of $ZnCl_2$ (eq 25).

$$
\underset{H}{\overset{Et}{}}\!\!C=C\!\!\underset{Al(i\text{-}Bu)_2}{\overset{Et}{}} + I\!-\!\!\bigcirc\!\!Me
$$

cat. Pd(PPh$_3$)$_4$

1 week $\longrightarrow$ no reaction   (25)

cat. Pd(PPh$_3$)$_4$

ZnCl$_2$, 1 hr $\longrightarrow$ $\underset{H}{\overset{Et}{}}\!\!C=C\!\!\overset{Et}{}\!\!\bigcirc\!\!Me$

88%

Also effective were Cd salts, such as CdCl$_2$. However, CuI was only marginally effective, and InI$_3$, SnCl$_4$ and TiCl$_4$ were totally ineffective.

If the three-step mechanism in Scheme I indeed operates, and if ZnCl$_2$ participates in the transmetallation reactions with both alkenylalanes and Pd catalysts, only a catalytic amount of ZnCl$_2$ would be required. This point has been borne out by the results summarized in equation 26.

$$
\underset{H}{\overset{Et}{}}\!\!C=C\!\!\underset{Al(i\text{-}Bu)_2}{\overset{Et}{}} + \underset{H}{\overset{Br}{}}\!\!C=C\!\!\underset{COOMe}{\overset{Me}{}} \xrightarrow[20\text{-}22^\circ]{\text{cat. Pd(PPh}_3)_4} \underset{H}{\overset{Et}{}}\!\!C=C\!\!\underset{\underset{H}{\overset{}{|}}}{\overset{Et}{}}C=C\!\!\underset{COOMe}{\overset{Me}{}}
$$

(26)

ZnCl$_2$ (equiv)	Time (1 hr)	Yield (%)
0	6	<2
0.2	2	72
0.5	1	85
1.0	1	82

The probable intermediacy of alkenylzinc derivatives has been further evidenced by the fact that the reaction of (E)-2-methyl-1-heptenylzinc chloride (9), preformed by treating the corresponding alkenyl iodide first with n-butyllithium and then with ZnCl$_2$, reacts with 1-iodo-1-hexyne at least as fast as the ZnCl$_2$-promoted reaction of (E)-(2-methyl-1-heptenyl)dimethylalane with the same iodoalkyne (Scheme IV). Although the scope of the double-metal-catalyzed alkenyl-aryl, alkenyl-alkenyl and alkenyl-alkynyl coupling reactions is unclear, we have not so far encountered any major difficulty.

We are now ready to apply this new procedure to the problem of the stereoselective conversion of β-ionone into vitamin A.

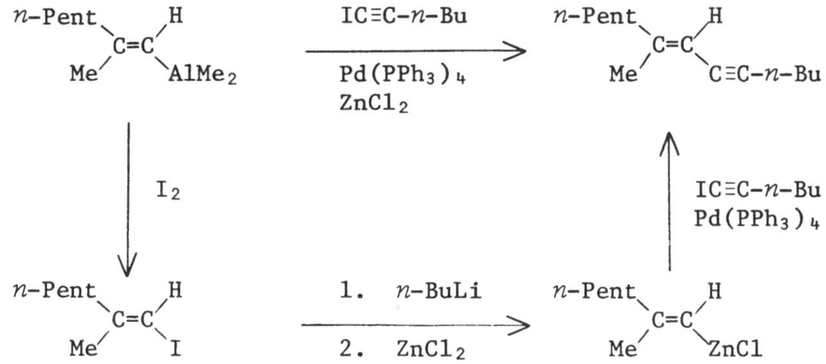

9

Scheme IV

Quite disappointingly, none of the known procedures for the conversion of methyl ketones into terminal acetylenes which we have tried so far has produced the desired dienyne 10 in high yield. After many trials and errors, we have found that the following one-pot procedure cleanly produces 10 in 80-85% isolated yield[19] (eq 27).

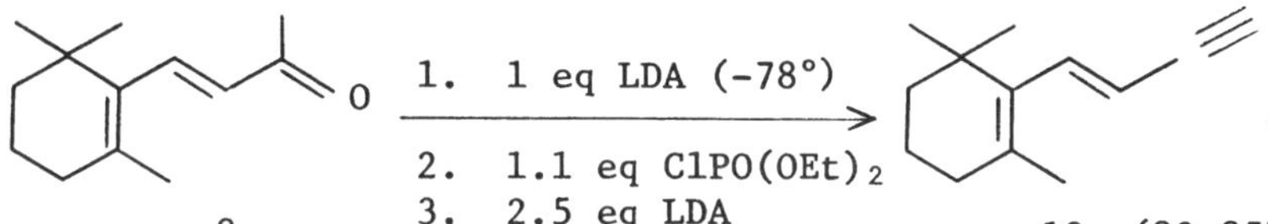

8   10 (80-85%)   (27)

No difficulty was encountered in converting 10 into dehydrovitamin A(11) in 50-60% yield by the double-metal-catalyzed cross coupling (eq 28). The required trimethylsilyl ether of (E)-5-iodo-3-methyl-2,4-pentenyn-1-ol was readily prepared from commercially available (E)-3-methyl-2,4-pentenyn-1-ol *via* silylation-iodination.

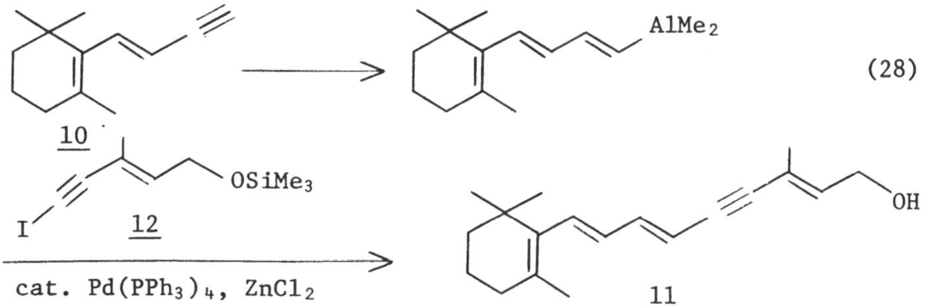

10   12   11   (28)

We are currently engaged in refining the cross-coupling step and converting 11 into our final product, vitamin A.

Although we have not dealt with the Ni- or Pd-catalyzed cyanation of aryl halides,[49] it has been shown to produce aryl cyanides in high yields. The reaction appears to proceed by the mechanism shown in Scheme I.

$$\text{ArX} + \text{MCN} \xrightarrow{\text{cat. NiL}_n \text{ or PdL}_n} \text{ArCN} \qquad (29)$$

M = K or Na. X = I or Br

A BRIEF SURVEY OF THE Ni- AND Pd-PROMOTED CROSS COUPLING WITH sp³ HYBRIDIZED REAGENTS

We have so far dealt only with unsaturated organometallics and organic halides, i.e., aryl, alkenyl and alkynyl derivatives. Although our own study of the Ni- and Pd-catalyzed cross coupling with the sp^3 hybridized reagents has been very limited, a brief survey of their cross-coupling reactions might be appropriate to put our discussion of the Ni- and Pd-catalyzed cross coupling into proper perspective. The sp^3 hybridized reagents are loosely defined in this review as those containing primary, secondary and tertiary alkyl, allyl, propargyl, benzyl and various α-carbonyl and α-hetero-substituted alkyl groups.

A.   Organometallic Reagents Containing Allylic, Propargylic and Benzylic Groups

The stoichiometric reaction of π-allylnickel derivatives with organic halides[2] (Corey-Semmelhack-Hegedus reaction) mentioned earlier provides one of the most versatile methods of allylation. In view of two extensive reviews[2] on this reaction, it will suffice to mention that the reaction generally proceeds well with primary and secondary alkyl, benzyl, aryl and alkenyl halides. On the other hand, the product yields are usually poor with tertiary alkyl halides and moderate with α-haloketones. Allylic halides induce an extensive cross-homo scrambling. Little appears to be known about the reaction of propargyl and alkynyl halides.

The stoichiometric reaction of π-allylpalladium derivatives with enolates[5] (Tsuji-Trost reaction) would be too expensive to be of wide synthetic utility. On the other hand, the Pd-catalyzed reaction of allylic acetates with certain enolates[6] appears promising. At present, however, its scope is rather limited, since only certain "activated" enolates have been found to participate in this reaction.

Benzylzinc derivatives are readily obtainable by the direct metallation of benzylic bromides with 2-2.5 equiv of Zn powder.[50] In general, only trace amounts of bibenzyls are formed, if a sufficient amount of Zn powder is added. This is in sharp contrast with the corresponding Li reaction which is complicated by the competitive Wurtz coupling. Even the corresponding Mg reaction is usually accompanied by the formation of 5-10% of bibenzyls. We have found that the benzylzinc halides react with aryl[35] and alkenyl[32] iodides and bromides in the presence of a catalytic amount of a Ni-phosphine or Pd-phosphine complex (eqs 30 and 31).

$$PhCH_2ZnBr \ + \ ArX \ \xrightarrow[25°]{\text{cat. } NiL_n \text{ or } PdL_n} \ PhCH_2Ar \qquad (30)$$

$$PhCH_2ZnBr \ + \ \underset{R^2}{\overset{R^1}{>}}C=C\underset{X}{\overset{R^3}{<}} \ \xrightarrow[25°]{\text{cat. } PdL_n} \ \underset{R^2}{\overset{R^1}{>}}C=C\underset{CH_2Ph}{\overset{R^3}{<}} \qquad (31)$$

X = I or Br. NiL_n = Ni(PPh₃)₄, etc., PdLn = Pd(PPh₃)₄, etc.

Benzylmagnesium halides are less satisfactory but acceptable in most cases. Both Ni and Pd complexes are satisfactory in most of the benzyl-aryl coupling reactions. However, the Pd complexes are clearly superior to the corresponding Ni complexes in the benzyl-alkenyl coupling, as the latter catalysts tend to induce double-bond migration to form conjugated arylalkenes.[32]

At present, only aryl and alkenyl halides have been shown to participate in the reaction. Alkynyl and alkyl halides which do not contain any β-hydrogen atom bound to an sp³ carbon atom may also undergo this reaction. On the other hand, those alkyl halides which contain any sp³ carbon-bound β-hydrogen atoms are likely to cause a serious difficulty as discussed later.

It should be pointed out that, although a few arylcopper compounds have been reacted with benzyl halides to form the corresponsing diarylmethanes in relatively low yields,[1] little is known about the cross-coupling reaction of benzylcopper compounds.

B. Organometallic Reagents Containing Primary, Secondary and Tertiary Alkyl Groups

There is a major difficulty associated with the Ni- and Pd-catalyzed cross coupling. As is well known, alkylnickel and alkylpalladium compounds containing sp³ carbon-bound β-hydrogen

atoms have a strong tendency to undergo dehydrometallation.   Therefore, in order to observe a successful cross coupling with such
compounds, the desired reaction must be much faster than the dehydrometallation reaction. Assuming that the desired cross
coupling proceeds *via* oxidative addition-transmetallation-reductive
elimination, there are two distinct cases represented by Schemes
V and VI. Of course, some reactions combine the features shown
in both of these Schemes.

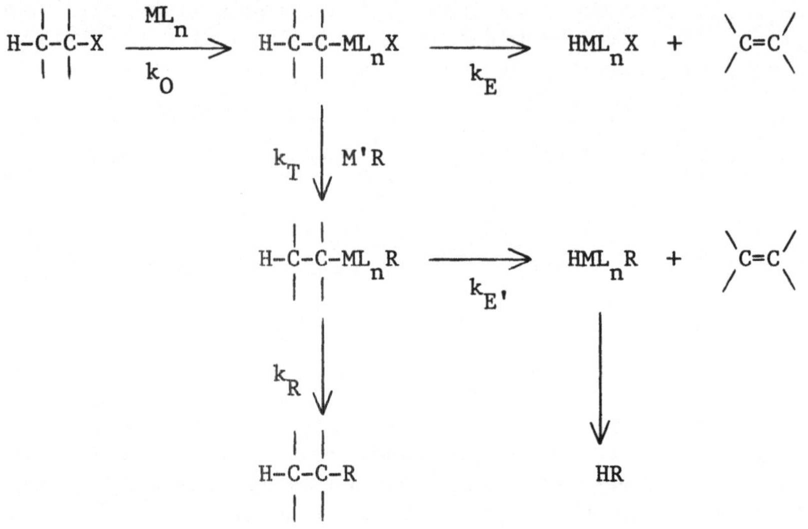

$$
\begin{array}{ccccc}
\text{H}-\overset{|}{\underset{|}{\text{C}}}-\overset{|}{\underset{|}{\text{C}}}-\text{X} & \xrightarrow[k_O]{ML_n} & \text{H}-\overset{|}{\underset{|}{\text{C}}}-\overset{|}{\underset{|}{\text{C}}}-ML_nX & \xrightarrow{k_E} & HML_nX \ + \ \underset{/}{\overset{\backslash}{\text{C}}}=\underset{\backslash}{\overset{/}{\text{C}}}
\end{array}
$$

$k_T \Big\downarrow M'R$

$$
\begin{array}{ccc}
\text{H}-\overset{|}{\underset{|}{\text{C}}}-\overset{|}{\underset{|}{\text{C}}}-ML_nR & \xrightarrow{k_{E'}} & HML_nR \ + \ \underset{/}{\overset{\backslash}{\text{C}}}=\underset{\backslash}{\overset{/}{\text{C}}}
\end{array}
$$

$k_R \Big\downarrow \qquad\qquad\qquad\qquad \Big\downarrow$

$$
\text{H}-\overset{|}{\underset{|}{\text{C}}}-\overset{|}{\underset{|}{\text{C}}}-\text{R} \qquad\qquad\qquad \text{HR}
$$

Scheme V

$$
RX \quad \xrightarrow[k_O]{ML_n} \quad RML_nX
$$

$k_T \Big\downarrow M'-\overset{|}{\underset{|}{\text{C}}}-\overset{|}{\underset{|}{\text{C}}}-\text{H}$

$$
RML_n-\overset{|}{\underset{|}{\text{C}}}-\overset{|}{\underset{|}{\text{C}}}-\text{H} \xrightarrow{k_{E'}} RML_nH \ + \ \underset{/}{\overset{\backslash}{\text{C}}}=\underset{\backslash}{\overset{/}{\text{C}}}
$$

$k_R \Big\downarrow \qquad\qquad\qquad\qquad \Big\downarrow$

$$
\text{R}-\overset{|}{\underset{|}{\text{C}}}-\overset{|}{\underset{|}{\text{C}}}-\text{H} \qquad\qquad\qquad \text{RH}
$$

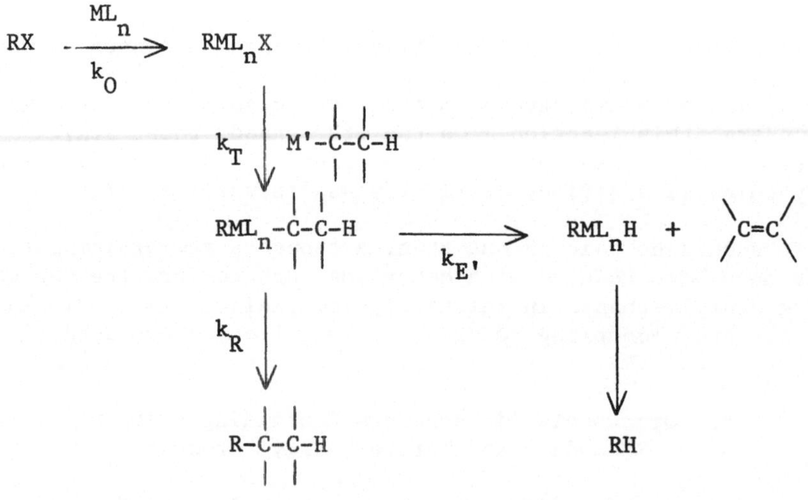

Scheme VI

A tentative but seemingly general conclusion of significance based on the available information is that there has been essentially no successful example of the Ni- or Pd-catalyzed cross coupling with alkyl halides containing sp^3 carbon-bound β-hydrogens. It is likely, although not clear, that the oxidative addition step is complicated by dehydrometallation (Scheme V). Irrespective of the precise reason, this represents one of the most serious limitations of the Ni- or Pd-catalyzed cross coupling. Clearly, the Cu-promoted cross coupling[1] is distinctly superior to the Ni- or Pd-catalyzed reaction in such cases.

One way of circumventing the difficulty mentioned above is to first convert the β-hydrogen-containing alkyl halides into the corresponding organometallic reagents and then employ the latter in the cross coupling, which presumably proceeds *via* Scheme VI.

The Ni-catalyzed reaction of alkylmagnesium halides has been extensively studied by Kumada and co-workers.[51] The available results indicate the following. If certain Ni-phosphine complexes containing bidentate ligands, such as $Ph_2P(CH_2)_2PPh_2$ (dpe) and $Ph_2P(CH_2)_3PPh_2$ (dpp), are used as catalysts, primary and secondary alkylmagnesium halides can be coupled with aryl and alkenyl halides in high yields, whereas those Ni complexes which contain monodentate ligands, such as PBu_3 and PPh_3, induce a competitive reduction of organic halides[51] (eq 32).

$Cl_2Ni(dpp)$	85%	4%	0%
$Cl_2Ni(PPh_3)_2$	6%	13%	24%

Furthermore, the reduction is usually accompanied by the formation of isomeric cross-coupled products.

Although some unique applications of the Ni-catalyzed alkylation, such as asymmetric alkylation[51d,e,i] and a one-step synthesis of cyclophanes and heterophanes,[51f] have been reported, most of the compounds prepared by the Ni-catalyzed alkylation can often be better prepared by the organocuprate reaction[1] or by other conventional methods. Further investigation is therefore necessary to delineate more clearly the advantages and disadvantages of the Ni-catalyzed alkylation over the previously developed methodologies.

Much less is known at present about the Pd-catalyzed alkylation. Murahashi[13] reported in 1975 a stoichiometric reaction of
alkenylpalladium derivatives with methyllithium and n-butyllithium.
Their attempts to induce a Pd-catalyzed cross-coupling reaction of
alkenyl halides with alkyllithiums resulted in the formation of
acetylenes as the main products. On the other hand, the reaction
of methylmagnesium iodide with (E)- or (Z)-bromostyrene in the
presence of a catalytic amount of Pd(PPh$_3$)$_4$ produced (E)- or (Z)-
propenylbenzene in quantitative yield. More recently,
Linstrumelle[44] reacted (E)- or (Z)-1-octenyl iodide with ethylmagnesium bromide and obtained the desired cross-coupled product
in ca. 85% yield. All of these reactions are nearly completely
stereospecific. The Pd-catalyzed alkylation of aryl halides was
first reported by Fauvarque.[52] However, only methylmagnesium
bromide was used as the alkylating agent. Ishikawa[38] reacted
ethylmagnesium bromide and n-butyllithium with phenyl iodide in
the presence of a catalytic amount of PhPdI(PPh$_3$)$_2$ and obtained
ethylbenzene and n-butylbenzene in 32 and 45% yields, respectively.
While these results are interesting, their unique synthetic
capability is at present unclear.

C. Other Organometallics

In a limited number of cases, the Ni-catalyzed cross-coupling
reaction of lithium enolates with aryl halides have been
studied[53,54] and applied to the synthesis of a Cephalotaxus
alkaloid.[53b] However, the reaction is reported to be inferior
to the photo-induced free-radical α-arylation of enolates.[55]
More recently, Fauvarque reported a Ni- or Pd-catalyzed α-aryl-
ation of Reformatsky reagents[56] (eq 33).

$$ArX \; + \; BrZnCH_2COOEt \xrightarrow{\text{cat. NiL}_n \text{ or PdL}_n} ArCH_2COOEt \quad (33)$$

An indirect approach to the α-arylation of ketones and aldehydes
has been reported by Kumada[51h] (eq 34).

$$\underset{Br}{\overset{R^1}{\diagup}}C=C\underset{OSiMe_3}{\overset{R^2}{\diagdown}} \; + \; ArMgX \xrightarrow{\text{Ni(dpp)Cl}_2} \underset{Ar}{\overset{R^1}{\diagup}}C=C\underset{OSiMe_3}{\overset{R^2}{\diagdown}}$$

$$\xrightarrow{H_3O^+} \underset{Ar}{\overset{R^1}{\diagup}}CHCR^2 \quad \overset{O}{\overset{\|}{}} \qquad (34)$$

The reaction is also adaptable to the α-alkylation of carbonyl
compounds.

Migita[57] reported recently that allyltributyltin reacted at 100° with aryl halides in the presence of $Pd(PPh_3)_4$ (eq 35).

$$ArBr \ + \ Bu_3Sn \diagup\!\!\diagdown\!\!\diagup \quad \xrightarrow[\ 100°\ ]{cat. \ Pd(PPh_3)_4} \quad Ar \diagup\!\!\diagdown\!\!\diagup \qquad (35)$$

$$72\text{-}100\%$$

Although the precise mechanism of the reaction is unknown, it may involve either the oxidative addition-transmetallation-reductive elimination sequence (Scheme I) or the oxidative addition-carbo-palladation-elimination involving destannopalladation (Scheme III).

Finally, a few Pd-catalyzed reactions of acyl halides with organomercury[58] and organotin[59] reagents have been reported as new routes to ketones (eq 36).

$$R^1COX \ + \ R^2MX_n \quad \xrightarrow{\ cat. \ PdL_n\ } \quad R^1COR^2 \qquad (36)$$

$$M = Hg \ or \ Sn$$

CONCLUSIONS

The results and discussions presented in this review seem to permit the following tentative conclusions.

1. The Ni- or Pd-catalyzed cross coupling represents the second major innovation of cross coupling by the introduction of transition metals. With respect to the $Csp^2\text{-}Csp^2$ and $Csp^2\text{-}Csp$ cross coupling, the Ni- or Pd-catalyzed cross coupling often provides some distinct advantages over the corresponding Cu method. On the other hand, the Ni- or Pd-catalyzed cross coupling of those alkyl reagents which contain sp^3 carbon-bound $\underline{\beta}$-hydrogen(s) tends to be complicated by dehydrometal-lation. In such a case, the organocopper method appears generally more satisfactory than the Ni- or Pd-catalyzed reactions.

2. In many cases, both Ni-phosphine and Pd-phosphine complexes are highly satisfactory. However, the Pd-phosphine complexes are superior to the Ni complexes in the following cases: (a) alkenyl-alkenyl cross coupling, (b) cross-coupling reactions involving alkynyl groups, and (c) reaction of nitro-containing compounds. Although Pd complexes are considerably more expensive than the corresponding Ni complexes, the former complexes do not seem to have any other serious disadvantages in comparison to the latter complexes, at least in the cross

coupling of two unsaturated groups. This overall superiority
of Pd complexes represents one of the significant findings in
recent studies by us and others.

3. Another finding of major significance in our study is that,
    whereas highly basic organometallics, such as those containing
    Li and Na, are associated with a number of problems, less
    basic organometallics, such as those containing Mg, Zn, Al,
    Sn and Zr, tend to provide far more favorable results. First,
    the rate of the Ni- or Pd-catalyzed cross coupling is often
    much higher with these organometallics than with organoalkali
    metals. Second, the extent of undesirable side reactions, such
    as halogen-metal exchange and elimination reactions, is much
    less than those observed with organoalkali metals. Third,
    the reactions of organometallics containing Zn, Al, Sn and
    Zr are generally far more chemoselective than those of organo-
    alkali metals and Grignard reagents. When all of these
    factors are taken into consideration, organozinc compounds
    appear to be the reagents of choice.

4. Many stereo-defined alkenylmetals containing Al and Zr are
    readily obtainable *via* hydrometallation and carbometallation
    of acetylenes. These alkenyl derivatives can be directly
    utilized in the cross coupling, thereby providing a number of
    simple and selective routes to olefins.

5. In cases where the reactivity of organometallics containing
    Al and Zr is not sufficiently high, it can be significantly
    augmented by the addition of Zn and Cd salts, such as $ZnCl_2$.
    This double-metal-catalyzed cross coupling appears to be of
    wide applicability.

6. The current scope of the Ni- or Pd-catalyzed cross coupling
    with respect to the two organic groups to be coupled is in-
    dicated in Tables shown below.

Other Ni- or Pd-promoted cross-coupling reactions which have been
shown to produce the expected products, at least in moderate yields,
include the following. The first group represents R^1 of R^1M, and
the second group R^2 of R^2X.

Ni-Promoted reactions

allyl-benzyl, allyl-1° alkyl, allyl-2° alkyl, allyl-$\underline{\alpha}$-halocarbonyl,
enolate-aryl, cyano-aryl.

Pd-Promoted reactions

1° alkyl-acyl, enolate-allyl, cyano-aryl.

Table 8

Scope of the Ni- or Pd-Catalyzed Cross
Coupling of Two Unsaturated Groups

R^1M \ R^2 of R^2X		Aryl	Alkenyl	Alkynyl
Aryl	Ni	+	+	$-^a$
	Pd	+	+	+
Alkenyl	Ni	+	$-^b$	$-^a$
	Pd	+	+	+
Alkynyl	Ni	$-^a$	$-^a$	$-^{a,c}$
	Pd	+	+	$-^c$

[a] The product yield is generally low. [b] The reaction tends to be complicated by stereochemical scrambling. [c] Statistical mixtures of three diynes are formed.

Table 9

Scope of the Ni- or Pd-Promoted Cross
Coupling Involving sp^3 Hybridized Groups

R^1M \ R^2 of R^2X		Aryl	Alkenyl	Alkynyl
Allyl	Ni	+	+	?
	Pd	+	?	?
Benzyl	Ni	+	$-^a$	?
	Pd	+	+	?
1° & 2° Alkyl	Ni	+	+	?
	Pd	+	+	?
3° Alkyl	Ni	$-^b$	$-^b$	$-$ (?)
	Pd	$-$ (?)	$-$ (?)	$-$ (?)

[a] An extensive double bond migration takes place.
[b] Low yield.

Except with π-allylnickel reagents, primary, secondary and tertiary alkyl halides containing sp^3 carbon-bound β-hydrogen(s) do not appear to participate in the Ni- or Pd-promoted cross coupling. Organometallics containing tertiary alkyl group(s) are also quite reluctant to participate in the Ni- or Pd-promoted cross coupling.

ACKNOWLEDGMENTS

I wish to thank Professor J. H. Brewster and the other members of the organizing committee of the H. C. Brown symposium for inviting me to present a lecture on which this review is based.

I am deeply indebted to my co-workers whose names have appeared in our papers cited in this review, especially Shigeru Baba for his initial explorations in 1974-1976, and Anthony O. King, Nobuhisa Okukado and David E. Van Horn, for their subsequent developments of the Pd- and Ni-catalyzed cross coupling. Financial support by the National Science Foundation, the donors of the Petroleum Research Funds administered by the American Chemical Society, Matthey Bishop, Inc. and, last but not least, Syracuse University are gratefully acknowledged.

REFERENCES

1. For an extensive review, see G. H. Posner, *Org. React.*, <u>22</u>, 253 (1975).
2. For reviews, see (a) M. F. Semmelhack, *Org. React.*, <u>19</u>, 115 (1972); (b) L. S. Hegedus, *J. Organometal. Chem. Library*, <u>1</u>, 329 (1976).
3. (a) M. F. Semmelhack, P. M. Helquist, and L. D. Jones, *J. Am. Chem. Soc.*, <u>93</u>, 5908 (1971); (b) M. F. Semmelhack, P. M. Helquist, and J. P. Gorzynski, *J. Am. Chem. Soc.*, <u>94</u>, 9234 (1972).
4. J. Tsuji and H. Takahasi, *Tetrahedron Lett.*, 4387 (1965).
5. For an extensive review, see B. M. Trost, *Tetrahedron*, <u>33</u>, 2615 (1977).
6. B. M. Trost and T. R. Vanhoeven, *J. Am. Chem. Soc.*, <u>98</u>, 630 (1976).
7. For a review, see R. F. Heck, *R. A. Welch Found. Conf. Chem. Res.*, <u>17</u>, 53 (1974).
8. K. Tamao, K. Sumitani, and M. Kumada, *J. Am. Chem. Soc.*, <u>94</u>, 4374 (1972).
9. R. J. P. Corriu and J. P. Masse, *J. Chem. Soc., Chem. Comm.*, 144 (1972).
10. M. S. Kharasch and O. Reinmuth, "Grignard Reactions of Nonmetallic Substances," Chap. 16, Prentice-Hall, New York, 1954.
11. D. G. Morrell and J. K. Kochi, *J. Am. Chem. Soc.*, <u>97</u>, 7262 (1975).
12. G. W. Parshall, *J. Am. Chem. Soc.*, <u>96</u>, 2360 (1974).
13. M. Yamamura, I. Moritani, and S.-I. Murahashi, *J. Organometal. Chem.*, <u>91</u>, C39 (1975).
14. L. Cassar, *J. Organometal. Chem.*, <u>93</u>, 253 (1975).
15. K. Sonogashira, Y. Tohda, and N. Hagihara, *Tetrahedron Lett.*, 4467 (1975).

16. E. Negishi and S. Baba, *J. Chem. Soc., Chem. Comm.*, 596 (1976).

17. S. Baba and E. Negishi, *J. Am. Chem. Soc.*, 98, 6729 (1976).

18. A. O. King, E. Negishi, F. J. Villani, Jr., and A. Silveira, Jr., *J. Org. Chem.*, 43, 358 (1978).

19. Unpublished results by E. Negishi and A. O. King.

20. (a) J. C. Ware, and T. G. Traylor, *J. Am. Chem. Soc.*, 89, 2304 (1967); (b) A. Hosomi and T. G. Traylor, *J. Am. Chem. Soc.*, 97, 3682 (1975); (c) G. D. Hartman and T. G. Traylor, *J. Am. Chem. Soc.*, 97, 6147 (1975); (d) For a review of pertinent studies by Eaborn and co-workers, see C. Eaborn, *J. Organometal. Chem.*, 100, 53 (1975).

21. (a) C. E. Castro and R. D. Stephens, *J. Org. Chem.*, 28, 2163 (1963); (b) R. D. Stephens and C. E. Castro, *J. Org. Chem.*, 28, 3313 (1963); (c) C. E. Castro, R. Havlin, V. K. Honwad, A. Malte, and S. Mojé, *J. Am. Chem. Soc.*, 91, 6464 (1969).

22. H. A. Dieck and R. F. Heck, *J. Organometal. Chem.*, 93, 259 (1975).

23. For some indirect syntheses of terminal arylalkynes involving organocopper reagents, see (a) R. E. Atkinson, R. F. Curtis, D. M. Jones and J. A. Taylor, *J. Chem. Soc. (C)*, 2173 (1969); (b) R. Oliver and D. R. M. Walton, *Tetrahedron Lett.*, 5209 (1972); (c) I. Barrow and A. E. Pedler, *Tetrahedron*, 32, 1829 (1976); (d) J. S. Kiely, P. Boudjouk, and L. L. Nelson, *J. Org. Chem.*, 42, 2626 (1977).

24. (a) A. Butenandt, E. Hecker, M. Hopp, and W. Koch, *Annalen*, 658, 39 (1962); (b) G. Zweifel and N. L. Polston, *J. Am. Chem. Soc.*, 92, 4068 (1970); (c) F. Naf, R. Decorzant, W. Thommen, B. William, and G. Ohloff, *Helv. Chem. Acta*, 58, 1016 (1975).

25. For other stereoselective syntheses of unsymmetrical enynes, see Footnote 4 in Ref. 26.

26. A. O. King, N. Okukado and E. Negishi, *J. Chem. Soc., Chem. Comm.*, 683 (1977).

27. For a review, see G. Eglinton and W. McCrae, *Advan. Org. Chem.*, 4, 225 (1963).

28. See, for example, R. Eastmond and D. R. M. Walton, *J. Chem. Soc., Chem. Comm.*, 204 (1968).

29. J. A. Sinclair and H. C. Brown, *J. Org. Chem.*, 41, 1078 (1976).

30. See, for example, T. K. Devon and A. I. Scott, "Handbook of Naturally Occurring Compounds," 2 Vols., Academic Press, New York, 1972 and 1975.

31. For an organocopper procedure for the synthesis of enynes, see J. F. Normant, A. Commercon, and J. Villieras, *Tetradron Lett.*, 1465 (1975).

32. Unpublished results by E. Negishi and N. Okukado.

33. H. Van de Walle and A. Henne, *Bull. Sci. Acad. Roy. Belg.*, 11, 360 (1925).

34. For a review, see J. Mathieu and J. Weill-Raynal, "Formation of Carbon-Carbon Bonds," Vol. 2, Thieme, Stuttgart, 1975.

35. E. Negishi, A. O. King, and N. Okukado, *J. Org. Chem.*, 42, 1821 (1977).

36. K. Tamao, A. Minato, N. Miyake, T. Matsuda, Y. Kiso, and M. Kumada, *Chem. Letters*, 133 (1975).

37. R. L. Clough, P. Mison, and J. D. Roberts, *J. Org. Chem.*, 41, 2252 (1976).

38. A. Sekiya and N. Ishikawa, *J. Organometal. Chem.*, 118, 349 (1976).

39. See, for example, P. W. Collins, E. Z. Dajani, M. S. Bruhn, C. H. Brown, J. R. Palmer, and R. Pappo, *Tetrahedron Lett.*, 4217 (1975).

40. For a review, see J. Schwartz, *J. Organometal. Chem. Library*, 1, 461, (1976).

41. C. A. Bertelo and J. Schwartz, *J. Am. Chem. Soc.*, 97, 228 (1975).

42. E. Negishi and D. E. Van Horn, *J. Am. Chem. Soc.*, 99, 3168 (1977).

43. N. Okukado, D. E. Van Horn, W. L. Klima, and E. Negishi, *Tetrahedron Lett.*, 1027 (1978).

44. H. P. Dang and G. Linstrumelle, *Tetrahedron Lett.*, 191 (1978).

45. D. E. Van Horn and E. Negishi, *J. Am. Chem. Soc.*, 100, 2252 (1978).

46. For a review, see J. F. Normant, *J. Organometal. Chem. Library*, 1, 219 (1976).

47. For a review, see F. Kienzle, *Pure Appl. Chem.*, 47, 183 (1976).

48. E. Negishi, N. Okukado, A. O. King, D. E. Van Horn, and B. I. Spiegel, *J. Am. Chem. Soc.*, 100, 2254 (1978).

49. (a) L. Cassar, *J. Organometal. Chem.*, 54, C57 (1973); (b) A. Sekiya and N. Ishikawa, *Chem. Letters*, 277 (1975); (c) K. Takagi, T. Okamoto, Y. Sakakibara, A. Ohno, S. Oka, and N. Hayama, *Bull. Chem. Soc. Japan*, 48, 3298 (1975).

50. M. Gaudemar, *Bull. soc. chim. Fr.*, 974 (1962).

51. (a) For a review, see M. Kumada in Y. Ishii and M. Tsutsui, eds., "Organotransition-Metal Chemistry," Plenum, New York, 1975, p. 211; (b) K. Tamao, Y. Kiso, K. Sumitani, and M. Kumada, *J. Am. Chem. Soc.*, 94, 9268 (1972); (c) Y. Kiso, K. Tamao, and M. Kumada, *J. Organometal. Chem.*, 50, C12 (1973); (d) Y. Kiso, K. Tamao, N. Miyake, K. Yamamoto, and M. Kumada, *Tetrahedron Lett.*, 3 (1974); (e) T. Hayashi, M. Tajika, K. Tamao, and M. Kumada, *J. Am. Chem. Soc.*, 98, 3718 (1976); (f) K. Tamao, S. Kodama, T. Nakatsuka, Y. Kiso, and M. Kumada, *J. Am. Chem. Soc.*, 97, 4405 (1975); (g) K. Tamao, K. Sumitani, Y. Kiso, M. Zembayashi, A. Fujioka, S. Kodama, I. Nakajima, A. Minato, and M. Kumada, *Bull. Chem. Soc. Japan*, 49, 1958 (1976); (h) K. Tamao, M. Zembayashi, and M. Kumada, *Chem. Letters*, 1239 (1976); (i) M. Zembayashi, K. Tamao, T. Hayashi, T. Mise, and M. Kumada, *Tetrahedron Lett.*, 1799 (1977).

52. J. F. Fauvarque and A. Jutand, *Bull. soc. chim. Fr.*, 765 (1976).

53. (a) M. F. Semmelhack, R. D. Stauffer, and T. D. Rogerson, *Tetrahedron Lett.*, 4519 (1973); (b) M. F. Semmelhack, B. P. Chong, R. D. Stauffer, T. D. Rogerson, A. Chong, and L. D. Jones, *J. Am. Chem. Soc.*, 97, 2507 (1975).

54. A. A. Millard and M. W. Rathke, *J. Am. Chem. Soc.*, 99, 4833 (1977).

55. (a) J. K. Kim and J. F. Bunnett, *J. Am. Chem. Soc.*, 92, 7464 (1970); (b) R. Rossi and J. F. Bunnett, *J. Am. Chem. Soc.*, 94, 683 (1972); R. Rossi and J. F. Bunnett, *J. Org. Chem.*, 38, 1407 (1973).

56. J. F. Fauvarque and A. Jutand, *J. Organometal. Chem.*, 132, C17 (1977).

57. M. Kosugi, K. Sasazawa, Y. Shimizu and T. Migita, *Chem. Letters*, 301 (1977).

58. K. Takagi, T. Okamoto, Y. Sakakibara, A. Ohno, S. Oka, and N. Hayama, *Chem. Letters*, 951 (1975).

59. D. Milstein and J. K. Stille, *J. Am. Chem. Soc.*, 100, 3636 (1978).

HERBERT C. BROWN - A BIOGRAPHICAL NOTE

James H. Brewster

Department of Chemistry, Purdue University

West Lafayette, Indiana 47907

This is a good time to look back on the life and accomplishments of H. C. Brown. He is, formally, retired but his mind is as quick, tenacious and inventive as ever and the passage of years, each leaving it residue of experience, has served only to consolidate his skills in research. We expect him to be productive for many years and can look to the past for clues to the future.

THE EARLY YEARS[1,2,3]

Herbert C. Brown was born in London on May 22, 1912. His father was a cabinetmaker and the family lived in an apartment project that the Rothschilds had built to house refugees from the pogroms of the Czar. They left England in 1914 and emigrated to Chicago to join other members of the family. His father ran a hardware store but died when Herb was 14, making it necessary for him to leave school and go to work to support his mother and three sisters. In 1929 he was able to return to Englewood High School while continuing to work in the hardware store on evenings and weekends. He completed two years of schooling in one year and wrote for the school publications as well. After the family business failed and the store was sold, Herb worked as a shoe salesman and held jobs packing notebook paper and men's belts. This experience persuaded him to resume his education.

He enrolled in Crane Junior College in February of 1933 and there took a freshman chemistry course that kindled a flame of enthusiasm that still burns strong. It was at Crane that he met Sarah Baylen, now his wife of more than 40 years. She was his classmate in the slide-rule course and she graded his papers in the freshman

319

chemistry course and has described her ambivalent response to his
performance in each class.[2] When city colleges closed for lack of
funds at the depth of the depression, Herb and Sarah were allowed to
continue their training in the home laboratory of Dr. Nicholas D.
Cheronis. At this time Herb also took evening courses at Armour
Institute (now Illinois Institute of Technology) and a home-study
course in quantitative analysis from the University of Chicago. The
colleges opened again in the fall of 1934; they attended Wright Junior
College and then entered the University of Chicago on scholarship as
Juniors majoring in chemistry. Again, Herb completed two years of
work in one. He graduated in 1936 and Professor Julius Stieglitz
encouraged him to go to graduate school. Sarah's graduation gift -
Alfred Stock's Baker lectures on "Hydrides of Boron and Silicon"-
persuaded him to choose Professor H. I. Schlesinger as his research
adviser. His thesis dealt with the reduction of carbonyl compounds
with diborane;[4] he obtained the Ph.D. in 1938. There followed a
year in postdoctoral work with Professor Morris S. Kharasch, exploring
free-radical chemistry, chlorination and sulfonation with sulfonyl
chloride and carboxylation with oxalyl chloride and phosgene.[5] He
was made instructor in 1939 and assistant to Schlesinger, who was
investigating the metal borohydrides. This work was supported as a
defense project during the early years of World War II and produced
new ways of generating diborane in quantity as well as methods for
making sodium borohydride by reaction of sodium hydride with boron
halides or esters of boric acid.[6] The experience gained during
these years formed the basis of his later work with boron compounds
as Lewis acids, as selective reducing agents and as versatile reagents
for synthesis in organic chemistry. Beyond this, it probably influenced the style of his later research operations which, like Schlesinger's, featured close integration of the efforts of many persons
under the immediate supervision of seasoned postdoctoral research
assistants.

STERIC STRAINS[7,8]

     While he was an instructor at Chicago Brown was allowed to
direct the research of master's students. With them, he carried out
the first of those studies of steric effects which, throughout his
career, have provided a counterpoint of theory to his more practical
studies of the use of boron compounds in synthesis. That was a time
when steric hindrance had become "the last refuge of the puzzled
organic chemist"[9] and was generally held in low esteem. The electronic theory of organic structure had first subsumed Kekule's classical model and then advanced beyond it with the non-classical concept of delocalization, expressed then in the dominant resonance
model but also in the nascent molecular orbital model. Beyond that
it had opened the way to rational thought about detailed reaction

mechanism and had proven so successful in that regard that all else seemed superfluous.

By providing measurements of magnitude, Brown demonstrated that steric effects could be as large as the major electronic effects and so have both quantitative and qualitative influence on the course of reactions. Just as delocalization could slow down reactions by stabilizing the ground state or accelerate them by stabilizing intermediates (and, thus, transition states) so steric effects could hinder the approach of reagents or accelerate the departure of a leaving group.

Front (or F-) strain, operating directly between two entities, either as they approached or separated, subsumed most of the earlier concepts of steric hindrance. The magnitudes of such effects could be determined through studies of equilibria and heats of formation of the molecular addition compounds formed by boron Lewis acids and amines (eq 1); they were, indeed, large enough to matter, rather routinely larger than 2 or 3 kcal/mol. Back (or B-) strain was an

$$R_3B \;+\; :NR_3' \;\rightleftharpoons\; \underset{R}{\overset{R}{\underset{}{}}}B\!-\!\underset{R'}{\overset{R'}{N}}\!\!-\!R' \tag{1}$$

internal effect-crowding among groups attached to the same atom, made manifest when the bond angle changes. Typically, it would come into play when hybridization at the central atom changed, as from sp^3 (bond angle $ca.$ 109.5°) to sp^2 (bond angle $ca.$ 120°). Thus, relief of B-strain could promote the ionization of a highly ramified tertiary halide (eq 2). Generation of B-strain could be one of the factors hindering the reverse sort of process. Internal (or I-) strains were

$$\underset{R}{\overset{R}{\underset{R}{}}}C\!-\!X \;\rightleftharpoons\; \underset{R}{\overset{R}{\underset{R}{}}}C\!\oplus \;:X^{\ominus} \tag{2}$$

(tetrahedral)   (planar)

those non-bonded interactions which are especially important in ring compounds. They would include strains due to abnormal bond angles, to the eclipsing of groups on adjacent atoms or to transannular crowding in medium-sized rings. These are, in essence, effects which must be significant if the others are and they provide convenient rationalizations of trends in reactivity in alicyclic compounds.

It would be expected that compounds of similar shape - homomorphs - would show similar strains, as o-t-butyltoluene (1) and
the adduct (2) of trimethylborane and 2-picoline:

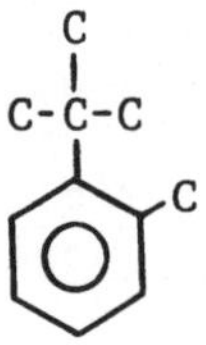

5.6 kcal/mol   5.9 kcal/mol
      1   2

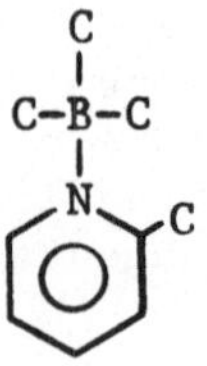

Similar strains should also be present in the transition states for
the formation of other homomorphs, as the ammonium salts 3 and 4,
where energies of activation for formation of the bonds denoted by
broken lines are shown.

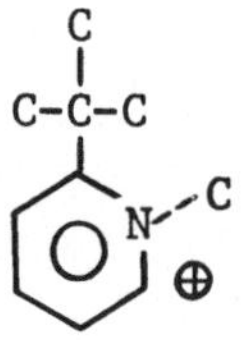

ca. 4 kcal/mol   5.9 kcal/mol
      3   4

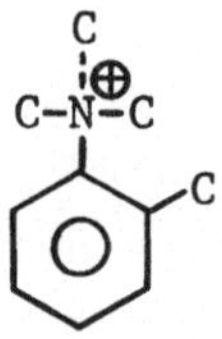

The idea that steric, as well as electronic, effects were important in the transition states of reactions provided useful insights
into the mechanisms of base-induced elimination. The conclusion
that steric effects controlled the Hofmann-Saytzeff regiochemistry
of such reactions brought Herb into controversy with C. K. Ingold,
one of the founding fathers of modern mechanistic chemistry. Although
this was no tea party, it was not so searing a dispute as the one
to come - but, rather, a good preliminary bout.

AROMATIC SUBSTITUTION[10,11]

A concern with the relative importance of steric and electronic
factors led to studies of the directive effects of substituents in

electrophilic aromatic substitution. This, in turn, led to more
detailed studies of particular reactions, notably those in which
Lewis acids act as catalysts by forming molecular addition compounds
with reagents. Kinetic studies suggested that in some cases the
aromatic nucleus acted as a nucleophile to displace the complexed
leaving group while in others it more passively accepted the atten-
tions of the reactive cation. It was found that the Hammett equation could be applied to reactions involving attack on the ring itself, but that new σ constants were required for those substituents
capable of resonant electron release to cationic centers. These
σ^+ constants[12] were later to prove useful in other studies of the
structures of carbonium ions. One of the most important concepts to
be developed and exploited in connection with this work was the
Selectivity Relationship[13]- the more reactive a reagent, the less
selective it will be. The absolute magnitude of the reaction constant, ρ^+ (which will be negative in these reactions), is then a
useful measure of the selectivity of a reaction proceeding through
a carbocationic intermediate.

CLASSICAL AND NONCLASSICAL CARBONIUM IONS[14-20]

More than ninety of Brown's publications[5] deal with the non-
classical carbonium ion problem. Some of them are summarizing state-
ments of his position as it developed over the years;[14-19] an almost
final summary is to be found in his recent book,[20] which contains
commentary by Paul Schleyer. Most of these publications are in communication form, as is appropriate for reports on continuing studies
of a subtle and difficult problem. Such communications invited
replies and the public discourse sometimes became heated and abra-
sive - the Great Debate of recent organic chemistry, with Brown
standing tenaciously alone against the massed mechanistic establish-
ment.

It is difficult to recall how abundant, exuberant and extrava-
gant was the use of the non-classical approach to reaction mechanisms
some 20 years ago. When Brown questioned the free use and easy ac-
ceptance of non-classical formulations in 1961[14] it was like the
child saying, "But the emperor has no clothes." Some still reject
his point of view on scientific ground and some have never forgiven
him for taking some of the joy out of organic chemistry - but his
sobering and chilling corrective was one that was much needed for,
indeed, an elaborate mechanistic edifice was a-building on a quick-
sand of *ad libitum* dotted line.

We may illustrate the nature of the problem and of Brown's contribution to its solution by considering the case of that Lorelei,
the 2-norbornyl cation.

Winstein and Trifan[21] reported that *exo*- norbornylbrosylate
acetolyzed 350 times as fast as the *endo* isomer. (eq 3). Chiral
exo gave almost fully racemic product; indeed it racemized (by internal return of the ions) faster than it solvolyzed. The product
was almost wholly *exo*.

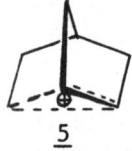

$$\text{OBs} \xrightarrow[\text{fast}]{\text{HOAc}} \quad + \quad \longleftarrow \tag{3}$$

exo   *endo*

99.5+% racemic from *exo*
7% chiral from *endo*

The non-classical symmetrically bridged ion (5) was suggested

<u>5</u>

to account for:

    a) the high *exo-endo* rate ratio
    b) the rapid and essentially complete racemization in the
    cation (even before it is free of the counter-ion) (the
    endo compound was assumed to give some direct displacement),
    c) the predominating formation of *exo* product.

The possibility that the non-classical structure was simply the
transition state for a pair of enantiomeric equilibrating ions (eq 4)

$$\rightleftharpoons \qquad \equiv \tag{4}$$

was dismissed as not accounting for either the *exo-endo* rate ratio or the predominating formation of *exo* products.

This was essentially an electronic answer, one that neglected steric effects that Brown saw as being potentially important. Ionization of the *endo* isomer would tend to force the departing anion back against the other side of the six-membered ring (6).

6

This effect would clearly be more important in the *endo* isomer than in the *exo* and would produce a decrease in the rate of solvolysis of the former. Studies of U-shaped molecules showed that such effects could be significant; studies of model monocyclic compounds tended to support the view that the *endo* rate was unusually slow rather than that the *exo* rate was unusually high. A careful consideration of potential energy curves indicated that it was the same difference in transition state energies that produced both the high *exo-endo* rate ratios and, in the reverse (bond-forming) process the high proportion of *exo* product. Studies of other reactions of norbornane derivatives, as, for example, the hydroboration of norbornene, showed a strong general preference for *exo* attack, again presumably due to differences in the energies of *exo* and *endo* transition states.

But this did no more than establish that a steric alternative to the electronic interpretation was viable. More was required. Trapping experiments indicated that racemization was not always complete as would be required of an obligatory symmetrically bridged first intermediate. It is possible, however, that tight ion-pairing or the intervention of small amounts of processes other than cation formation could produce small biases for retention of chirality. The "probe of increasing electron demand" (p. 1) showed that the *exo-endo* rate ratio in the 2-arylnorbornyl series (7) is not affected by substituents on the aromatic ring. This is powerful evidence against the existence of any significant stabilization of the carbonium center other than that provided by the aryl group. Where participation does occur (as in π systems) rate ratios are profoundly affected by substituents on the aromatic ring. This result indicates

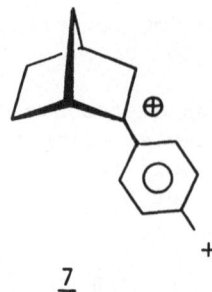

<u>7</u>

that whatever delocalization effects operate in the ionization of the
exo isomer operate with similar strength in the ionization of the
endo isomer.

   Well, then, is the norbornyl ion itself classical or non-classical? <u>That</u> question may not be answerable.

   Sometimes we approach the Heisenberg limit, where Nature takes
the fifth amendment and answers our questions with paraphrases or
even parodies of our questions. Perhaps we can not phrase the questions correctly. Perhaps we try to put boundaries in the middle of
a continuum. To this observer, it appears that there is a continuum
between the extremes of well-defined resonance and well-defined tau-
tomerism. At one end we have a single, stabilized structure at the

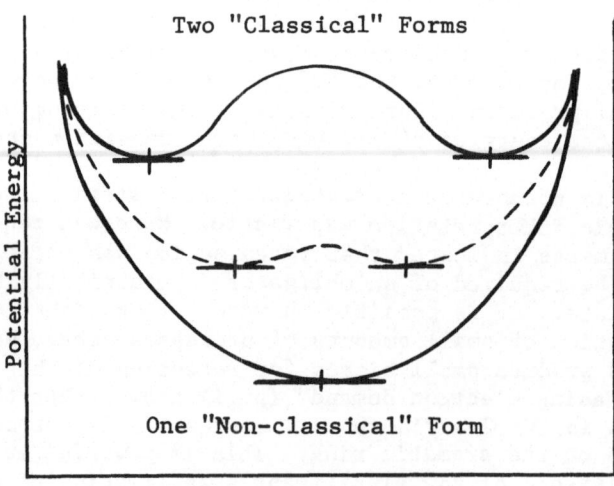

Figure 1.   The "Classical-nonclassical" continuum.

bottom of a "non-classical" well. At the other, two structures side-
by-side in "classical" wells, unstabilized and separated by enough
of a barrier that their interconversion is slow and observable. This
is shown diagrammatically in Figure 1. In between, as we reduce the
barrier, we find the two structures interconverting more and more
readily and therefore becoming more stabilized and less fully classical. (This point has been developed in detail by F. K. Fong.[22])
Brown has provided powerful evidence that the 2-norbornyl cation
lies in the middle region of this continuum in that it is best described as a rapidly interconverting pair of unsymmetrical ions;
that is a magnificent achievement. But we find ourselves in a seman-
tic quagmire when we try to put a label on such an ion if we insist
on the bipolarity: classical vs. non-classical. To say that the
existence of <u>any</u> stabilization means that the ion is non-classical,
or to say that only a totally symmetrical species is non-classical
is to deny significance to one end or the other of the continuum and
to seek victory by definition.

If controversy is good, *per se*, because it causes the participants to take care in advancing new postulates and to put forth their
best efforts to achieve a convincing solution to a problem, then
there should be no matter of deciding who won and who gets the palm
of victory. All participants deserve credit for the stimulation they
provided to one another and to the concerned on-lookers. There can
be no argument that H. C. Brown proved himself, in this arena, to
be one of the most stimulating chemists of our era.

ORGANOBORANES[23-30]

The discovery, with B. C. Subba Rao in 1956, that diborane would
add rapidly and quantitatively to olefins in ether solution at room
temperature has made the organoboranes readily available as chemical
intermediates. The boron atom adds to the less-substituted carbon
atom; depending upon steric factors, mono-, di-, or tri-alkylboranes
may be formed (eqs 5 - 7). All of these products give the

$$RCH{=}CH_2 \ + \ B_2H_6 \ \longrightarrow \ (RCH_2CH_2{-})_3B \qquad\qquad (5)$$

$$CH_3{-}\underset{\underset{CH_3}{|}}{C}{=}CHCH_3 \ + \ B_2H_6 \ \longrightarrow \ (CH_3{-}\underset{\underset{CH_3}{|}}{CH}{-}\underset{\underset{CH_3}{|}}{CH}{-})_2BH \qquad (6)$$

$$CH_3{-}\underset{\underset{CH_3}{|}}{C}{=}\underset{\underset{CH_3}{|}}{C}{-}CH_3 \ + \ B_2H_6 \ \longrightarrow \ CH_3{-}\underset{\underset{CH_3}{|}}{CH}{-}\underset{\underset{CH_3}{|}}{\overset{\overset{CH_3}{|}}{C}}{-}BH_2 \qquad (7)$$

synthetically useful reactions whereby the boron atom is replaced
(below) but the mono- and di-alkylboranes are also useful as reducing
or hydroborating agents, most notably the adducts from α-pinene
(di-isopinocampheylborane) (eq 8) and 1,5-cyclooctadiene (9-borabi-
cyclononane, or 9-BBN) (eq 9). The hydroboration reaction occurs

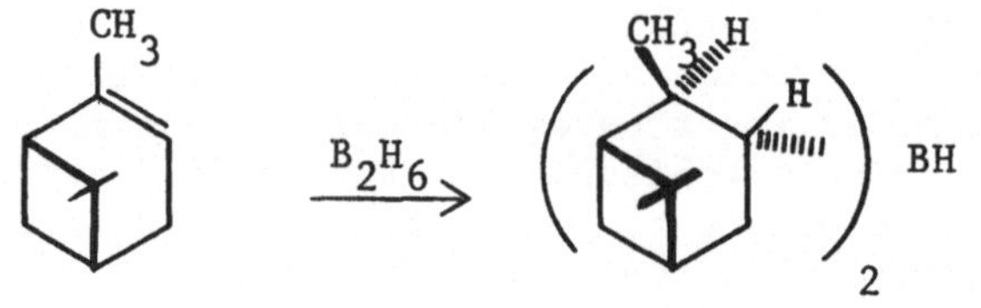

$$\text{(8)}$$

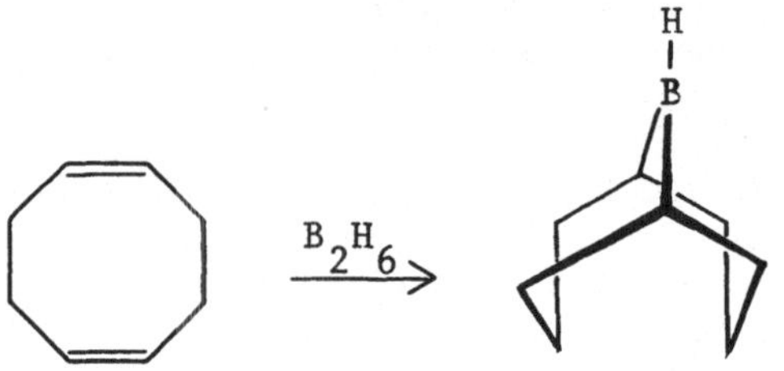

$$\text{(9)}$$

without skeletal rearrangement and gives clean *cis*- addition to the
less-hindered side of the double bond. Sterically crowded boranes
rearrange when heated in polyether solvents with migration of the
boron atom to the least crowded position on the carbon chain. This
furnishes organoboranes that would otherwise have to be made from
terminal olefins; indeed, the fact that the hydroboration is reversible makes these sometimes unstable or less accessible olefins
available from internal or endocyclic olefins (eq 10).

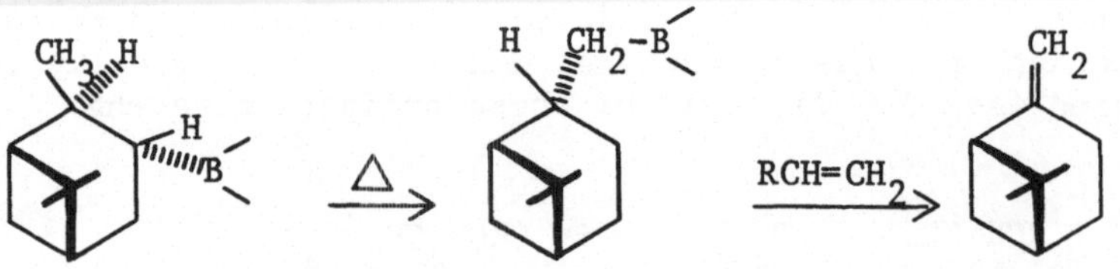

$$\text{(10)}$$

The boron atom can be replaced with complete retention of configuration by hydrogen, deuterium, hydroxy and amino groups (Scheme
I). The use of chiral boranes, as di-isopinocampheylborane, leads
to chiral products, often in high enantiomeric purity.

Valuable as these reactions are, those that lead to carbon-carbon
bond formation should prove to be even more so. The alkyl groups

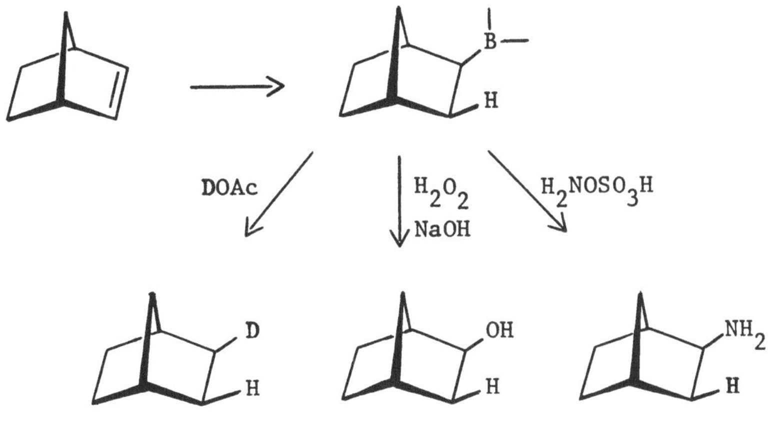

SCHEME I

of organoboranes are coupled with silver nitrate (eq 12); mixed boranes give mixed products (eq 13). Alkyl groups can be added to

$$CH_3\underset{CH_3}{\overset{CH_3}{\underset{|}{\overset{|}{C}}}}-CH_2\underset{CH_3}{\underset{|}{CH}}-CH_2B\langle \xrightarrow[NaOH]{AgNO_3} CH_3\underset{CH_3}{\overset{CH_3}{\underset{|}{\overset{|}{C}}}}-CH_2\underset{CH_3}{\underset{|}{CH}}-CH_2CH_2\underset{CH_3}{\underset{|}{CH}}-CH_2\underset{CH_3}{\overset{CH_3}{\underset{|}{\overset{|}{C}}}}-CH_3 \quad (12)$$

$$R_3B + R'_3B \longrightarrow RR + 2R'R + R'R' \quad (13)$$

α,β-unsaturated aldehydes and ketones under free radical catalysis, (eqs 14, 15) while α-bromoesters and α-bromoketones can be alkylated in the presence of strong bases (eqs 16, 17):

$$R_3B + CH_3-CH{=}CH-\overset{O}{\overset{\|}{C}}-H \longrightarrow R-\underset{CH_3}{\underset{|}{CH}}-CH_2-\overset{O}{\overset{\|}{C}}-H \quad (14)$$

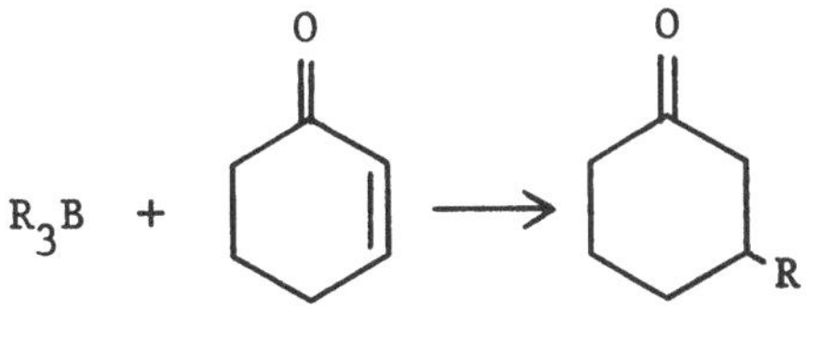

(15)

$$R_3B + BrCH_2CO_2Et \xrightarrow{KOtBu} RCH_2CO_2Et \qquad (16)$$

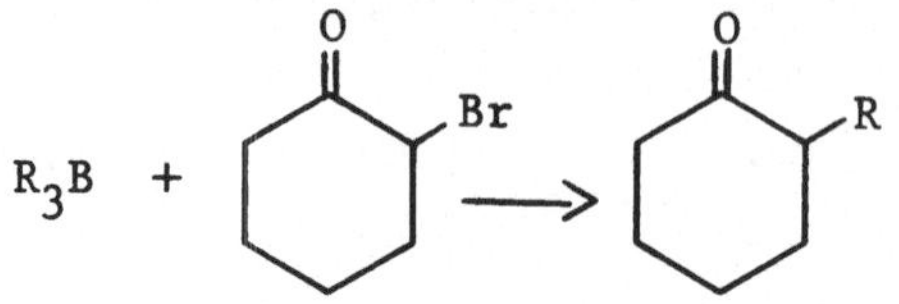

(17)

The use of 9-BBN derivatives is especially noteworthy because methyl and phenyl compounds can also be employed (eq 18):

(18)

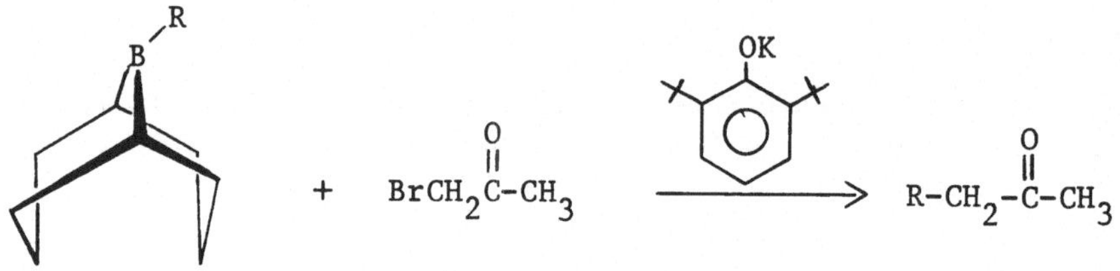

$R = -CH_3$, $-C_6H_5$, or alkyl from an olefin

New route to alcohols (eqs 19-21), aldehydes (eq 21) and ketones (eq 20) are provided by the reaction of organoboranes with carbon monoxide.

$$R_3B + CO \longrightarrow R_3C\text{-}BO \xrightarrow{(ox)} R_3C\text{-}OH \qquad (19)$$

$$R_3B + CO \xrightarrow{H_2O} R_2\overset{OH}{\overset{|}{C}}\text{—}\overset{OH}{\overset{|}{B}}\text{—}R \begin{cases} \xrightarrow{NaOH} R_2CH\text{-}OH \\ \xrightarrow{(ox)} R_2C{=}O \end{cases} \qquad (20)$$

$$R_3B + CO \xrightarrow{LiAlH(OCH_3)_3} R\text{-}\overset{OM}{\overset{|}{C}}H\text{-}BR_2 \begin{cases} \xrightarrow{NaOH} RCH_2OH \\ \xrightarrow{(ox)} R\text{-}\underset{H}{\underset{|}{C}}{=}O \end{cases} \qquad (21)$$

Annulation reactions based on this kind of process have been developed (eq 22).

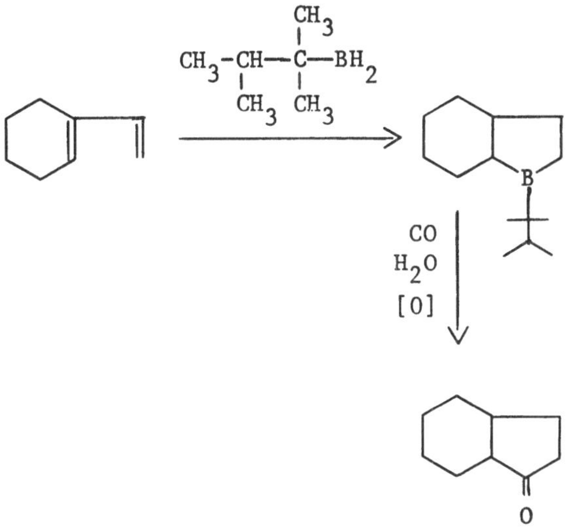

(22)

It has sometimes been said of H. C. Brown that, "he is not so much a scientist, as an inventor." This comment has some merit if one chooses to distinguish the detached, rarified and somewhat passive observer from the involved and active creator. Organoborane chemistry was not just discovered, it was also the product of a high order of creative imagination — an artistic triumph as well as a scientific one. Such creativity is rare indeed and when it is coupled with sharp powers of observation, outstanding technical skill, and superb management, it produces a man who is, in truth, a craftsman. It is fitting, then, that he has received many of the major recognitions of his profession — National Academy of Sciences (1957), the Nichols Medal (1959), the ACS Award for Creative Research in Synthetic Organic Chemistry (1960), American Academy of Arts and Sciences (1966), the Linus Pauling Medal (1968), an honorary doctorate from his alma mater (1968), the Medal of Science (1969) and the Roger Adams Medal and Award (1971).

And, as has been demonstrated at this symposium, the respect and affection of his students and colleagues.

REFERENCES

1. H. C. Brown, "Boranes in Organic Chemistry", Cornell University Press, Ithaca, N.Y., 1972, pp. 3-15.
2. S. B. Brown, in "Remembering HCB", Department of Chemistry, Purdue University, W. Lafayette, Indiana, 1978, pp. 3-6.
3. R. J. Britton, Purdue University, Department of Chemistry Newsletter, January 1978.

4. H. C. Brown, H. I. Schlesinger and A. Burg, *J. Am. Chem. Soc.*, <u>61</u>, 673 (1939).

5. For a full bibliography see, "Remembering HCB", reference 2, pp. 229-268.

6. H. I. Schlesinger, H. C. Brown, *et al.*, *J. Am. Chem. Soc.*, <u>75</u>, 186 (1953).

7. H. C. Brown, *J. Chem. Soc.*, 1248 (1956).

8. Ref. 1, pp. 53-128.

9. F. E. Ray, "Organic Chemistry", Lippincott, Philadelphia, 1941, p. 522.

10. K. L. Nelson and H. C. Brown, in "The Chemistry of Petroleum Hydrocarbons", D. T. Brooks, C. E. Boord, S. S. Kurtz and L. Schmerling, editors, Reinhold Publishing Corp. New York, 1955, Chapter 56.

11. L. M. Stock and H. C. Brown, in "Advances in Physical Organic Chemistry", V. Gold, editor, Academic Press, Inc., London, 1963, pp. 35-154.

12. H. C. Brown and Y. Okamoto, *J. Am. Chem. Soc.*, <u>80</u>, 4979 (1958).

13. This concept originated in earlier studies of free radical halogenation with Glen Russell.

14. H. C. Brown, "The Transition State", The Chemical Society, London, 1962, pp. 140-158.

15. H. C. Brown, *Chem. Brit.*, <u>2</u>, 199 (1966).

16. H. C. Brown, *Chem. Engr. News*, <u>45</u>, No. 7, 87 (1967).

17. Ref. 1, pp. 131-205.

18. H. C. Brown, *Acc. Chem. Res.*, <u>6</u>, 377 (1973).

19. H. C. Brown, *Tetrahedron*, 32, 179 (1976).

20. H. C. Brown (with P. v. R. Schleyer), "The Nonclassical Ion Problem", Plenum Press, New York, 1977.

21. S. Winstein and D. Trifan, *J. Am. Chem. Soc.*, <u>74</u>, 1147, 1154 (1952).

22. F. K. Fong, *J. Am. Chem. Soc.*, <u>96</u>, 7638 (1974).

23. H. C. Brown, *J. Chem. Ed.*, <u>38</u>, 173 (1961).

24. H. C. Brown, *Tetrahedron*, <u>12</u>, 117 (1961).

25. H. C. Brown, "Hydroboration", W. A. Benjamin, Inc., New York, (1962).

26. H. C. Brown, *Acc. Chem. Res.*, <u>2</u>, 65 (1969).

27. H. C. Brown, *Chem. Brit.*, <u>7</u>, 453 (1971).

28. Ref. 1, pp. 255-446.

29. H. C. Brown (with G. W. Kramer, A. B. Levy and M. !. Midland), "Organic Synthesis *via* Boranes", Wiley, New York, 1975.

30. H. C. Brown, *J. Pure Appl. Chem.*, <u>47</u>, 49 (1976).

ABOUT THE AUTHORS WHO WERE ASSOCIATED WITH PROFESSOR BROWN

Shelton Bank was born in New York City in 1932 and obtained the B.S. at Brooklyn College in 1954. With two years out for service in the army, he received his Ph.D. with Professor Brown in 1960, with a study of "The Buttressing Effect in 2-Methyl-3-t-butylpyridine and 2,6-Dimethyl-3-t-butylpyridine." He spent a year in postdoctoral work with P. D. Bartlett at Harvard and then several years at Esso Research. He joined the faculty of the State University of New York at Albany in 1966 and became professor in 1972. His research interests include: aromatic radical anions, base-catalyzed isomerization of olefins, and the thermodynamic and kinetic properties of cyclic and bicyclic compounds.

Frederick R. Jensen was born in Yerington, Nevada, in 1925. After service with the navy he obtained the B.S. from the University of Nevada in 1952 and the M.S. a year later. At Purdue, with Professor Brown, he did research on aromatic substitution and wrote one of the early massive theses, entitled "A Study of the Friedel-Crafts Acylation and Sulfonylation Reactions." The Ph.D. was awarded in 1955. He is now professor at the University of California, Berkeley, with wide research interests in electrophilic reactions of aliphatics, cyclopropanes and aromatics, the chemistry of organometallics, conformational analysis and the measurement of fast rates by NMR.

George W. Kabalka was born in Wyandotte, Michigan in 1943 and graduated from the University of Michigan in 1965. He did his Ph.D. thesis with Professor Brown on "The 1,4-Addition Reaction of Trialkylboranes with α,β-Unsaturated Carbonyl Compounds" and received the doctorate in 1970. He is now at the University of Tennessee, with research interests in the free radical reactions of organometallics, mechanisms of organometallic reactions and the preparation of physiologically active compounds containing radionuclides.

Clinton F. Lane was born in Iowa City, Iowa in 1944 and grew up in rural areas near Muscatine. He attended Iowa State University and then came to Purdue. His research with Professor Brown and

333

his thesis on "Bromination of Organoboranes" earned him the Ph.D. in 1972. He was a research associate at Cornell University and then joined Aldrich-Boranes, where he has been active in developing commercial production and applications of hydride reducing agents and boranes.

Richard C. Larock was born in Berkeley, California in 1944 and attended the University of California at Davis, where he did undergraduate research with George Zweifel. Graduating in 1967, he came to Purdue to work with Professor Brown. With his thesis, "The Mercuration of Organoboranes," he obtained the Ph.D. in 1972. A year's postdoctorate at Harvard was followed by appointment to the staff at Iowa State University, where he does research in the broad area of application of organometallics to synthesis.

M. Mark Midland was born in Fort Dodge, Iowa, in 1946 and graduated from Iowa State University in 1968. He was influenced by Glen Russell to come to Purdue to work with Professor Brown. "The Autoxidation and other Free-radical Reactions of Organoboranes" won him the Ph.D. in 1972. He stayed on several years as a research associate and then joined the faculty at the University of California, Riverside in 1975. There, he continued research in organoborane chemistry with special emphasis on asymmetric syntheses giving large excess of one enantiomer.

Ei-ichi Negishi was born in Shinkyo City, Japan, in 1935. He obtained the B.S. from the University of Tokyo in 1958 and then worked at the Research Institute of Teijin, Ltd. for a few years. He did graduate work at the University of Pennsylvania and obtained the doctorate in 1963. After several more years at Teijin, he returned to this country and was a research associate with Professor Brown from 1966 to 1972. He then joined the faculty at Syracuse University where he has an active research program in organometallic and synthetic chemistry.

Edward N. Peters was born in Providence, R.I., in 1943 and graduated from the University of Rhode Island in 1967. He was one of the relatively few predoctoral students who worked on the non-classical carbonium ion problem with Dr. Brown. His thesis (1972) dealt with "The Effect of Substituents on the Rates of Solvolysis of 2-Aryl-2-Norbornenyl and Cyclopropylaryl Carbinyl Derivatives." He did an additional year of postdoctoral research at Purdue and then joined the research staff of Union Carbide where he is now.

M. Ravindranathan was born in Enakulam, Kerala State, India in 1944 and attended the Maharaja's College there. He obtained the Ph.D. from the Indian Institute of Science in Bangalore in 1966 and came to Purdue the following year to work with Professor Brown. He returned to India in the fall of 1977 and is now at the Central University of Hyderabad.

Milorad M. Rogić was born in Belgrade, Yugoslavia in 1931. He obtained both the B.S. (1956) and the Ph.D. (1961) from the University
  of Belgrade. He did steroid research at the Worcester Foundation
  for Experimental Biology and postdoctoral research at Notre Dame
  University and the University of Saskatchewan. He came to Purdue
  in 1966 and did research with Professor Brown until 1969, when he
  joined Allied Chemical Corporation, where he is a group leader.
  His research interests are largely in the field of reaction mechanism as it applies to commercially significant processes.

Glen A. Russell was born in Rensselaer County, New York, in 1925.
  He attended Rensselear Polytechnic Institute and obtained the
  BChE. in 1947 and the M.S. in 1948. One of Professor Brown's early
  Purdue Ph.D.'s, he worked on "The Effect of Structure on the Relative Reactivities of Hydrogen Atoms as Determined in Photochemical
  Halogenation" and was granted the doctorate in 1951. He did research at General Electric Company for about seven years and then
  joined the faculty of Iowa State University, where he is now
  professor. His research interests lie broadly in the area of physical organic chemistry, especially in the mechanism of free radical reactions, oxidations and the uses of ESR.

Leon M. Stock was born in Detroit, Michigan in 1930 and graduated
  from the University of Michigan in 1952. After two years in the
  army, he entered Purdue and obtained the Ph.D. with Professor
  Brown in 1959. His thesis dealt with "The Non-Catalytic Halogenation of Toluene, t-Butylbenzene, Anisole and Biphenyl." He joined
  the faculty of the University of Chicago and in 1958 and became
  professor in 1971. His research interests lie broadly in the area
  of the effects of structure and solvents on reactivity, electrophilic aromatic substitution, models for the evaluation of inductive effects and the study of organic radicals by EPR.

George Zweifel was born in Rapperswill, Switzerland in 1926 and received the degree, Dr. Sc. Tech., from the Swiss Federal Institute
  of Technology in 1955. He then worked as a research associate at
  Edinburgh and Birmingham Universities before coming to Purdue in
  1958. With Dr. Brown, he helped open up the field of hydroboration
  in five highly productive years. He went to the University of
  California, Davis in 1963 and became professor in 1972. There he
  has studied the synthetic chemistry of organoalanes, especially
  those available from acetylenes.